Encyclopedia of Agricultural Sciences

Volume I

Encyclopedia of Agricultural Sciences Volume I

Edited by **Laura Vivian**

R CALLISTO REFERENCE

New York

Published by Callisto Reference,
106 Park Avenue, Suite 200,
New York, NY 10016, USA
www.callistoreference.com

Encyclopedia of Agricultural Sciences: Volume I
Edited by Laura Vivian

International Standard Book Number: 978-1-63239-171-1 (Hardback)

Printed in the United States of America.

Contents

Preface

The human race has been dependent on agriculture since its early days. Agriculture has provided for the basic needs of food and nutrition and has been a key occupation of human civilization until the industrial revolution. The science of agriculture has evolved in manifold ways. It varies as a consequence of climates, cultures and technologies. Agronomy and agricultural science are intertwined as they both focus on the research and development of new techniques for improving current agricultural practices for the best possible results.

Agricultural science is a multidisciplinary field of study which encompasses the diversified practices of agriculture and their implementation in varied fields of science and technology. It comprises advanced production techniques, pest control, minimizing adverse environmental effects, research and development of food production techniques to meet global requirements and many other topics. The agricultural science prospered during the eighteenth century when the study of fertilizers and plant physiology gained pace. In the twenty-first century, the technological advancements have revolutionized the science of agriculture. Biotechnology, genetic engineering, precision farming and such fields have given a new direction to agriculture science.

This book is dedicated to such advancements and their applications. But, this progress comes at a cost. The effort to meet the worldwide demand has also affected the quality of crops which is said to have a negative impact on human health and environment. Such aspect of agriculture science is often debated and discussed by experts and researchers all over the world.

I would like to thank all the contributors of this book. I would also like to thank the publishing house for their endless efforts. This book is a result of the collaboration of veterans across the globe and I feel blessed to have this opportunity to work with them.

Editor

Short-term influence of anaerobically-digested and conventional swine manure, and N fertilizer on organic C and N, and available nutrients in two contrasting soils

Sukhdev S. Malhi[1*], R. L. Lemke[2], M. Stumborg[3], F. Selles[4]

[1]Agriculture and Agri-Food Canada, Melfort, Canada; [*]Corresponding Author
[2]Agriculture and Agri-Food Canada, Saskatoon, Canada
[3]Agriculture and Agri-Food Canada, Swift Current, Canada
[4]Agriculture and Agri-Food Canada, Brandon, Canada

ABSTRACT

A three-year (2006-2008) field experiment was conducted at Swift Current and Star City in Saskatchewan to determine the short-term influence of land-applied anaerobically digested swine manure (ADSM), conventionally treated swine manure (CTSM) and N fertilizer on total organic C (TOC), total organic N (TON), light fraction organic C (LFOC), light fraction organic N (LFON) and pH in the 0 - 7.5 and 7.5 - 15 cm soil layers, and ammonium-N, nitrate-N, extractable P, exchangeable K and sulphate-S in the 0 - 15, 15 - 30, 30 - 60, 60 - 90 and 90 - 120 cm soil layers. Treatments included spring and autumn applications of CTSM and ADSM at a 1x rate (10,000 and 7150 $L \cdot ha^{-1}$, respectively) applied each year, a 3x rate (30,000 and 21,450 $L \cdot ha^{-1}$, respectively) applied once at the beginning of the experiment, plus a treatment receiving commercial fertilizer (UAN at 60 $kg \cdot N \cdot ha^{-1} \cdot yr^{-1}$) and a zero-N control. There was no effect of swine manure rate, type and application time on soil pH. Mass of TOC and TON in the 15 cm soil layer increased significantly with swine manure application compared to the control, mainly at the Swift Current site, with greater increases from 3x rate than 1x rate (by 2.21 $Mg \cdot C \cdot ha^{-1}$ and 0.167 $Mg \cdot N \cdot ha^{-1}$). Compared to the control, mass of LFOC and LFON in the 15 cm soil layer increased with swine manure application at both sites, with greater increases from 3x rate than 1x rate (by 287 $kg \cdot C \cdot ha^{-1}$ and 26 $kg \cdot N \cdot ha^{-1}$ at Star City, and by 194 $kg \cdot C \cdot ha^{-1}$ and 19 $kg \cdot N \cdot ha^{-1}$ at Swift Current). Mass of TOC and TON in soil layer was tended to be greater with ADSM than CTSM, but mass of LFOC and LFON in soil was greater with CTSM than ADSM. Mass of TOC, TON, LFOC and LFON in soil also increased with annual N fertilizer application compared to the control (by 3.2 $Mg \cdot C \cdot ha^{-1}$ for TOC, 0.195 $Mg \cdot N \cdot ha^{-1}$ for TON, 708 $kg \cdot C \cdot ha^{-1}$ for LFOC and 45 $kg \cdot N \cdot ha^{-1}$ for LFON). In conclusion, our findings suggest that the quantity and quality of organic C and N in soil can be affected by swine manure rate and type, and N fertilization even after three years, most likely by influencing inputs of C and N through crop residue, and improve soil quality.

Keywords: Anaerobic Digestion; Available N; P, K and S; Organic C and N; Soil; Swine Manure

1. INTRODUCTION

Of the approximately 30 million hogs marketed in Canada, nearly one-half of that industry is located in the Canadian prairie region, and approximately 90% of intensive livestock operations (ILOs) store manure in liquid form in a holding tank or lagoon until it can be land-applied. Land application of liquid swine manure (LSM) is an effective source of nutrients for crop production [1-3]. Economically feasible, environmentally friendly, and socially acceptable management of LSM from ILOs is a key element for the future viability of this industry. In LSM, there is usually less than 2% solid material [4] and most of the nutrients are in plant-available inorganic form. Thus, LSM can potentially increase soil organic C (SOC) mainly by supplying nutrients to crops [5,6] and increasing above and below ground plant biomass thereby adding organic matter to the soil. In the Prairie Provinces of Canada, previous research has documented the agronomic benefits of LSM application on enhancing crop yields [1]. Increased soil fertility is an important benefit of LSM application that substantially increases the concentration of N, P, K and micronutrients in soil [1,3].

Anaerobic digestion is a promising technology that may reduce greenhouse gas (GHG, CH_4 and N_2O) emissions by utilizing the biogas produced during digestion to displace fossil fuels and by reducing emissions during lagoon storage. The effects of land-applied anaerobically digested swine manure (ADSM) versus conventionally treated swine manure (CTSM) or N fertilizer on crop yields and GHG emissions in the Canadian prairies are presented in our previous report [7]. However, the research information on the impact of ADSM versus CTSM or N fertilizer on soil biochemical and chemical properties is lacking in the Canadian prairies, especially in the Parkland region. The objective of this study was to compare relative effects of land-applied ADSM, CTSM, or N fertilizer on quantity and quality of soil organic C and N (TOC, TON, LFOC and LFON), and some soil chemical properties (pH, ammonium-N, nitrate-N, extractable P, exchangeable K and sulphate-S).

2. MATERIALS AND METHODS

A field experiment was conducted over three years from 2006 to 2008 at two field sites in Saskatchewan [Star City (Dark Gray Luvisol soil) and Swift Current (Brown Chernozem soil)], having contrasting soil and climatic conditions. Precipitation in the growing season (May, June, July and August) at the two sites from 2006 to 2008, and long-term (30-year) average of precipitation in May to August at the nearest Environment Canada Meteorological Station (AAFC Melfort and AAFC Swift Current) are presented in **Table 1**. Precipitation in the 2006 growing season was slightly below average at both sites. In 2007, the growing season precipitation was much below long-term average at Swift Current (with particularly limited precipitation in July), but was slightly above average at Star City. In 2008, the growing season precipitation was much higher than average (especially in June) at Swift Current, but much below average (especially in May during seeding) at Star City. Treatments included autumn and spring applications of CTSM and ADSM at a 1x rate (10,000 and 7150 $L \cdot ha^{-1}$ respectively) applied each year, and a 3x rate (30,000 and 21,450 $L \cdot ha^{-1}$ respectively) applied once at the beginning of the study. A treatment receiving commercial fertilizer urea-ammonium nitrate (UAN) solution and a check (no N) were also included. Eleven treatments (**Table 2**) were arranged in a randomized complete block design with four replications. Liquid swine manures were applied by the Prairie Agricultural Machinery Institute (PAMI) using a customized applicator, which injected the material to 10 cm. All plots were seeded to barley (*Hordeum vulgare* L.) in each of the three years, and harvested for seed and straw yield, and total N uptake. In the autumn of 2008, soil in each plot was sampled to 0 - 7.5, 7.5 - 15 and 15 - 20 cm depths for TOC, TON, LFOC, LFON and pH, and to 0 - 15, 15 - 30, 30 - 60 and 60 - 90 cm depths for ammonium-N, nitrate-N, extractable P, exchangeable K and sulphate-S.

For TOC, TON, LFOC, LFON and pH, soil cores at 10 locations in each plot were collected using a 2.4 cm diameter coring tube. Bulk density of soil was determined by the core method using soil weight and core volume [8]. The soil samples were air dried at room temperature after removing coarse roots and easily detectable crop

Table 1. Monthly cumulative precipitation in the growing season during 2006, 2007 and 2008 at Star City and Swift Current, Saskatchewan.

Location/Year	Precipitation (mm)				
	May	June	July	August	Total
Star City					
2006	63	73	39	46	221
2007	71	119	47	40	277
2008	6	32	117	22	177
30-year mean	46	66	76	57	245
Swift Current					
2006	35	96	31	21	183
2007	26	48	10	19	103
2008	27	152	64	69	312
30-year mean	50	66	52	40	208

Short-term influence of anaerobically-digested and conventional swine manure, and N fertilizer on organic C and N, and available nutrients in two contrasting soils

3

Table 2. List of treatments and the corresponding total amount of N applied and input of C from crop residue returned and land-applied liquid swine manure (LSM) during a three-year (2006-2008) field study at Star City and Swift Current, Saskatchewan.

Time of application	Product applied[z]	Application rate of LSM or N fertilizer	Total amount of N applied in 3 years ($kg \cdot N \cdot ha^{-1}$)	Input of C from crop residue plus LSM in 3 years at Star City ($kg \cdot C \cdot ha^{-1}$)	Input of C from crop residue plus LSM in 3 years at Swift Current ($kg \cdot C \cdot ha^{-1}$)
	Control	No manure or N fert	0	4037	3710
Autumn	ADSM-3x	$21,450 \, L \cdot ha^{-1}$	214	6665	5753
	ADSM-1x	$7150 \, L \cdot ha^{-1}$	205	6446	4719
	CTSM-3x	$30,000 \, L \cdot ha^{-1}$	403	7603	5919
	CTSM-1x	$10,000 \, L \cdot ha^{-1}$	360	8765	4872
Spring	ADSM-3x	$21,450 \, L \cdot ha^{-1}$	257	6745	5270
	ADSM-1x	$7150 \, L \cdot ha^{-1}$	255	7601	4444
	CTSM-3x	$30,000 \, L \cdot ha^{-1}$	343	7405	5701
	CTSM-1x	$10,000 \, L \cdot ha^{-1}$	326	7906	5210
	UAN	$60 \, kg \cdot N \cdot ha^{-1}$	180	6158	5278

[z]ADSM = anaerobically digested swine manure, CTSM = conventionally treated swine manure, UAN = urea ammonium nitrate (liquid), 3x = once in 3 years, 1x = annual application.

residues, and ground to pass a 2-mm sieve. Sub-samples were pulverized in a vibrating-ball mill (Retsch, Type MM2, Brinkman Instruments Co., Toronto, Ontario) for determination of TOC, TON, LFOC and LFON in soil. Soil samples used for organic C and N analyses were tested for the presence of inorganic C (carbonates) using dilute HCl, and none was detected in any soil sample. Therefore, C in soil associated with each fraction was considered to be of organic origin. Total organic C in soil was measured by Dumas combustion using a Carlo Erba instrument (Model NA 1500, Carlo Erba Strumentazione, Italy), and Technicon Industrial Systems [9] method was used to determine TON in the soil. Light fraction organic matter (LFOM) was separated using a NaI solution of 1.7 $Mg \cdot m^{-3}$ specific gravity, as described by Janzen et al. [10] and modified by Izaurralde et al. [11]. The C and N in LFOM (LFOC, LFON) were measured by Dumas combustion.

Soil samples (ground to pass a 2-mm sieve) taken for organic C and N from the 0 - 15 cm layer were also monitored for pH in 0.01 M $CaCl_2$ solution with a pH meter. For other chemical properties, soil cores (using a 4 cm diameter coring tube) were collected at 4 locations in each plot from the 0 - 15, 15 - 30, 30 - 60, 60 - 90 and 90 - 120 cm layers. The bulk density of each depth was calculated using soil weight and core volume [8]. The soil samples were air dried at room temperature, ground to pass a 2-mm sieve, and analyzed for ammonium-N [12] and nitrate-N [13] by extracting soil in a 1:5 soil: 2M KCl solution; extractable P [9] by extracting soil in Kelowna extract, exchangeable K [14] and sulphate-S [15].

The data on each parameter were subjected to analysis of variance (ANOVA) using GLM procedure in SAS [16]. For each ANOVA, the least significant difference at $P \leq$ 0.05 ($LSD_{0.05}$) was used to determine significant differences between treatment means, and standard error of the mean (SEM) and significance are also reported.

3. RESULTS

3.1. Soil Biochemical Properties

At Star City, there was no significant beneficial effect of swine manure or UAN fertilizer application on TOC and TON mass in soil compared to the zero-N control treatment (**Table 3**). At Swift Current, mass of TOC and TON in soil increased with application of swine manure at 3x rate compared to control in the 0 - 7.5 and also in the total 0 - 15 cm depth, with the greatest increase from 3x rate of ADSM applied in spring (**Table 4**). On average, TOC and TON in soil was greater with 3x rate (once in 3 years) than 1x rate (annual application) of swine manure, and greater with ADSM than CTSM in some cases.

At Star City, mass of LFOC and LFON in soil increased with increasing rate of swine manure and also with UAN application compared to the zero-N control treatment in the 0 - 7.5 cm layer (**Table 5**). On average, mass of LFOC and LFON was greater with the 3x rate (once in 3 years) than the 1x rate (annual application) of swine manure in the 0 - 7.5 cm soil layer, but there was little or no effect of timing and type of swine manure application on these parameters. At Swift Current, there was a significant effect of swine manure and N fertilizer treatments on mass of LFOC and LFON in the 0 - 7.5 cm

Table 3. Effect of land-applied anaerobically digested swine manure (ADSM), conventionally treated swine manure (CTSM) and urea-ammonium-nitrate (UAN) solution fertilizer over three years from 2006 to 2008 on mass of total organic C (TOC) and total organic N (TON) in soil in autumn 2008 at Star City, Saskatchewan, Canada (Gray Luvisol soil).

Treatments	TOC mass ($Mg \cdot C \cdot ha^{-1}$) in soil layers (cm)			TON mass ($Mg \cdot N \cdot ha^{-1}$) in soil layers (cm)		
	0 - 7.5	7.5 - 15	0 - 15	0 - 7.5	7.5 - 15	0 - 15
Control	22.27	16.58	38.85	2.024	1.519	3.543
ADSM-3x Autumn	22.57	17.73	40.31	2.037	1.636	3.673
ADSM-1x Autumn	22.26	15.52	37.78	1.999	1.476	3.474
CTSM-3x Autumn	21.86	17.49	39.34	2.082	1.751	3.833
CTSM-1x Autumn	21.31	16.21	37.52	1.926	1.463	3.388
ADSM-3x Spring	22.52	15.99	38.51	2.106	1.588	3.694
ADSM-1x Spring	22.54	17.94	40.48	2.045	1.723	3.768
CTSM-3x Spring	21.96	16.43	38.39	2.001	1.556	3.557
CTSM-1x Spring	21.55	17.90	39.45	1.924	1.673	3.597
UAN Spring	23.64	19.13	42.76	2.106	1.774	3.880
$LSD_{0.05}$	ns	ns	ns	ns	ns	ns
SEM (Probability)	0.733^{ns}	1.249^{ns}	1.620^{ns}	0.0731^{ns}	0.1116^{ns}	0.1505^{ns}
Manure rate						
1x	21.91	16.89	38.80	1.973	1.584	3.557
3x	22.23	16.91	39.14	2.057	1.633	3.690
$LSD_{0.05}$	ns	ns	ns	0.112	ns	ns
SEM (Probability)	0.359^{ns}	0.653^{ns}	0.837^{ns}	0.0385^{*}	0.0592^{ns}	0.0816^{ns}
Manure type						
ADSM	22.47	16.80	39.27	2.047	1.606	3.653
CTSM	21.67	17.01	38.68	1.983	1.611	3.594
$LSD_{0.05}$	1.04	ns	ns	ns	ns	ns
SEM (Probability)	0.359^{*}	0.653^{ns}	0.837^{ns}	0.0385^{ns}	0.0592^{ns}	0.0816^{ns}
Manure application time						
Autumn	22.00	16.74	38.74	2.011	1.581	3.592
Spring	22.14	17.07	39.21	2.019	1.635	3.654
$LSD_{0.05}$	ns	ns	ns	ns	ns	ns
SEM (Probability)	0.359^{ns}	0.653^{ns}	0.837^{ns}	0.0385^{ns}	0.0592^{ns}	0.0816^{ns}

* and ^{ns} refer to significant treatment effects in ANOVA at $P \leq 0.10$ and not significant, respectively.

soil layer (**Table 6**). On average, mass of LFOC and LFON in soil was greater with the 3x rate (once in 3 years) than the 1x rate (annual application) of swine manure, but there was little effect of timing and type of swine manure application on these parameters.

At both sites, the correlation coefficients among the TOC, TON, LFOC and LFON fractions in soil were strong, and were highly significant between TOC and TON, and between LFOC and LFON (**Table 7**). At Swift Current, the correlation between TOC and LFOC or LFON was significant at $P = 0.12$ or 0.15. The correlation coefficients between crop residue C input over 3 growing seasons (**Table 1**) and TOC, TON, LFOC or LFON were not significant in any case at Star City, but

Short-term influence of anaerobically-digested and conventional swine manure, and N fertilizer on organic C and N, and available nutrients in two contrasting soils

5

Table 4. Effect of land-applied anaerobically digested swine manure (ADSM), conventionally treated swine manure (CTSM) and urea-ammonium-nitrate (UAN) solution fertilizer over three years from 2006 to 2008 on mass of total organic C (TOC) and total organic N (TON) in soil in autumn 2008 at Swift Current, Saskatchewan, Canada (Dark Brown Chernozem soil).

Treatments	TOC mass ($Mg \cdot C \cdot ha^{-1}$) in soil layers (cm)			TON mass ($Mg \cdot N \cdot ha^{-1}$) in soil layers (cm)		
	0 - 7.5	7.5 - 15	0 - 15	0 - 7.5	7.5 - 15	0 - 15
Control	19.50	15.60	35.10	1.948	1.723	3.671
ADSM-3x Autumn	21.59	16.69	38.28	2.096	1.760	3.856
ADSM-1x Autumn	20.50	17.57	38.07	2.065	1.814	3.879
CTSM-3x Autumn	22.02	15.88	37.90	2.126	1.711	3.837
CTSM-1x Autumn	20.50	16.17	36.67	2.062	1.714	3.776
ADSM-3x Spring	23.31	17.52	40.83	2.347	1.853	4.200
ADSM-1x Spring	20.41	16.17	36.58	2.016	1.694	3.710
CTSM-3x Spring	22.08	16.69	38.77	2.121	1.726	3.847
CTSM-1x Spring	20.67	14.97	35.64	2.084	1.623	3.707
UAN Spring	20.68	16.90	37.58	2.023	1.701	3.724
$LSD_{0.05}$	2.14	ns	3.48	0.178	ns	0.290
SEM (Probability)	0.736^{*}	0.845^{ns}	1.201^{*}	0.0613^{*}	0.0682^{ns}	0.0999^{*}
Manure rate						
1x	20.52	16.22	36.74	2.057	1.711	3.768
3x	22.25	16.70	38.95	2.172	1.762	3.935
$LSD_{0.05}$	1.14	ns	1.87	0.107	ns	0.167
SEM (Probability)	0.392^{**}	0.439^{ns}	0.644^{*}	0.0366^{*}	0.0343^{ns}	0.0575^{*}
Manure type						
ADSM	21.45	16.99	38.44	2.131	1.780	3.911
CTSM	21.32	15.93	37.25	2.098	1.693	3.792
$LSD_{0.05}$	ns	1.28	ns	ns	0.100	0.167
SEM (Probability)	0.392^{ns}	0.439^{*}	0.644^{ns}	0.0366^{ns}	0.0343^{*}	0.0575^{*}
Manure application time						
Autumn	21.15	16.58	37.73	2.087	1.750	3.837
Spring	21.62	16.34	37.96	2.142	1.724	3.866
$LSD_{0.05}$	ns	ns	ns	ns	ns	0.167
SEM (Probability)	0.392^{ns}	0.439^{ns}	0.644^{ns}	0.0366^{ns}	0.0343^{ns}	0.0575^{ns}

*, ** and ^{ns} refer to significant treatment effects in ANOVA at $P \leq 0.10$, $P \leq 0.05$, $P \leq 0.01$ and not significant, respectively.

was significant for LFOC and LFON at Swift Current. For linear regressions between crop residue C input and TOC, TON, LFOC or LFON, the R^2 values were not significant in any case at Star City, but highly significant for LFOC and LFON at Swift Current (**Table 8**).

3.2. Soil Chemical Properties and Distribution of Available N, P, K and S in the Soil Profile

There was no significant effect of swine manure (fre-quency, type and application time) or N fertilizer application after three years on soil pH in the 0 - 15 cm layer at either site (data not shown). The soil pH ranged from 6.4 to 6.7 at Star City and from 5.8 to 6.5 at Swift Current among different treatments. There was also no effect of swine manure or N fertilizer treatments on ammonium-N and exchangeable K in soil at both sites, and sulphate-S in soil at Swift Current (data not shown). The amount of nitrate-N increased with the 3x rate of swine manure application in the 30 - 60, 60 - 90 and 90 - 120 cm soil layers at Star City (**Table 9**), and in all soil layers

Table 5. Effect of land-applied anaerobically digested swine manure (ADSM), conventionally treated swine manure (CTSM) and urea-ammonium-nitrate (UAN) solution fertilizer over three years from 2006 to 2008 on mass of light fraction organic C (LFOC) and light fraction organic N (LFON) in soil in autumn 2008 at Star City, Saskatchewan, Canada (Gray Luvisol soil).

Treatments	LFOC mass ($Mg \cdot C \cdot ha^{-1}$) in soil layers (cm)			LFON mass ($Mg \cdot N \cdot ha^{-1}$) in soil layers (cm)		
	0 - 7.5	7.5 - 15	0 - 7.5	7.5 - 15	0 - 7.5	7.5 - 15
Control	1931	649	2580	135	35	170
ADSM-3x Autumn	2263	603	2866	158	33	191
ADSM-1x Autumn	2116	749	2865	143	40	183
CTSM-3x Autumn	2286	1004	3290	154	52	206
CTSM-1x Autumn	2306	762	3068	149	41	190
ADSM-3x Spring	2333	817	3150	160	47	207
ADSM-1x Spring	2034	673	2707	134	36	170
CTSM-3x Spring	2277	957	3234	161	55	216
CTSM-1x Spring	1878	875	2753	127	48	175
UAN Spring	2203	952	3155	150	55	205
$LSD_{0.05}$	ns	ns	ns	ns	ns	ns
SEM (Probability)	206.8^{ns}	122.1^{ns}	241.7^{ns}	13.3^{ns}	7.2^{ns}	14.6^{ns}
Manure rate						
1x	2083	765	2848	138	41	179
3x	2290	845	3135	158	47	205
$LSD_{0.05}$	296	ns	346	19	ns	21
SEM (Probability)	101.5^{ns}	55.0^{ns}	118.7^{*}	6.5^{*}	3.1^{ns}	7.1^{*}
Manure type						
ADSM	2187	710	2897	149	39	188
CTSM	2187	900	3087	148	49	197
$LSD_{0.05}$	ns	160	ns	ns	9	ns
SEM (Probability)	101.5^{ns}	55.0^{*}	118.7^{ns}	6.5^{ns}	3.1^{*}	7.1^{ns}
Manure application time						
Autumn	2242	780	3022	151	41	192
Spring	2131	830	2961	145	47	192
$LSD_{0.05}$	ns	ns	ns	ns	ns	ns
SEM (Probability)	101.5^{ns}	55.0^{ns}	118.7^{ns}	6.5^{ns}	3.1^{ns}	7.1^{ns}

*, * and ^{ns} refer to significant treatment effects in ANOVA at $P \leq 0.10$, $P \leq 0.05$ and not significant, respectively.

up to the 120 cm depth at Swift Current (**Table 10**). Application of UAN fertilizer had a significant effect on nitrate-N in soil at Swift Current, but no effect on soil nitrate-N at Star City. The increase in nitrate-N due to swine manure in the 120 cm soil profile was greater with CTSM than ADSM, and also greater with autumn application than spring application at Star City site. However, the opposite was true at Swift Current. The amounts of

extractable P in soil tended to increase in a few cases with swine manure application in the 0 - 15 and 15 - 30 cm layers at Star City (**Table 11**) and in the 0 - 15, 15 - 30 or 30 - 60 cm layers at Swift Current (**Table 12**). Application of UAN fertilizer had no significant effect on extractable P in soil at either site. On average, extractable P in soil tended to be greater with CTSM than ADSM at Swift Current, but there was no effect of swine manure

Short-term influence of anaerobically-digested and conventional swine manure, and N fertilizer on organic C and N, and available nutrients in two contrasting soils

7

Table 6. Effect of land-applied anaerobically digested swine manure (ADSM), conventionally treated swine manure (CTSM) and urea-ammonium-nitrate (UAN) solution fertilizer over three years from 2006 to 2008 on mass of light fraction organic C (LFOC) and light fraction organic N (LFON) in soil in autumn 2008 at Swift Current, Saskatchewan, Canada (Dark Brown Chernozem soil).

Treatments	LFOC mass ($Mg·C·ha^{-1}$) in soil layers (cm)			LFON mass ($Mg·N·ha^{-1}$) in soil layers (cm)		
	0 - 7.5	7.5 - 15	0 - 7.5	7.5 - 15	0 - 7.5	7.5 - 15
Control	1808	635	2443	123	36	159
ADSM-3x Autumn	2726	806	3532	186	46	231
ADSM-1x Autumn	2360	936	3296	159	53	212
CTSM-3x Autumn	2935	905	3740	201	47	249
CTSM-1x Autumn	2570	884	3454	172	51	223
ADSM-3x Spring	2783	712	3496	190	39	229
ADSM-1x Spring	2272	774	3046	151	43	193
CTSM-3x Spring	2677	777	3454	182	44	225
CTSM-1x Spring	2825	814	3640	189	45	234
UAN Spring	2397	886	3284	163	51	214
$LSD_{0.05}$	679	ns	ns	48	ns	ns
SEM (Probability)	233.8^{*}	94.8^{ns}	294.9^{ns}	16.5^{*}	5.5^{ns}	19.8^{ns}
Manure rate						
1x	2507	852	3359	167	48	215
3x	2780	775	3555	190	44	234
$LSD_{0.05}$	348	ns	ns	25	ns	ns
SEM (Probability)	119.4^{*}	45.4^{ns}	146.3^{ns}	8.4^{*}	2.7^{ns}	9.9^{ns}
Manure type						
ADSM	2535	807	3342	171	45	216
CTSM	2752	820	3572	186	47	233
$LSD_{0.05}$	ns	ns	ns	ns	ns	ns
SEM (Probability)	119.4^{ns}	45.4^{ns}	146.3^{ns}	8.4^{ns}	2.7^{ns}	9.9^{ns}
Manure application time						
Autumn	2648	857	3505	179	49	228
Spring	2639	769	3408	178	43	221
$LSD_{0.05}$	ns	ns	ns	ns	8	ns
SEM (Probability)	119.4^{ns}	45.4^{ns}	146.3^{ns}	8.4^{ns}	2.7^{*}	9.9^{ns}

* and ^{ns} refer to significant treatment effects in ANOVA at $P \leq 0.10$, $P \leq 0.05$, $P \leq 0.01$, $P \leq 0.001$ and not significant, respectively.

type, rate or application time on extractable P in soil at Star City. At Star City, the amount of sulphate-S in soil increased (but not significantly) with swine manure application mainly in the 30 - 60, 60 - 90 and 90 - 120 cm layers (**Tables 13**). Application of UAN fertilizer had no significant effect on sulphate-S in soil at Star City, and in fact sulphate-S in the surface 0 - 15 cm soil layer tended

to decrease compared to the zero-N control treatment. On average, sulphate-S in soil was greater with ADSM than CTSM, considerably greater with autumn application than spring application, and slightly greater with 1x rate than 3x rate of swine manure. There was no effect of any amendment treatment on sulphate-S in soil at Swift Current, and exchangeable K in soil at both sites (data not

Table 7. Relationships among organic C or N fractions (TOC, TON, LFOC, LFON) in the 0 - 15 cm soil, or between crop residue and/or swine manure C input from 2006 to 2008 growing seasons and organic C or N stored in the 0 - 15 cm soil sampled in autumn 2008 at Star City (Gray Luvisol) and Swift Current (Dark Brown Chernozem), Saskatchewan, Canada.

Soil	Parameter	Correlation coefficients (r)			
		TOC	TON	LFOC	LFON
Relationships among soil organic C or N fractions					
Star City	TOC		0.820^{**}	-0.004^{ns}	0.041^{ns}
	TON			0.279^{ns}	0.276^{ns}
	LFOC				0.950^{***}
	LFON				
Swift Current	TOC		0.913^{***}	0.492^{*}	0.527^{*}
	TON			0.406^{ns}	0.436^{ns}
	LFOC				0.994^{***}
	LFON				
Relationships between crop residue and/or swine manure C input and soil organic C or N fractions					
Star City		-0.196^{ns}	-0.083^{ns}	0.413^{ns}	0.226^{ns}
Swift Current		0.587^{*}	0.386^{ns}	0.891^{***}	0.921^{***}

*, **, *** and ^{ns} refer to significant treatment effects in ANOVA at $P \leq 0.10$, $P \leq 0.01$, $P \leq 0.001$ and not significant, respectively.

Table 8. Linear regressions for relationships between crop residue and swine manure C input from 2006 to 2008 growing seasons and organic C or N (TOC, TON, LFOC, LFON) stored in the 0 - 15 cm soil sampled in autumn 2008 at Star City (Gray Luvisol) and Swift Current (Dark Brown Chernozem), Saskatchewan, Canada.

Soil	Crop parameter (X)	Soil C or N parameter (Y)	^{z}Linear regression (Y = a + bX)	R^2
Star City	Crop residue C input	TOC	$Y = 40.97 - 0.0002X$	0.038^{ns}
		TON	$Y = 3.721 - 0.00001X$	0.009^{ns}
		LFOC	$Y = 2420 + 0.079X$	0.170^{ns}
		LFON	$Y = 170.5 + 0.003X$	0.052^{ns}
Swift Current	Crop residue C input	TOC	$Y = 30.21 + 0.001X$	0.345^{ns}
		TON	$Y = 3.378 + 0.00009X$	0.149^{ns}
		LFOC	$Y = 848.9 + 0.489X$	0.794^{**}
		LFON	$Y = 41.41 + 0.035X$	0.847^{**}

^{z}Y = Soil organic C or N fraction (TOC and TON as Mg C or N·ha^{-1}; and LFOC, LFON as kg C or N·ha^{-1}; a = Intercept on Y, origin of the line; b = Regression coefficient of Y on X, slope of line; X = Crop residue and/or swine manure C input (Mg·ha^{-1}); ** and ^{ns} refer to significant treatment effects in ANOVA at $P \leq 0.01$ and not significant, respectively.

shown).

3.3. Amounts of N Uptake in Crop, Nitrate-N in Soil, N Balance Sheets, and Recovery of Applied N

The N balance over the 2006 to 2008 period for the 10 treatments included the amount of nitrate-N recovered in the 0 - 90 cm soil in autumn 2008 and in seed yield (which was removed from the land/field), and N applied as UAN or swine manure, plus N added in seed at seeding over 3 years, and the estimated amount of N balance and unaccounted N (**Tables 14** and **15**). At Star City, the estimated amounts of nitrate-N recovered in soil in autumn 2008 plus N recovered (removed) in seed in 3 years in various treatments ranged from 139 to 357 kg·N·ha^{-1}. The corresponding values of N applied as UAN fertilizer or manure plus N added in seed at seeding in 3 years ranged from 7 to 410 kg·N·ha^{-1}. The amounts of N that

Short-term influence of anaerobically-digested and conventional swine manure, and N fertilizer on organic C and N, and available nutrients in two contrasting soils

9

Table 9. Effect of land-applied anaerobically digested swine manure (ADSM), conventionally treated swine manure (CTSM) and urea-ammonium-nitrate (UAN) solution fertilizer over three years from 2006 to 2008 on the amount of residual nitrate-N in soil in autumn 2008 at Star City (Gray Luvisol), Saskatchewan, Canada.

Treatments	Amount of nitrate-N ($kg \cdot N \cdot ha^{-1}$) in soil layers (cm)					
	0 - 15	15 - 30	30 - 60	60 - 90	90 - 120	0 - 120
Control	6.7	1.2	2.4	4.2	6.2	20.7
ADSM-3x Autumn	5.2	1.4	6.6	14.1	16.6	43.9
ADSM-1x Autumn	5.4	1.7	3.6	7.9	8.2	26.8
CTSM-3x Autumn	5.6	2.0	8.5	37.1	33.8	87.0
CTSM-1x Autumn	3.8	1.5	3.9	7.3	9.0	25.5
ADSM-3x Spring	7.2	1.6	4.8	11.8	13.2	38.6
ADSM-1x Spring	4.9	0.8	1.9	4.5	6.3	18.4
CTSM-3x Spring	5.8	1.3	7.5	25.8	21.2	61.6
CTSM-1x Spring	3.9	1.0	3.6	7.1	8.1	23.7
UAN Spring	4.8	1.1	4.3	7.3	8.3	25.8
$LSD_{0.05}$	ns	0.7	3.9	10.6	7.3	18.8
SEM (Probability)	0.88^{ns}	0.23^{*}	1.35^{*}	3.67^{***}	2.52^{***}	6.49^{***}
Manure rate						
1x	4.5	1.2	3.2	6.7	7.9	23.5
3x	5.9	1.6	6.9	22.2	21.2	57.8
$LSD_{0.05}$	1.3	0.4	2.0	6.8	4.7	12.1
SEM (Probability)	0.43^{*}	0.13^{*}	0.69^{**}	2.34^{***}	1.63^{***}	4.16^{***}
Manure type						
ADSM	5.7	1.4	4.2	9.6	11.1	32.0
CTSM	4.8	1.4	5.9	19.3	18.0	49.4
$LSD_{0.05}$	1.3	ns	2.0	6.8	4.7	12.1
SEM (Probability)	0.43^{*}	0.13^{ns}	0.69^{*}	2.34^{**}	1.63^{**}	4.16^{**}
Manure application time						
Autumn	5.0	1.6	5.6	16.6	16.9	45.7
Spring	5.4	1.2	4.5	12.3	12.2	35.6
$LSD_{0.05}$	ns	0.4	ns	6.8	4.7	12.1
SEM (Probability)	0.43^{ns}	0.13^{*}	0.69^{ns}	2.34^{ns}	1.63^{*}	4.16^{*}

*, **, *** and ^{ns} refer to significant treatment effects in ANOVA at $P \leq 0.10$, $P \leq 0.05$, $P \leq 0.01$, $P \leq 0.001$ and not significant, respectively.

could not be accounted for ranged from −132 to 57 $kg \cdot N \cdot ha^{-1}$. The amounts of unaccounted N from N applied/added ranged from 91 to 192 $kg \cdot N \cdot ha^{-1}$. At Swift Current, the estimated amounts of nitrate-N recovered in soil in autumn 2008 plus N recovered (removed) in seed in 3 years in various treatments ranged from 170 to 399 $kg \cdot N \cdot ha^{-1}$. The corresponding values of N applied as UAN fertilizer or manure plus N added in seed at seed-

ing in 3 years ranged from 7 to 410 $kg \cdot N \cdot ha^{-1}$. The amounts of N that could not be accounted for ranged from −163 to 101 $kg \cdot N \cdot ha^{-1}$. The amounts of unaccounted N from N applied/added ranged from 35 to 207 $kg \cdot N \cdot ha^{-1}$. The percent recovery of applied N over 3 years ranged from 37.7% to 50.0% in seed and from 50.7% to 65.0% in seed + straw at Star City, and from 0.3% to 7.0% in seed and from 27.5 to 54.7 in seed + straw at

Table 10. Effect of land-applied anaerobically digested swine manure (ADSM), conventionally treated swine manure (CTSM) and urea-ammonium-nitrate (UAN) solution fertilizer over three years from 2006 to 2008 on the amount of residual nitrate-N in soil in autumn 2008 at Swift Current (Dark Brown Chernozem), Saskatchewan, Canada.

Treatments	Amount of nitrate-N (kg·N·ha^{-1}) in soil layers (cm)					
	0 - 15	15 - 30	30 - 60	60 - 90	90 - 120	0 - 120
Control	5.9	2.0	2.9	12.8	37.5	61.1
ADSM-3x Autumn	10.3	7.8	29.9	85.9	46.6	180.5
ADSM-1x Autumn	14.8	9.8	53.6	43.1	43.0	164.3
CTSM-3x Autumn	11.4	6.6	63.8	79.1	30.0	190.9
CTSM-1x Autumn	13.8	13.7	121.7	61.4	41.1	251.7
ADSM-3x Spring	12.5	8.5	45.8	122.1	83.5	272.4
ADSM-1x Spring	11.8	10.1	50.1	41.6	22.8	136.4
CTSM-3x Spring	13.5	5.7	69.2	89.5	46.0	223.9
CTSM-1x Spring	17.8	16.7	53.9	48.0	43.2	179.6
UAN Spring	10.0	7.1	22.5	68.0	63.8	171.4
LSD$_{0.05}$	5.7	7.4	56.1	45.5	31.2	91.4
SEM (Probability)	1.98[*]	2.56[*]	19.33[*]	15.69[**]	10.77[*]	31.51[**]
Manure rate						
1x	14.6	12.6	69.8	48.5	37.5	183.0
3x	11.9	7.1	52.2	94.1	51.5	216.8
LSD$_{0.05}$	3.0	4.1	ns	24.2	18.0	ns
SEM (Probability)	1.03[*]	1.41[*]	11.02[ns]	8.30[***]	6.19[*]	18.69[ns]
Manure type						
ADSM	12.3	9.0	44.9	73.2	49.0	188.4
CTSM	14.1	10.7	77.2	69.5	40.1	211.6
LSD$_{0.05}$	ns	ns	32.1	ns	ns	ns
SEM (Probability)	1.03[ns]	1.41[ns]	11.02[*]	8.30[ns]	6.19[ns]	18.69[ns]
Manure application time						
Autumn	12.6	9.4	67.3	67.4	40.2	196.9
Spring	13.9	10.2	54.8	75.3	48.9	203.1
LSD$_{0.05}$	ns	ns	ns	ns	ns	ns
SEM (Probability)	1.03[ns]	1.41[ns]	11.02[ns]	8.30[ns]	6.19[ns]	18.69[ns]

[*], [*], [**], [***] and [ns] refer to significant treatment effects in ANOVA at $P \leq 0.10$, $P \leq 0.05$, $P \leq 0.01$, $P \leq 0.001$ and not significant, respectively.

Swift Current in the various swine manure or N fertilizer treatments (**Tables 14** and **15**). The recovery of applied N in seed or seed + straw for swine manure was usually greater in the 3x than the 1x rate and also greater with the ADSM than the CTSM treatments.

4. DISCUSSION

Research has shown potential for improvement in organic C and/or N storage in soil and/or increase in soil fertility level from the application of LSM [1,5,6] and N fertilization [17-19]. Previous research has also suggested that long-term application of LSM can increase N, P, S and K fertility of soil, due to the return of these nutrients in manure and in crop residue to soil over years [1]. Similarly, in our study, applications of LSM and N fertilizer increased organic C and N, and amounts of

Short-term influence of anaerobically-digested and conventional swine manure, and N fertilizer on organic C and N,
and available nutrients in two contrasting soils

11

Table 11. Effect of land-applied anaerobically digested swine manure (ADSM), conventionally treated swine manure (CTSM) and urea-ammonium-nitrate (UAN) solution fertilizer over three years from 2006 to 2008 on the amount of residual extractable P in soil in autumn 2008 at Star City (Gray Luvisol), Saskatchewan, Canada.

Treatments	Amount of extractable P ($kg \cdot N \cdot ha^{-1}$) in soil layers (cm)					
	0 - 15	15 - 30	30 - 60	60 - 90	90 - 120	0 - 120
Control	21.4	8.6	12.6	6.0	4.5	53.1
ADSM-3x Autumn	22.0	11.5	7.1	5.7	4.3	50.6
ADSM-1x Autumn	21.7	10.5	9.1	3.9	3.3	48.5
CTSM-3x Autumn	18.4	8.6	10.9	3.3	3.6	44.8
CTSM-1x Autumn	27.2	11.9	7.4	4.7	5.0	56.2
ADSM-3x Spring	18.4	9.1	10.3	5.5	4.2	47.5
ADSM-1x Spring	34.2	16.8	8.7	4.8	4.1	68.6
CTSM-3x Spring	39.3	11.2	12.7	6.4	4.2	73.8
CTSM-1x Spring	24.0	9.8	10.8	5.0	7.5	57.1
UAN Spring	21.8	10.0	9.4	4.8	5.2	51.2
$LSD_{0.05}$	ns	ns	ns	ns	ns	ns
SEM (Probability)	5.88[ns]	2.29[ns]	3.42[ns]	1.06[ns]	1.43[ns]	8.28[ns]
Manure rate						
1x	26.8	12.3	9.0	4.6	5.0	57.5
3x	24.5	10.1	10.3	5.2	4.1	54.2
$LSD_{0.05}$	ns	ns	ns	ns	ns	ns
SEM (Probability)	3.24[ns]	1.21[ns]	1.73[ns]	0.53[ns]	0.76[ns]	4.39[ns]
Manure type						
ADSM	24.1	12.0	8.8	5.0	4.0	53.9
CTSM	27.2	10.4	10.5	4.9	5.1	58.1
$LSD_{0.05}$	ns	ns	ns	ns	ns	ns
SEM (Probability)	3.24[ns]	1.21[ns]	1.73[ns]	0.53[ns]	0.76[ns]	4.39[ns]
Manure application time						
Autumn	22.3	10.6	8.6	4.4	4.1	50.0
Spring	29.0	11.7	10.6	5.4	5.0	61.7
$LSD_{0.05}$	ns	ns	ns	ns	ns	12.8
SEM (Probability)	3.24[ns]	1.21[ns]	1.73[ns]	0.53[ns]	0.76[ns]	4.39[*]

[*] and [ns] refer to significant treatment effects in ANOVA at $P \leq 0.10$ and not significant, respectively.

plant-available N, P or S in soil in many cases, depending on soil type/site. The following sections discuss the short-term effects of LSM and N fertilization on soil biochemical and chemical properties.

4.1. Soil Biochemical Properties

Earlier research has shown positive effects of swine manure or N fertilizer application on crop yield, and soil organic matter and fertility [1,5,6]. Similarly, we found increase in TOC and TON from swine manure application due to its dual effect by directly contributing to organic C and N, plus additional indirect contribution of C from increased crop residue (roots, stubble, straw, chaff/fallen leaves) returned to the land/soil, as evidenced by greatest increase in straw yield in this treatment [7]. Inorganic fertilizers supply specific nutrients, but do not

Table 12. Effect of land-applied anaerobically digested swine manure (ADSM), conventionally treated swine manure (CTSM) and urea-ammonium-nitrate (UAN) solution fertilizer over three years from 2006 to 2008 on the amount of residual extractable P in soil in autumn 2008 at Swift Current (Dark Brown Chernozem), Saskatchewan, Canada.

Treatments	Amount of extractable P (kg$\cdot$P$\cdot$ha^{-1}) in soil layers (cm)					
	0 - 15	15 - 30	30 - 60	60 - 90	90 - 120	0 - 120
Control	53.0	8.4	7.7	7.3	13.4	89.7
ADSM-3x Autumn	55.0	9.1	8.7	8.2	13.9	94.9
ADSM-1x Autumn	51.6	8.4	11.0	8.0	19.6	98.6
CTSM-3x Autumn	48.4	8.1	6.9	7.3	25.9	96.6
CTSM-1x Autumn	56.8	10.6	7.1	9.1	24.7	108.3
ADSM-3x Spring	48.1	11.0	11.6	5.4	16.6	92.7
ADSM-1x Spring	52.4	8.4	8.5	4.6	16.7	90.6
CTSM-3x Spring	66.6	13.7	8.1	6.1	15.7	110.2
CTSM-1x Spring	59.6	8.6	7.2	6.2	18.4	100.0
UAN Spring	54.3	10.4	9.1	5.3	14.6	93.7
LSD$_{0.05}$	ns	ns	ns	ns	ns	ns
SEM (Probability)	7.25^{ns}	2.01^{ns}	1.68^{ns}	1.80^{ns}	3.33^{ns}	8.64^{ns}
Manure rate						
1x	55.1	9.0	8.5	7.0	19.8	99.4
3x	54.5	10.5	8.8	6.8	18.0	98.6
LSD$_{0.05}$	ns	ns	ns	ns	ns	ns
SEM (Probability)	3.71^{ns}	1.02^{ns}	0.83^{ns}	0.87^{ns}	1.77^{ns}	4.06^{ns}
Manure type						
ADSM	51.8	9.2	10.0	6.5	16.7	94.2
CTSM	57.9	10.3	7.4	7.2	21.2	104.0
LSD$_{0.05}$	ns	ns	2.4	ns	5.2	11.8
SEM (Probability)	3.71^{ns}	1.02^{ns}	0.83*	0.87^{ns}	1.77$^{.}$	4.06$^{.}$
Manure application time						
Autumn	53.0	9.1	8.4	8.2	21.0	99.7
Spring	56.7	10.5	8.9	5.6	16.8	98.5
LSD$_{0.05}$	ns	ns	ns	2.5	5.2	ns
SEM (Probability)	3.71^{ns}	1.02^{ns}	0.83^{ns}	0.87*	1.77$^{.}$	4.06^{ns}

$^{.}$, * and ^{ns} refer to significant treatment effects in ANOVA at $P \leq 0.10$, $P \leq 0.05$ and not significant, respectively.

contribute directly to soil organic matter, and thus may result in much less contribution to soil organic C and N. However, in our study, there was relatively greater storage of organic C and N from N fertilizer application than swine manure, at least at Star City site. The smaller storage of TOC or TON from swine manure or UAN fertilizer applications at Star City than Swift Current was probably due to the differences in soil type (Gray Luvisol loam soil at Star City versus Brown Chernozem silt loam at Swift Current) and climatic conditions (relatively moister soils at Star City than Swift Current) at the two sites, resulting in greater turn over of organic matter at Star City compared to Swift Current.

In our study, the changes in LFOC and LFON due to LSM application and N fertilization were more pronounced than TOC and TON in both soils. For example,

Short-term influence of anaerobically-digested and conventional swine manure, and N fertilizer on organic C and N, and available nutrients in two contrasting soils

13

Table 13. Effect of land-applied anaerobically digested swine manure (ADSM), conventionally treated swine manure (CTSM) and urea-ammonium-nitrate (UAN) solution fertilizer over three years from 2006 to 2008 on the amount of residual sulphate-S in soil in autumn 2008 at Star City (Gray Luvisol), Saskatchewan, Canada.

Treatments	Amount of sulphate-S ($kg \cdot S \cdot ha^{-1}$) in soil layers (cm)					
	0 - 15	15 - 30	30 - 60	60 - 90	90 - 120	0 - 120
Control	27.9	8.8	12.7	11.2	13.8	74.4
ADSM-3x Autumn	23.3	9.7	20.5	44.5	45.1	143.1
ADSM-1x Autumn	27.8	10.4	13.2	19.0	30.1	100.5
CTSM-3x Autumn	18.9	10.1	13.4	10.7	9.9	63.0
CTSM-1x Autumn	22.7	11.1	16.5	38.2	42.6	131.1
ADSM-3x Spring	30.2	10.1	11.6	12.1	11.2	75.2
ADSM-1x Spring	25.8	9.1	10.4	9.5	8.8	63.6
CTSM-3x Spring	18.1	9.8	13.2	9.5	8.6	59.4
CTSM-1x Spring	17.0	8.1	12.7	13.9	22.2	73.9
UAN Spring	18.6	10.3	12.0	12.4	13.7	67.0
$LSD_{0.05}$	ns	ns	ns	ns	ns	ns
SEM (Probability)	6.48^{ns}	0.99^{ns}	4.24^{ns}	12.68^{ns}	11.77^{ns}	28.29^{ns}
Manure rate						
1x	23.3	9.7	13.2	20.1	25.9	92.2
3x	22.7	9.9	14.7	19.2	18.8	85.3
$LSD_{0.05}$	ns	ns	ns	ns	ns	ns
SEM (Probability)	3.15^{ns}	0.51^{ns}	2.23^{ns}	7.04^{ns}	6.62^{ns}	15.42^{ns}
Manure type						
ADSM	26.8	9.8	13.9	21.3	23.8	95.6
CTSM	19.2	9.8	13.9	18.1	20.9	81.9
$LSD_{0.05}$	9.2	ns	ns	ns	ns	ns
SEM (Probability)	3.15^{*}	0.51^{ns}	2.23^{ns}	7.04^{ns}	6.62^{ns}	15.42^{ns}
Manure application time						
Autumn	23.2	10.3	15.9	28.1	31.9	109.4
Spring	22.8	9.3	12.0	11.2	12.8	68.1
$LSD_{0.05}$	ns	ns	ns	20.5	19.3	44.9
SEM (Probability)	3.15^{ns}	0.51^{ns}	2.23^{ns}	7.04^{*}	6.62^{*}	15.42^{*}

$\cdot$, * and ^{ns} refer to significant treatment effects in ANOVA at $P \leq 0.10$, $P \leq 0.05$ and not significant, respectively.

in the 0 - 15 cm soil layer after 3 years, and compared to the zero-N control treatment, the manure and N fertilizer treatments, respectively, increased TOC by 10.1% and 3.2%, TON by 9.5% and 2.3%, LFOC by 22.3% and 16.0% and LFON by 20.6% and 12.9% at Star City. The corresponding increases at Swift Current were 7.1% and 7.8% for TOC, 1.4% and 4.9% for TON, 34.4% and 41.5% for LFOC and 34.6% and 41.5% for LFON, re-

spectively. Other researchers have also observed greater responses of LFOC and LFON to N fertilization and other management practices than TOC and TON [18-20]. Our findings confirm that the changes in LFOC and LFON can be considered good indicators of changes of organic C and N in soil as a result of manure addition or appropriate fertilization. This also suggests that monitoring the changes in LFON and LFOC in the surface soil

Table 14. Balance sheets of land-applied anaerobically digested swine manure (ADSM), conventionally treated swine manure (CTSM) and urea-ammonium-nitrate (UAN) solution fertilizer over three years from 2006 to 2008 at Star City (Gray Luvisol), Saskatchewan, Canada.

Parameters	Control	UAN spring	Fall application				Spring application			
			ADSM-3x	ADSM-1x	CTSM-3x	CTSM-1x	ADSM-3x	ADSM-1x	CTSM-3x	CTSM-1x
Nitrate-N recovered in soil (0 - 90 cm) after 3 years in fall 2008 (kg·N·ha^{-1})	21	26	44	27	87	26	39	18	62	24
N recovered in seed in 3 years kg·N·ha^{-1})	118	199	225	204	270	284	235	242	253	262
N recovered in soil after 3 years + N recovered in seed in 3 years (kg·N·ha^{-1})	139	225	269	231	357	310	274	260	315	286
Total N applied in UAN or in SM in 3 years (kg·N·ha^{-1})	0	180	214	205	403	360	257	255	343	326
Organic N added in seed in 3 years (kg·N·ha^{-1})	7	7	7	7	7	7	7	7	7	7
Total N added in UAN + SM + seed in 3 years (kg·N·ha^{-1})	7	187	221	212	410	367	264	262	350	333
N balance (N applied in UAN/SM/seed − N recovered in seed) (kg·N·ha^{-1})	−111	−12	−4	8	140	83	29	20	97	71
Unaccounted N (N applied in UAN/SM/seed − N recovered in soil + seed) (kg·N·ha^{-1})	−132	−38	−48	−19	53	57	−10	2	35	47
N recovered in seed in 3 years from applied N (kg·N·ha^{-1})		81	107	86	152	166	117	124	135	144
N recovered in soil after years + seed in 3 years from applied N (kg·N·ha^{-1})		86	130	92	218	171	134	121	176	147
N balance (N applied in UAN/SM/seed − N recovered in seed from applied N) (kg·N·ha^{-1})		106	114	126	258	201	147	138	215	189
Unaccounted N (N applied in UAN/SM/seed − N recovered in soil + seed from applied N) (kg·N·ha^{-1})		101	91	120	192	196	130	141	174	186
Recovery of applied N in seed over 3 years (%)		45.0	50.0	42.0	37.7	46.1	45.5	48.6	39.4	44.2
N recovered in seed + straw in 3 years (kg·N·ha^{-1})	159	264	298	263	365	371	313	311	346	341
Recovery of applied N in seed + straw over 3 years (%)		58.3	65.0	50.7	51.1	58.9	59.9	59.6	54.5	55.8

could be a good strategy to determine the potential for N supplying power, and improvement in soil quality/health. The trends of higher organic C and N in light organic fractions than total organic fractions in the manure and N fertilizer treatments were most likely associated with greater inputs of C and N to soil through manure, and also straw, chaff [17] and roots [21,22].

The relative greater increases in C or N for LFOC or LFON than TOC or TON in our study are in agreement with other research, where light organic fraction was also more responsive to management practices than total organic fraction [18-20]. Unlike TOC and TON, there was a greater build-up of light fraction organic C or N at Swift Current than at Star City, in spite of greater input

of C from crop residues plus LSM in 3 years at Star City than Swift Current. We do not have any real explanation for this unusual trend for the greater build-up of light organic fraction under relatively warmer temperature conditions at Swift Current than Star City, but this may be possibly due to relatively drier conditions which may have resulted in relatively slower decomposition of freshly added crop residues at swift Current than Star City.

Earlier long-term research studies have shown strong and highly significant correlations among TOC, TON, LFOC and LFON fractions in soil due to management practices [18-20]. However, in our study, the strong positive correlations were found only between TOC and TON,

Short-term influence of anaerobically-digested and conventional swine manure, and N fertilizer on organic C and N, and available nutrients in two contrasting soils

15

Table 15. Balance sheets of land-applied anaerobically digested swine manure (ADSM), conventionally treated swine manure (CTSM) and urea-ammonium-nitrate (UAN) solution fertilizer over three years from 2006 to 2008 at Swift Current (Dark Brown Chernozem), Saskatchewan, Canada.

Parameters	Control	UAN spring	Fall application				Spring application			
			ADSM-3x	ADSM-1x	CTSM-3x	CTSM-1x	ADSM-3x	ADSM-1x	CTSM-3x	CTSM-1x
Nitrate-N recovered in soil (0 - 90 cm) after 3 years in autumn 2008 (kg·N·ha^{-1})	61	171	181	164	191	252	272	136	224	180
N recovered in seed in 3 years (kg·N·ha^{-1})	109	110	123	114	118	110	127	112	132	116
N recovered in soil after 3 years + N recovered in seed in 3 years kg·N·ha^{-1})	170	281	304	278	309	362	399	248	356	296
Total N applied in UAN or in SM in 3 years (kg·N·ha^{-1})	0	180	214	205	403	360	257	255	343	326
Organic N added in seed in 3 years (kg·N·ha^{-1})	7	7	7	7	7	7	7	7	7	7
Total N added in UAN + SM + seed in 3 years (kg·N·ha^{-1})	7	187	221	212	410	367	264	262	350	333
N balance (N applied in UAN/SM/seed – N recovered in seed) (kg·N·ha^{-1})	−102	77	98	98	292	257	137	150	218	217
Unaccounted N (N applied in UAN/SM/seed – N recovered in soil + seed) (kg·N·ha^{-1})	−163	−94	−83	−66	101	5	−135	14	−6	37
N recovered in seed in 3 years from applied N kg·N·ha^{-1})		1	14	5	9	1	18	3	23	7
N recovered in soil after years + seed in 3 years from applied N (kg·N·ha^{-1})		111	134	108	139	192	229	78	186	126
N balance (N applied in UAN/SM/ seed – N recovered in seed from applied N) (kg·N·ha^{-1})		186	207	207	401	366	249	259	327	326
Unaccounted N (N applied in UAN/SM/seed – N recovered in soil + seed from applied N) (kg·N·ha^{-1})		76	87	104	271	175	35	184	164	207
Recovery of applied N in seed over 3 years (%)		0.6	6.5	2.4	2.2	0.3	7.0	1.2	6.7	2.1
N recovered in seed + straw in 3 years (kg·N·ha^{-1})	162	242	279	240	297	261	290	233	299	259
Recovery of applied N in seed + straw over 3 years (%)		44.4	54.7	38.0	35.5	27.5	49.8	27.8	39.9	29.8

and between LFOC and LFON in both soils. Previous long-term studies have shown positive relationships between the input of increased amounts of manure and/or crop residue C or N and TOC, TON, LFOC or LFON, especially in the labile/light organic fractions [18-20, 23,24]. However, in our study after 3 years, the significant linear regressions between the amounts of C or N input and mass of organic C or N in the 0 - 15 cm soil layer in various organic fractions were found only for LFOC and LFON and only at Swift Current. This lack of significant relationships between C or N input and mass of organic C or N stored in soil was probably due to short duration of our study.

4.2. Soil Chemical Properties and Distribution of Available N, P, K and S in the Soil Profile

Slow acidification of soil from N fertilization has been earlier reported after long-term annual applications of moderate rates of N fertilizer to annual crops in North America [25-27]. However, in our study, there was no effect of manure or N fertilization on soil pH, and this was probably due to the shorter duration of our present study. In a study in Quebec, Canada, Ndayegamiye and Cote [5] also found no effect of pig slurry application on soil pH even after 10 annual applications.

There was no build-up of residual ammonium-N in

soil after three annual applications of swine manure or N fertilizer, no doubt due to the rapid nitrification of any ammonium-N released during mineralization of organic matter. The amount of residual nitrate-N in soil increased with increasing rate of swine manure in the 30 - 60, 60 - 90 and 90 - 120 cm layers in the 0 - 120 cm soil profile, particularly at Swift Current. This suggests potential risk of nitrate leaching below the root zone, even within the short duration of our study (only three years), as other long-term studies in China have shown a great potential of underground water contamination with nitrate-N from annual applications of farmyard manure (FYM) at relatively high rates [28-30]. Our findings also suggest the need for deep soil sampling, as soils in our study were sampled only to the 120 cm depth. In our study at Star City, there was a significant increase in nitrate-N in the soil profile with 3x LSM rate while there was only little increase in residual nitrate-N in the 0 - 120 cm soil profile due to fertilizer N application. The rate of fertilizer-N applied in our study was below the rate needed for optimum yield in this soil-climatic region [31], and the amount of N removed in the grain closely matched the amount of fertilizer-N added. This would have minimized the amount of surplus N available for leaching or other losses. However, a portion of the applied N may have been immobilized into the soil organic N pool, especially when straw was retained [20]. It is also possible that a portion of the residual soil nitrate-N may have been lost as gaseous N over the winter and especially in early spring after snow melting [32,33]. It is unlikely that much of the applied N at Star City leached below the 120 cm depth, as evidenced by little residual nitrate-N recovered in the 30 - 60, 60 - 90 and 90 - 120 cm soil layers in autumn 2008 sampling

At Swift Current, the amounts of fertilizer and manure N applied exceeded the amounts of N removed in the grain, and based on the moderate amounts of residual nitrate-N recovered in the 30 - 60, 60 - 90 and 90 - 120 cm soil layers in autumn 2008 at Swift Current it may well that a portion of the applied N had leached below the 120 cm depth, particularly at the high rate of manure. Previous research in Saskatchewan where soil samples were taken to 240 cm depth after 12 growing seasons, Malhi *et al.* [34] observed large amounts of residual nitrate-N accumulation in the 210 - 240 cm layer for treatments where N applications had exceeded N removals. It should be noted that at Swift Current, crops were drought stressed during grain filling during both 2006 and 2007 and final grain yields were greatly depressed, while in 2008 the study suffered severe hail damage prior to grain filling. Minimal grain N uptake at Swift Current in all three years no doubt influenced the amount of nitrate N accumulating in the soil profile. Regardless, the results also emphasizes the need for deep soil sampling (maybe up to 3 or 4 m depth) in future research in order to make valid conclusions related to nitrate leaching losses in the soil profile.

Earlier research in China has shown substantial increase in extractable P and total P in soil with long-term annual applications of FYM [35]. In our study, there was a tendency towards increased extractable P in the surface 0 - 5 cm soil with swine manure in some treatments even after three annual applications, probably due to fairly high concentration of P in swine manure. The increase of extractable P with swine manure only in the 0 - 15 cm soil layer suggests that P is relatively immobile, but the slow build-up of P in the surface soil, especially after repeated applications to increase crop production, may subsequently increase the potential risk of contamination of surface waters with P from surface run-off of water after snow melt in early spring and/or after heavy rainfall events which often occur in this region during summer.

Sulphate-S in soil tended to increase with swine manure at Star City. This suggests that swine manure either contained sulphate-S or possibly increased sulphate-S through mineralization of organic matter. Sulphate-S increased with increasing rate of swine manure. It is possible that a portion of the sulphate-S may have leached below the 120 cm depth, as evidenced by large amounts of sulphate-S in the 30 - 60, 60 - 90 and 90 - 120 cm layers, although no soil samples were obtained below 120 cm to verify this in our study. This suggests the need for future soil sampling to greater depths in order to make valid conclusions related to sulphate-S leaching. Earlier research in Saskatchewan has suggested that long-term application of LSM can increase K fertility of soil, due to the return of these nutrients in manure and in crop residue to soil over years [1]. However, in our present study, there was no increase in extractable K in soil from LSM or UAN application over three years at both sites.

4.3. Amounts of N Uptake in Crop, Nitrate-N in Soil, N Balance Sheets, and Recovery of Applied N

The amounts of unaccounted N increased with application of swine manure or N fertilizer compared to zero-N control. This unaccounted N reflects a portion of the applied N which did not become available to the crop, and may have been lost from the soil mineral N pool and/or from the soil-plant system. At Star City, it is unlikely that a portion of the applied N was leached down below 120 cm soil depth, because there was little nitrate-N recovered in the deeper soil layers in autumn 2008. At Swift Current, it is possible that a portion of the applied N may have leached down below 120 cm soil depth, because there were large amounts of nitrate-N recovered in the 30 - 60, 60 - 90 and 90 - 120 cm soil

Short-term influence of anaerobically-digested and conventional swine manure, and N fertilizer on organic C and N, and available nutrients in two contrasting soils

17

layers in autumn 2008 in many cases for swine manure and UAN treatments. Other researchers have reported an increase in the concentration of residual nitrate-N in the soil profile at high N fertilizer rates [36-39], and any soil nitrate-N below the effective root zone of crops is susceptible to leaching, The loss of nitrate-N through leaching can result in N contamination of groundwater, and thus represents a potential risk to groundwater quality and soil health [40]. Our N balance results suggest that a portion of the applied N in the N treatments may have been immobilized in soil organic N, as evidenced by higher amount of soil organic N, especially in LFON even after 3 years in autumn 2008 (**Tables 3-6**). At Star City, the amount of applied N recovered in LFON in soil ranged from -155 to 290 kg·N·ha^{-1} in various swine manure and UAN treatments. The corresponding values for the amounts of applied N recovered in total organic N in soil at Swift Current ranged from 36 to 529 kg·N·ha^{-1}. In addition, it is possible that a portion of the applied N may have been lost from the soil-plant system through denitrification (e.g., nitrous oxide and other N gases) due to wet surface soil conditions which temporarily exist in the present study area in most years in early spring after snow melt, or after occasional heavy rainfalls during summer and/or autumn [32,33,41]. It is also possible that a small portion of the applied N may have leached below the 120 cm soil depth profile, as suggested by Malhi *et al.* [34] who found large amounts of nitrate-N accumulation in the 120 to 240 cm soil profile in a long-term study in Saskatchewan with high input of N fertilizer and low crop intensity. This suggests the need for deep soil sampling below the 120 cm depth in future in our present long-term experiments.

Overall, the amount of residual soil nitrate-N recovered in the 0 - 120 cm soil profile was relatively small in the Gray Luvisol soil at Star City. This indicates low accumulation of nitrate-N in the soil profile. However, large amounts of unaccounted N from applied N suggest a great potential for gaseous N loss, especially in early spring after snow thawing when the surface soil is very wet (conducive to denitrification), and N immobilization, and possibility of some nitrate-N leaching below the 120 cm depth soil profile in the Brown Chernozem soil at Swift Current. However, as noted previously, grain N uptake was limited due to environmental conditions which no doubt influenced the amount of nitrate N accumulating in the soil profile. There were large amounts of N balance and unaccounted N in the zero-N treatments and also in the swine manure or N fertilizer treatments, especially at Swift Current. The implication of these large negative values for N balance and unaccounted N in the zero-N treatments is that large amounts of N became available to the crops in the growing seasons through mineralization of soil organic matter. However,

the large negative values for N balance and unaccounted N in the N fertilizer treatments at Swift Current suggest that the soil at this site may be gaining some N by wet/dry deposition through precipitation (rain/snow) and possibly by non-symbiotic N fixation. The Swift Current site is not close to any large city or industry, we don't know if soil at this site gained any N deposited through dry (snow) and wet (rainfall) precipitation. This supports the need for future research to obtain information on the contribution of N from rain/snow and non-symbiotic N fixation, or other outside sources, in order to optimize the use and accounting of N resources, and their effects on greenhouse gas (GHG) emissions to the atmosphere.

The percent recovery of applied N over 3 years ranged from 37.7% to 50.0% in seed and from 50.7% to 65.0% in seed + straw at Star City, and from 0.3% to 7.0% in seed and from 27.5 to 54.7 in seed + straw at Swift Current in various swine manure or N fertilizer treatments (**Tables 13** and **14**). The recovery of applied N in seed or seed + straw for swine manure was usually greater in the 3x rate than the 1x rate and also greater with the ADSM than the CTSM treatments. The greater recovery of applied N from swine manure in seed or seed + straw with the 3x rate than the 1x rate was possibly due to greater mineralization of any organic N because of much longer time of contact with soil microorganisms. The poor recovery of applied N in LSM or UAN fertilizer at Swift Current was most likely due to lack of crop response to these amendments, thus low input of organic C or N from crop residue which probably is the main/major source of C or N input to soil.

5. CONCLUSION

Our findings suggest that the quantity and quality of organic C and N in soil can be affected by swine manure rate and type, and N fertilization, most likely influencing inputs of C and N through crop residue, and improve soil quality.

6. ACKNOWLEDGEMENTS

We are grateful to Environmental Technologies Assessment for Agriculture (ETAA) program for funding, Prairie Agricultural Machinery Institute (PAMI) for excellent collaboration, Cudworth Pork Investors Group Ltd. and Clear-Green Environmental Inc. for providing access to raw and digested material, and D. Leach, D. Hahn and D. James for technical help.

REFERENCES

[1] Mooleki, S.P., Schoenau, J.J., Hultgreen, G., Wen, G. and Charles, J.L. (2002) Effect of rate, frequency and method of liquid hog manure application on soil nitrogen availability, crop performance and N use efficiency in east-

central Saskatchewan. *Canadian Journal of Soil Science*, **82**, 457-467.

[2] Mooleki, S.P., Schoenau, J.J., Hultgreen, G., Malhi, S.S. and Brandt, S. (2003) Soil and crop response to injected liquid hog manure on two Gray Luvisols. *Proceedings of Soils and Crops Workshop*, University of Saskatchewan, Saskatoon.

[3] Schoenau, J.J., Mooleki, S.P., Qian, P. and Malhi, S.S. (2003) Balancing availability of nutrients in manured soils. *Proceedings of Soils and Crops Workshop*, University of Saskatchewan, Saskatoon.

[4] Saskatchewan Agriculture and Food (SAF) (2003) Tri-Provincial manure application and use guidelines. Government of Saskatchewan, Regina Saskatchewan.

[5] Ndayegamiye, A. and Cote, D. (1989) Effect of long-term pig slurry and solid cattle manure application on soil chemical and biochemical properties. *Canadian Journal of Soil Science*, **69**, 39-47.

[6] Schoenau, J.J. and Assefa, B. (2004) Land application and handling of manure. In: M. Amrani, Ed., *Manure Research Findings and Technologies*, Alberta Agriculture, Food and Rural Development, Technical Press, Edmonton, 97-140.

[7] Lemke, R.L., Malhi, S.S., Selles, F. and Stumborg, M. (2012) Agronomic and greenhouse gas assessment of land-applied anaerobically-digested swine manure. *Agricultural Sciences*, In Review.

[8] Culley, J.L.B. (1993) Density and compressibility. In: Carter, M.R., Ed., *Soil Sampling and Methods of Analysis*, Lewis Publishers, Boca Raton, 529-549.

[9] Technicon Industrial Systems (1977) Industrial/simultaneous determination of nitrogen and/or phosphorus in BD acid digests. Industrial Method No. 334-74W/Bt, New York.

[10] Janzen, H.H., Campbell, C.A., Brandt, S.A., Lafond, G.P. and Townley-Smith, L. (1992) Light-fraction organic matter in soil from long term rotations. *Soil Science Society of America Journal*, **56**, 1799-1806.

[11] Izaurralde, R.C., Nyborg, M., Solberg, E.D., Janzen, H.H., Arshad, M.A., Malhi, S.S. and Molina-Ayala, M. (1997) Carbon storage in eroded soils after five years of reclamation techniques. In: Lal, R., Kimble, J.M., Follett, R.F. and Stewart, B.A., Eds., *Management of Carbon Sequestration in Soil*, Advances in Soil Science, CRC Press, Boca Raton, 369-385.

[12] Technicon Industrial Systems (1973) Ammonium in water and waste water. Industrial Method No. 90-70W-B. Revised January 1978. Technicon Industrial Systems, New York.

[13] Technicon Industrial Systems (1973) Nitrate in water and waste water. Industrial Method No. 100-70W-B. Revised January 1978. Technicon Industrial Systems, New York.

[14] Qian, P., Schoenau, J.J. and Karamanos, R.E. (1994) Simultaneous extraction of phosphorus and potassium with a new soil test: A modified Kelowna extraction. *Communications in Soil Science and Plant Analysis*, **25**, 627-635.

[15] Combs, S.M., Denning, J.L. and Frank, K.D. (1998) Sulfate-sulfur. Recommended Chemical Soil Test Procedures for the North Central Region. Missouri Agriculture Experiment Station Publication No. 221 (revised). Extension and Agricultural Information, I-98 Agricultural Building, University of Missouri, Columbia, 35-39.

[16] SAS Institute Inc. (2004) SAS product documentation. Version 8.

[17] Malhi, S.S. and Lemke, R. (2007) Tillage, crop residue and N fertilizer effects on crop yield, nutrient uptake, soil quality and greenhouse gas emissions in the second 4-yr rotation cycle. *Soil and Tillage Research*, **96**, 269-283.

[18] Malhi, S.S., Nyborg, M., Goddard, T. and Puurveen, D. (2011) Long-term tillage, straw management and N fertilization effects on quantity and quality of organic C and N in a Black Chernozem soil. *Nutrient Cycling in Agroecosystems*, **90**, 227-241.

[19] Malhi, S.S., Nyborg, M., Solberg, E.D., McConkey, B., Dyck, M. and Puurveen, D. (2011) Long-term straw management and N fertilizer rate effects on quantity and quality of organic C and N, and some chemical properties in two contrasting soils in western Canada. *Biology and Fertility of Soils*, **47**, 785-800.

[20] Malhi, S.S., Nyborg, M., Goddard, T. and Puurveen, D. (2010) Long-term tillage, straw and N rate effects on quantity and quality of organic C and N in a Gray Luvisol soil. *Nutrient Cycling in Agroecosystems*, **90**, 1-20.

[21] Lorenz, R.J. (1977) Changes in root weight and distribution in response to fertilization and harvest treatment of mixed prairie. In: Marshall, J.K., Ed., *The Belowground Ecosystem*, Range Science Department, Science Series, Colorado State University, Fort Collins, 63-71.

[22] Malhi, S.S. and Gill, K.S. (2002) Fertilizer N and P effects on root mass of bromegrass, alfalfa and barley. *Journal of Sustainable Agriculture*, **19**, 51-63.

[23] Bremer, E., Janzen, H.H. and Johnston, A.M. (1994) Sensitivity of total, light fraction and mineralizable organic matter to management practices in a Lethbridge soil. *Canadian Journal of Soil Science*, **74**, 131-138.

[24] Campbell, C.A., Janzen, H.H. and Juma, N.G. (1997) Case studies of soil quality in the Canadian Prairies: Long-term field experiments. In: Gregorih, E.G. and Carter, M.R., Eds., *Soil Quality for Crop Production and Ecosystem Health*, Chapter 17, Elsevier, New York, 351-397.

[25] Hoyt, P.B. and Hennig, A.M. (1982) Soil acidification by fertilizers and longevity of lime applications in the Peace River region. *Canadian Journal of Soil Science*, **62**, 155-163.

[26] Campbell, C.A. and Zentner, R.P. (1984) Effect of fertilizer on soil pH after seventeen years of continuous cropping in southwestern Saskatchewan. *Canadian Journal of Soil Science*, **64**, 750-710.

[27] Schwab, A.P., Owensby, C.E. and Kulyingyoung, S.

Short-term influence of anaerobically-digested and conventional swine manure, and N fertilizer on organic C and N, and available nutrients in two contrasting soils

19

(1990) Changes in soil chemical properties due to 40 years of fertilization. *Soil Science*, **149**, 35-43.

[28] Yuan, X., Tong, Y., Yang, X., Li, X. and Zhang, F. (2000) Effect of organic manure on soil nitrate accumulation. *Soil Environmental Science*, **9**, 197-200.

[29] Yang, S., Li, F., Malhi, S.S., Wang, P., Suo, D. and Wang, J. (2004) Long-term fertilization effects on crop yield and nitrate-N accumulation in soil in northwest China. *Agronomy Journal*, **96**, 1039-1049.

[30] Yang, S., Malhi, S.S., Song, J.R., Yue, W.Y., Wang, J.G. and Guo, T.W. (2006) Crop yield, N uptake and nitrate-N accumulation in soil as affected by 23 annual applications of fertilizers and manure on in the rainfd region of northwestern China. *Nutrient Cycling in Agroecosystems*, **76**, 81-94.

[31] Mooleki, S.P., Malhi, S.S., Lemke, R., Schoenau, J.J., Lafond, G., Brandt, S., Hultgreen, G., Wang, H. and May, W.E. (2010) Effect of nitrogen management, and phosphorus placement on wheat production in Saskatchewan. *Canadian Journal of Soil Science*, **90**, 319-337.

[32] Heaney, D.J., Nyborg, M., Solberg, E.D., Malhi, S.S. and Ashworth, J. (1992) Overwinter nitrate loss and denitrification potential of cultivated soils in Alberta. *Soil Biology and Biochemistry*, **24**, 877-884.

[33] Nyborg, M., Laidlaw, J.W., Solberg, E.D. and Malhi, S.S. (1997) Denitrification and nitrous oxide emissions from soil during spring thaw in a Malmo loam, Alberta. *Canadian Journal of Soil Science*, **77**, 53-160.

[34] Malhi, S.S., Brandt, S.A., Lemke, R., Moulin, A.P. and Zentner, R.P. (2009) Effects of input level and crop diversity on soil nitrate-N, extractable P, aggregation, organic

C and N, and N and P balance in the Canadian Prairie. *Nutrient Cycling in Agroecosystems*, **84**, 1-22.

[35] Yang, S., Malhi, S.S., Li, F., Suo, D., Jia, Y. and Wang, J. (2007) Long-term effects of manure and fertilization on soil organic matter and quality parameters of a calcareous soil in northwestern China. *Journal of Plant Nutrition and Soil Science*, **170**, 234-243.

[36] Benbi, D.K., Biswas, C.R. and Kalkat, J.S. (1991) Nitrate distribution and accumulation in an Ustochrept soil profile in a long-term fertilizer experiment. *Fertilizer Research*, **28**, 173-177.

[37] Guillard, K., Griffin, G.F., Allinson, D.W., Yamartino, W.R., Rafey, M.M. and Pietryzk, S.W. (1995) Nitrogen utilization of selected cropping systems in the US Northeast. II. Soil profile nitrate distribution and accumulation. *Agronomy Journal*, **87**, 199-207.

[38] Fan, J., Hao, M.D. and Shao, M.A. (2003) Nitrate accumulation in soil profile of dry land farming in Northwest China. *Pedosphere*, **13**, 367-374.

[39] Zhang, S.X., Li, X.Y., Li, X.P., Yuan, F.M., Yao, Z.H., Sun, Y.L. and Zhang, F.D. (2004) Crop yield, N uptake and nitrates in a fluvo-aquic soil profile. *Pedosphere*, **14**, 131-136.

[40] Zhang, W.L., Tian, Z.X., Zhang, N. and Li, X.O. (1996) Nitrate pollution of groundwater in northern China. *Agriculture, Ecosystem and Environment*, **59**, 223-231.

[41] Lemke, R.L., Izaurralde, R.C., Malhi, S.S., Arshad, M.A. and Nyborg, M. (1998) Nitrous oxide emissions from agricultural soils of the Boreal and Parkland regions of Alberta. *Soil Science Society of America Journal*, **62**, 1096-1102.

Yield and uptake of bahiagrass under flooded environment as affected by nitrogen fertilization

Gilbert C. Sigua[1*], Mimi M. Williams[2], Chad C. Chase Jr.[3], Joseph Albano[4], Manoch Kongchum[5]

[1]United States Department of Agriculture-Agricultural Research Service, Florence, SC, USA
[*]Corresponding Author
[2]United States Department of Agriculture-Natural Resources and Conservation Service, Brooksville, FL, USA
[3]United States Department of Agriculture-Agricultural Research Service, Clay Center, NE, USA
[4]United States Department of Agriculture-Agricultural Research Service, Fort Pierce, FL, USA
[5]School of Plant, Environmental, and Soil Sciences, Louisiana State University Agricultural Center, Baton Rouge, LA, USA

ABSTRACT

Bahiagrass (*Paspalum notatum*) is one of the most important forage grasses in subtropical region of USA and other tropical regions of the world. Although tolerant to short term flooding, bahiagrass is classified as a facultative upland (FACU+) species that suggest yield and plant persistence might be reduced under periods of extended waterlogging. The objective of this greenhouse study (2008-2009) was to determine the effect of nitrogen fertilization (0, 100, and 200 $kg \cdot N \cdot ha^{-1}$) on yield (DMY), crude protein content (CPC), and nitrogen uptake (NUP) of bahiagrass under varying flooded conditions (0, 14, 28, 56, and 84 days). Results disclosed an overwhelming effect of N application on yield and uptake component of bahiagrass. Averaged across flooding duration, results showed that DMY (R^2 = 0.91[**]), CPC (R^2 = 0.96[**]), and NUP (R^2 = 0.99[**]) were linearly related to increasing levels of N fertilization. Plants without N fertilization that were submerged between 14 to 84 days had significantly lower amount of DMY when compared with plants that were fertilized with 100 or 200 $kg \cdot N \cdot ha^{-1}$. Comparable DMY and NUP were obtained between plants fertilized with 200 $kg \cdot N \cdot ha^{-1}$ at 0 day of flooding ((11.7 ± 5.0) $ton \cdot ha^{-1}$) and plants fertilized with 200 $kg \cdot N \cdot ha^{-1}$ at 84 days of flooding ((9.8 ± 2.7) $ton \cdot ha^{-1}$). The practical implication of this study is that waterlogging may hamper yield and uptake while nitrogen fertilization could improve yield and uptake of bahiagrass under waterlogged condition.

Keywords: Bahiagrass; Flooding; Nitrogen Uptake; Dry Matter Yield; Nitrogen

1. INTRODUCTION

Flooding can have catastrophic impacts on the productivity of bahiagrass, as most crops including many forage species that are intolerant to excess water. Particularly little is known about the response of bahiagrass to the combined effect of waterlogging and the addition of nutrients (e.g., N and P). The use of N fertilizer prior to flooding may alleviate N deficiency because waterlogging causes a significant decrease in N content and rate of N accumulation in plants due to reduced root activity. Net assimilation rates and photosynthetic rates decline in plants experiencing root anaerobiosis, in part due to stomatal closure, and in part due to biochemical modifications [1,2]. Anaerobic conditions inhibit almost immediately the transport of nutrient ions by roots [3].

The cow-calf (*Bos taurus*) industry in subtropical United States especially in Gulf Coast states and other parts of the world depends almost totally on grazed pastures. Bahiagrass (*Paspalum notatum*) is one of the most important forage grasses grown in these regions of the world. Establishment of complete and uniform stand of bahiagrass in a short time period is important economically. Failure to obtain a good stand of bahiagrass means loss of not only the initial investment costs, but also production and its cash value [4]. In grasslands, waterlogging is frequently associated with other stresses, such as grazing, which may require specific and very different adaptive strategies and management [5,6]. Different plant species, including bahiagrass vary in their ability to withstand flooding and this undoubtedly contributes to their ecological distribution [7]. These adaptive strategies are not well understood and may still warrant extended investigations.

Relatively few studies and reports are available on the ability of N fertilizer to counteract the deleterious effects of waterlogging on terrestrial plants [8,9]. If additional fertilizer applications can improve the tolerance of forage grasses to waterlogging, then market value of environmental service of water storage will need to be adjusted to include additional fertilizer costs. Our hypothesis in this study is that bahiagrass will be negatively affected by extended flooding and N application could offset the potential negative effect of flooding on its yield, crude protein content, and nutrient uptake. The objective of this greenhouse study (2008-2009) was to determine the effect of nitrogen fertilization (0, 100, and 200 kg·N·ha^{-1}) on yield, crude protein content, and nitrogen uptake of bahiagrass under varying flooded conditions (0, 14, 28, 56, and 84 days).

2. MATERIALS AND METHODS

2.1. Preparation of Soils Media and Plant Materials

Soils needed for this two-year study (2008-2009) were obtained using a back-hoe from a Blichton soil series (loamy, siliceous, hyperthermic, Arenic Paleaudults) at a depth of 0 - 20 cm from a pasture at the USDA-ARS Subtropical Agriculture Research Station (STARS), Brooksville, FL in 2008 and 2009. The soil was air dried outside on an impervious surface at USDA-NRCS Plant Material Center (PMC), Brooksville, FL. Prior to drying, eight to 10 random samples were collected to determine some selected soil physical and initial soil chemical properties (**Table 1**). Blichton soils are poorly drained soils with water table at a depth of less than 25 cm for cumulative periods of 1 to 4 months during most years. These soils, like many of the soils in south Florida, have argillic and spodic horizons.

The introduced species of bahiagrass (cv. Tifton-9) was used in the study. All the plant materials were excavated from established stands of bahiagrass at STARS pasture. Approximately 15×10 cm plugs consisting of crowns, rhizomes, and roots (trimmed off to approx. 10-cm stubble height) were transplanted from the field 12 weeks prior to the initiation of the study each year. The plugs were planted into 15×60 cm planting columns that have been filled to within approximately 15 cm of the surface with the air dried, screened (1×1 cm screen) soils from the A horizon of Blichton series. The planting columns were sealed at the bottom to control water movement through the column. A hole was drilled at the side of the column and fitted with a drain tube with a stopcock to allow draining and sampling of the soil solution.

2.2. Experimental Treatments (Greenhouse Condition)

Immediately after planting, the columns with open drain tubes were moved into the greenhouse (22°/32°C,

Table 1. Selected physical and chemical properties of soils used in the study.

Soil Properties	Unit	Value
Particle Size		
Sand	g·kg^{-1}	868
Silt	g·kg^{-1}	75
Clay	g·kg^{-1}	57
Hydraulic Conductivity	cm·hr^{-1}	7.0
Bulk Density	kg·m^{-3}	1.4
Cation Exchange Capacity (CEC)	meq·100·g^{-1}	11.2
Ca	mg·kg^{-1}	2.9
Mg	mg·kg^{-1}	0.5
Na	mg·kg^{-1}	0.1
K	mg·kg^{-1}	0.1
pH		5.5

69% direct light) at the USDA-NRCS Plant Material Center greenhouse and allowed to recover and grow during the 12 wk pre-trial period. During the first four weeks of the pre-trial period, the plants were fertilized with a soluble complete fertilizer equivalent to 23 kg total of N, P, and K, and then fertilization discontinued. During the remainder of the adjustment periods (12 weeks), the plants were watered as needed to maintain the approximate soil field capacity.

Treatments were replicated five times using a 5×3 split plot arrangement in completely randomized block design. Flooding duration was the main treatment effect while N fertilization was the sub-plot feature. All columns received 40 kg·ha^{-1} of p as triple super phosphate granular fertilizer and appropriate N fertility treatments consisted of 0, 100, and 200 kg·ha^{-1} of N as NH_4NO_3 were applied to the appropriate columns. The soil flooding duration was consisted of 0-, 14-, 28-, 56-, and 84-days to mimic flooding occurrences in south Florida that may be associated with the need to temporarily store rainfall on pastureland in the summer. Flooding treatments were staggered such that termination of all flooding duration times coincided with the maximum flooding time of 84 days. For plants not receiving flood treatment, soil moisture was maintained at soil field capacity limit. Until a flooding treatment was started, all drain tubes remained open and the treatments were watered.

2.3. Plant Yield and Tissue Analysis

All treatments were destructively sampled at the end of maximum flooding time treatment of 84 days. Freshly cut aboveground growth was oven-dried at 60°C for 24 hours at the USDA-ARS Laboratory in Brooksville, FL. Plant samples were ground to pass through a 1-mm mesh screen in a Wiley mill. Ground forage was analyzed for total Kjeldahl nitrogen concentration [10] at the University of Florida Analytical Research Laboratory, Gainesville, FL. Crude protein content (%) and total nitrogen uptake ($kg \cdot N \cdot ha^{-1}$) were calculated using the equations below:

$$\text{Crude Protein Content} = \text{Nitrogen Concentration} \times 6.25 \quad (1)$$

$$\text{Total Nitrogen Uptake} = \text{DMY} \times \text{Nitrogen Concentration} \quad (2)$$

2.4. Redox Potential Measurements

Reduction-oxidation (Redox) potential of the soil solution was determined in 2008 (June 22 to September 4) and 2009 (June 23 to September 5) to monitor soil oxygen condition. Soil redox potential measurements have been used to characterize the intensity of reduction and oxidation and relate this to biological processes occurring in flooded soils. Measurements consist of three pieces of equipment: platinum electrode, reference electrode, and voltmeter. The platinum electrode and the reference electrode (calomel) were both buried into the soil column (7 - 10 cm depths) to be in contact with the soil solution. Wires from both the platinum electrode and reference electrode were connected to a voltmeter. The redox potential was reported based on standard hydrogen electrode (SHE) by adding value of +245 to the reading from the meter.

2.5. Statistical Analysis

The effects of flooding and N application on DMY, CPC, and NUP of bahiagrass were analyzed statistically following the PROC GLM procedures [11]. Where the F-test indicated a significant ($p \leq 0.05$) effect, means were separated, following the method of LSD test and Duncan Multiple Range test, using appropriate mean squares [11].

3. RESULTS AND DISCUSSION

3.1. Effects on Dry Matter Yield

Dry matter yield of bahiagrass varied significantly ($p \leq 0.0001$) with N fertilization, but was neither affected by flooding duration nor by the interaction of flooding duration and N fertilization (**Table 2**). The greatest DMY for bahiagrass of 12.3 $ton \cdot ha^{-1}$ was from tube with 14 days of flooding that was fertilized with 200 $kg \cdot N \cdot ha^{-1}$ while the least amount of DMY (2.7 $ton \cdot ha^{-1}$) was from tube that was flooded for 84 days and fertilized with 0 $kg \cdot N \cdot ha^{-1}$. Results disclosed an overwhelming effect of N application on DMY. This claim was exhibited by a much higher DMY under any flooding duration of plants that received higher N application as opposed to those plants without N fertilization.

Quite crucial to the interpretation of our results is the limited amount of oxygen that may be present following flooding. Oxygen level was indirectly measured using

Table 2. Yield, crude protein content, and nitrogen uptake of bahiagrass as affected by different flooding duration at different levels of nitrogen fertilization.

Flooding Duration (days)	Nitrogen Level ($kg \cdot N \cdot ha^{-1}$)	Dry Matter Yield ($ton \cdot ha^{-1}$)	Crude Protein Content (%)	Total Nitrogen Uptake ($kg \cdot ha^{-1}$)
	0	5.6 ± 2.4	5.6 ± 0.4	50.0 ± 21.7
0	100	9.8 ± 2.4	5.5 ± 0.9	84.5 ± 18.9
	200	11.7 ± 5.0	10.1 ± 2.4	187.6 ± 85.2
	0	5.9 ± 2.9	4.8 ± 1.0	46.6 ± 24.1
14	100	11.8 ± 2.7	6.1 ± 1.7	112.8 ± 27.9
	200	12.3 ± 2.8	9.5 ± 1.6	181.6 ± 14.8
	0	6.2 ± 2.4	5.1 ± 0.6	51.9 ± 24.8
28	100	7.1 ± 2.6	9.7 ± 2.5	110.6 ± 49.6
	200	11.3 ± 3.2	9.2 ± 1.7	160.6 ± 23.9
	0	6.2 ± 2.4	5.7 ± 1.6	59.5 ± 35.0
56	100	11.9 ± 3.2	5.7 ± 2.6	82.7 ± 40.3
	200	9.5 ± 4.9	8.9 ± 1.3	128.6 ± 52.9
	0	2.7 ± 0.8	4.5 ± 0.8	23.0 ± 4.6
84	100	7.8 ± 1.1	6.4 ± 1.7	79.5 ± 21.1
	200	9.8 ± 2.7	6.8 ± 1.1	105.7 ± 27.2
Sources of Variation			**F-values**	
Flooding (F)		2.04^{ns}	2.2^{ns}	2.36^{ns}
Nitrogen (N)		19.46^{***}	25.3^{***}	33.80^{***}
F × N		0.90^{ns}	3.03^{**}	0.96^{ns}

[ns]—not significant; [**]—significant at $p \leq 0.001$; [***]—significant at $p \leq 0.0001$.

the reduction-oxidation potentials (redox). Redox potential is an electrical measurement that shows the tendency of a soil solution to transfer electrons to or from a reference electrode. This measurement can estimate whether the soil is fully or partly aerobic or in anaerobic condition. **Figure 1** shows the different levels of redox potential readings from soils that were flooded from 0 to 84 days. Redox potentials (Eh values, mV) of the soils that were flooded from 0 to 84 days ranged from 400 mV to −150 mV. Except for the non-flooded (control) soils, all the soils attained the stage of anaerobic conditions between 8 - 10 days and became fully anaerobic thereafter. An Eh reading below 0 mV would mean limited supply of oxygen in the soil (**Figure 1**). It is generally accepted that energy deficit is one of the most severe problems encountered by plants when subjected to flooding. Oxygen is the terminal acceptor of electrons in the oxidative phosphorylation that indirectly provides the plant with ATP [12].

The effect of N fertilization on DMY is shown in **Figure 2(a)**. The amount of DMY increased linearly with increasing amount of N fertilization when averaged across flooding duration. The relationship between DMY and levels of N fertilization was described by the regression model given below:

$$DMY_{Bahiagrass} = 2.81x + 3.04$$
$$R^2 = 0.91^{**}$$ (3)

Our results show that bahiagrass having the ability to tolerate waterlogging is a tool not only to survive in such environment, but also to respond to growth stimulating factor, such as N fertilization. We can assume that soil N availability was quite similar in both waterlogged and non-waterlogged treatments. Nitrogen fertilization indi-

cated a positive result on DMY of bahiagrass [13,14]. Results are suggesting that waterlogging does not produce detrimental effect either in the growth of bahiagrass or its response capacity to stimulating growth factors, such as N fertilization.

As shown in **Figure 3(a)**, the DMY of bahiagrass was relatively comparable across flooding duration. This forage can be identified as somewhat tolerant to waterlogging. While extended period of flooding did not affect total biomass of bahiagrass, waterlogging slightly promoted their growth. For plants to be classified as tolerant to waterlogging can still maintain their growth rate in such condition [15-18]. Although flooding typically causes a reduction in the abundance of flood-sensitive plant species, it can also promote biomass growth in flood-tolerant species to exploit resources that otherwise would be shared with non-tolerant competitors [19,20]. Voesenek *et al.* [12] suggested that hormonal effects were involved in growth response for plants under waterlogged conditions because photosynthesis rates could be enhanced by increased leaf temperature at higher air vapor pressure deficit in most C_4 grasses (e.g., bahiagrass) while differences in photosynthetic activities between flooded and control plants may be accounted for by the differences in stomatal conductance.

Our visual observations during the experiment suggest that the aboveground biomass of bahiagrass under waterlogged conditions were similar with non-waterlogged forage species. Recent studies in grasslands of Argentina found that native grasses present plastic responses to flooding, such as aerenchyma tissue formation and increase in plant height [6,18]. Aerenchyma formation and leaf elongation are important for the recovery of contact with aerial environment and allow oxygen transport to

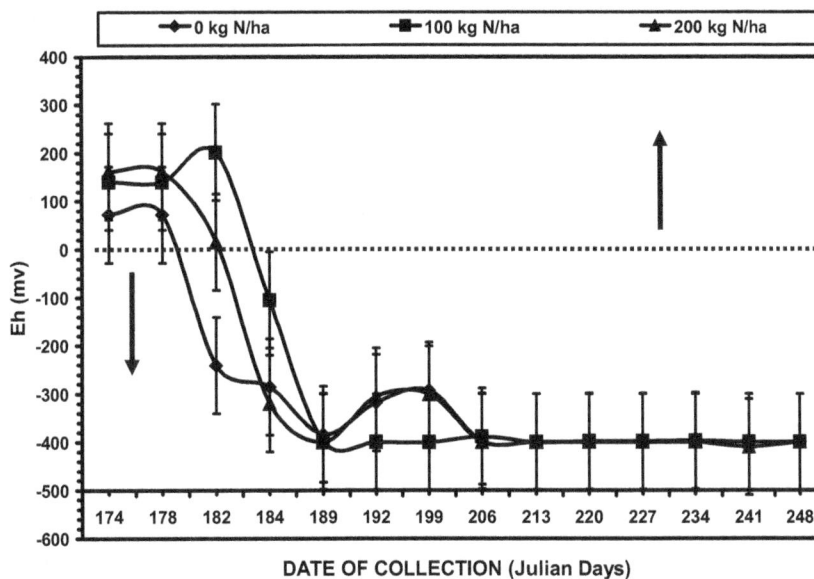

Figure 1. Average levels of reduction-oxidation (Eh) readings during the flooding experiment.

DRY MATTER YIELD (ton·ha⁻¹)

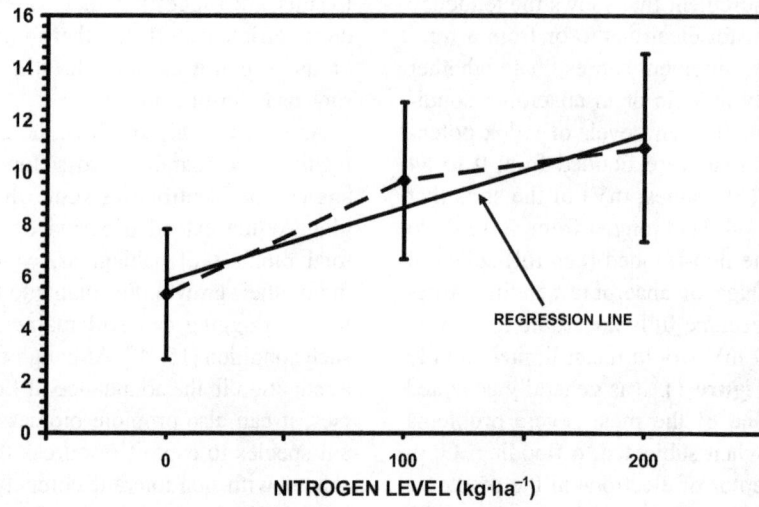

CRUDE PROTEIN CONTENT (%)

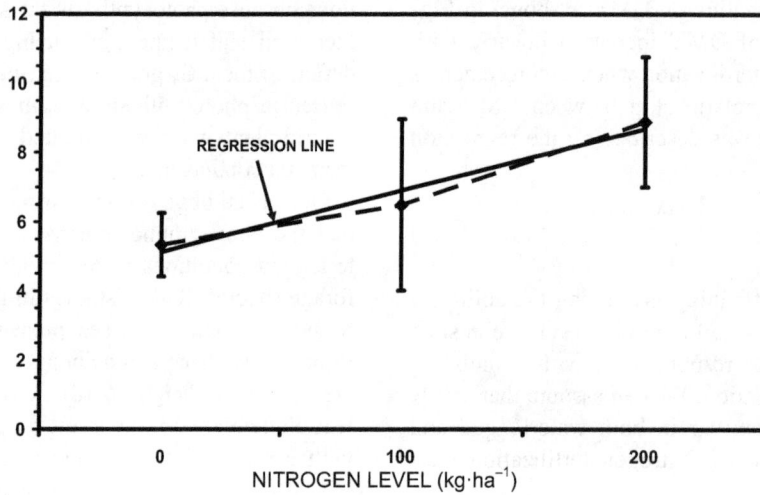

(a)

TOTAL NITROGEN CONCENTRATION IN TISSUES (%)

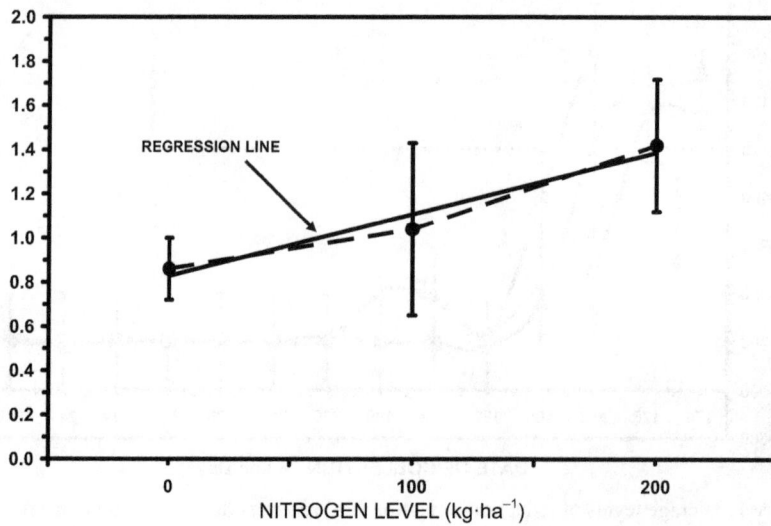

NITROGEN UPTAKE (kg·ha^{-1})

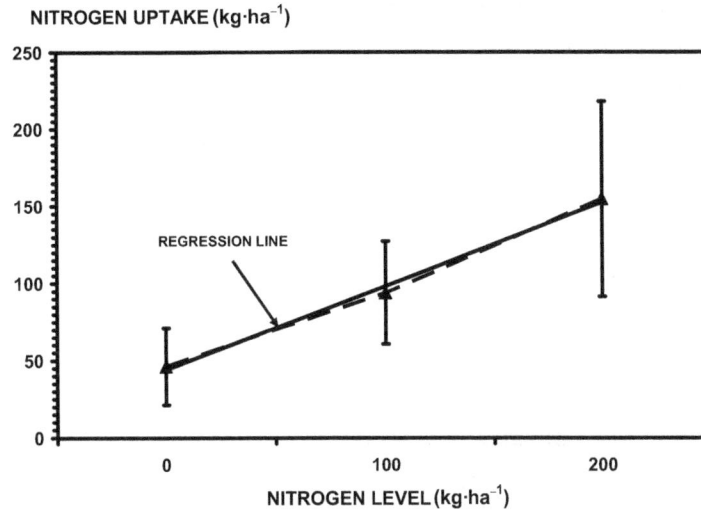

(b)

Figure 2. (a) Dry matter yield (2.81x + 3.04; $R^2 = 0.91^{**}$) and crude protein content (1.78x + 3.36; $R^2 = 0.96^{**}$) as affected by different levels of nitrogen fertilization; (b) Total nitrogen concentration (0.28x + 0.55; $R^2 = 0.96^{**}$) and nitrogen uptake (54.31x + 10.26; $R^2 = 0.99^{**}$) as affected by different levels of nitrogen fertilization.

DRY MATTER YIELD (ton·ha^{-1})

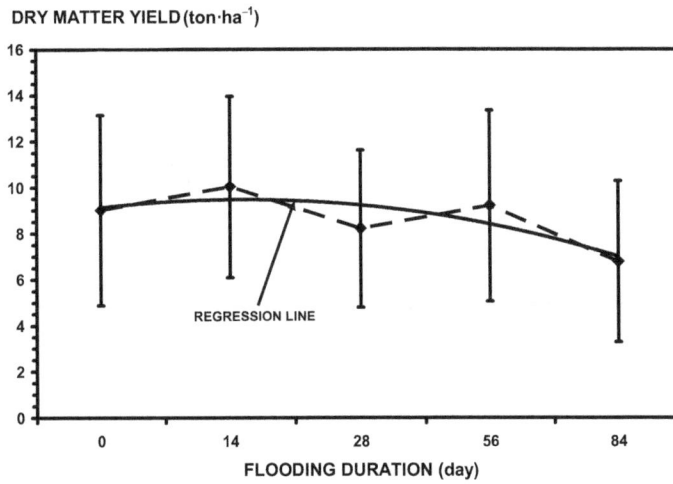

CRUDE PROTEIN CONTENT (%)

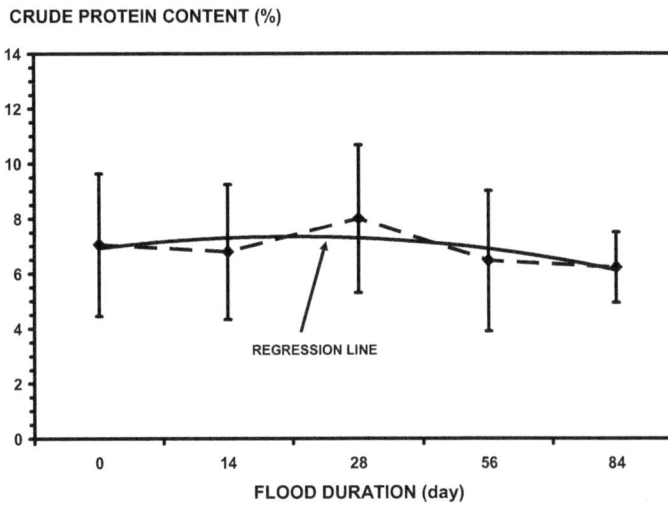

(a)

TOTAL NITROGEN CONCENTRATION (%)

NITROGEN UPTAKE (kg·ha^{-1})

(b)

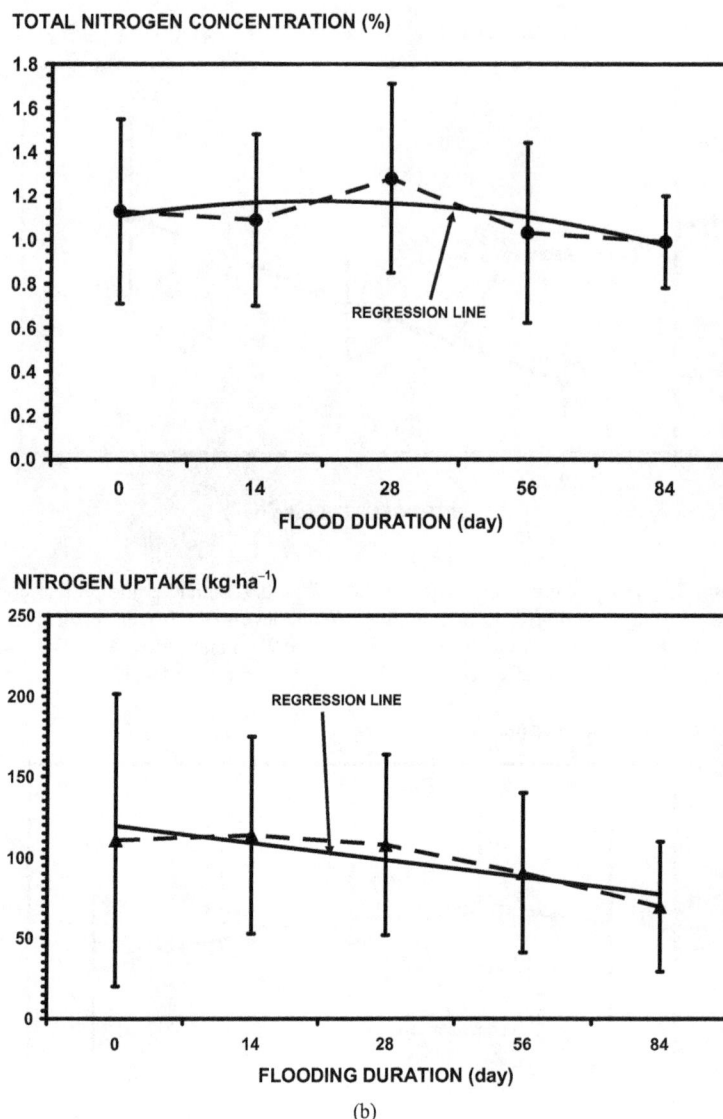

Figure 3. (a) Dry matter yield ($-0.29x^2 + 1.22x + 8.19$; $R^2 = 0.66^{**}$) and crude protein content ($-0.19x^2 + 0.95x + 6.15$; $R^2 = 0.49^{*}$) as affected by different flooding duration; (b) Total nitrogen concentration ($-0.03x^2 + 0.15x + 0.98$; $R^2 = 0.50^{*}$) and nitrogen uptake ($-10.61x + 130.17$; $R^2 = 0.82^{**}$) as affected by different flooding duration.

the submerged tissues of native grasses [21,22].

3.2. Effects on Crude Protein Content

The CPC of bahiagrass varied widely ($p \leq 0.001$) with the interaction effects of flooding duration and N fertilization. Crude protein contents of bahiagrass were also significantly affected by N fertilization ($p \leq 0.0001$). Interaction of flooding duration and levels of N on CPC are shown in **Table 2**. The greatest amount of CPC in bahiagrass (10.1%) was from the control tube (0 day flooding) with 200 kg·N·ha^{-1} while the least amount of CPC of 4.5% was from plants flooded with 84 days with 0 kg·N·ha^{-1}. It appears that bahiagrass that were fertilized

with higher amount of N have significantly higher amount of CPC when compared with those plants with no or less amount of applied N under any flooding duration (**Table 2**). These results may support the hypothesis of the study on the offsetting effect of N on the detrimental effect of flooding on CPC. The use of N fertilizer prior to flooding may alleviate N deficiency because waterlogging causes a significant decrease in N content and rate of N accumulation in plants due to reduced root activity. Net assimilation rates and photosynthetic rates decline in plants experiencing root anaerobiosis, in part due to stomatal closure, and in part due to biochemical modifications [1, 23].

The effect of N fertilization on CPC is shown in **Figure 2(a)**. The amount of CPC in bahiagrass increased linearly with increasing amount of N fertilization. Nitrogen fertilization for bahiagrass can be considered an adaptation to improve its nutrient uptake efficiency, and possibly may have had stimulated crude protein formation [13,14]. Nitrogen fertilization of pasture forages generally increases digestibility. Early reports [24-27] claimed that on bermudagrass pastures, CPC of the forage increased with each increment in N fertilization up to 504 kg·ha⁻¹. A regression model that described the relationship of CPC with levels of N (**x**) is given below:

$$CPC_{Bahiagrass} = 1.8x + 3.3$$
$$R^2 = 0.97^{**}$$
(4)

As observed, CPC of bahiagrass was positively affected by N fertilization despite of anoxic environment and these results could be somehow explained by sufficient nutrients that compensate for the negative effect of waterlogging. Earlier findings of Chapin [28] and Struik and Bray [29] showed that in nutrient-rich environments, root system can satisfy plant nutrient requirements resulting in normal metabolic activities of plants including crude protein formation. The beneficial effects of N fertilization on CPC of bahiagrass suggest the positive impact of N on keeping a good quality of forage even under waterlogged condition. Early published report of Sigua and Hudnall [30] confirmed the importance of gypsum and N fertilization on protein contents of four specie of wetland vegetation under saline environment. Their results disclosed highly significant protein content responses to gypsum and N additions. Increased growth, yield, and protein content were observed from plants that were fertilized with nitrogen.

3.3. Effects on Nitrogen Uptake

Total nitrogen uptake (NUP) of bahiagrass varied significantly (p ≤ 0.0001) with N fertilization, but was neither affected by flooding duration nor by the interaction of flooding duration and N fertilization (**Table 2**). It appears that bahiagrass that were fertilized with higher amount of N have significantly higher amount of NUP when compared with those plants with no or less amount of applied N under any flooding duration (**Table 2**). The greatest amount of NUP was from plants (187 kg·N·ha⁻¹) that were fertilized with 200 kg·N·ha⁻¹ with 0 day of flooding. Bahiagrass that were flooded for 84 days and did not receive N fertilization had the lowest NUP of 23 kg·N·ha⁻¹. These results demonstrated the mitigating effect of N fertilization in offsetting the negative impact of flooding could be mitigated by N fertilization.

The effect of N fertilization on NUP is shown in **Figure 2(b)**. The amount of NUP increased linearly with increasing amount of N fertilization when averaged across flooding duration. The relationship between NUP and levels of N fertilization (x) was described by a regression model given below:

$$NUP_{Bahiagrass} = 54.31x + 10.26$$
$$R^2 = 0.99^{**}$$
(5)

Results disclosed an overwhelming effect of N application on NUP. Averaged across flooding duration, bahiagrass that were fertilized at 200 kg·N·ha⁻¹ had an increase of about 235% NUP when compared with the control plants (0 kg·N·ha⁻¹). Bahiagrass that were fertilized with 100 kg·N·ha⁻¹ had an increase in NUP of about 103% when compared with the control plant. The increase in the amount of NUP between plants that were fertilized with 100 and 200 kg·N·ha⁻¹ was doubled (103% to 235%) significantly (**Table 2** and **Figure 2(b)**).

As shown in **Figure 3(b)**, the NUP of bahiagrass was relatively comparable across flooding duration. The average NUP of bahiagrass was negatively affected by increasing duration of flooding (x). This relationship was described by a regression model given below:

$$NUP_{Bahiagrass} = -10.61x + 130.17$$
$$R^2 = 0.82^{**}$$
(6)

Nitrogen uptake of bahiagrass being classified as a facultative (FACU+) upland species may have been reduced under periods of extended waterlogging. The lowest amount of NUP in bahiagrass was observed from plants that were flooded for 84 days. The greatest amount of NUP was observed from plants that were flooded between 0 and 14 days (**Table 2**). The lower NUP of bahiagrass could be related to anaerobic conditions that inhibit almost immediately the transport of nutrient ions by roots [3]. This may be due to insufficient energy to maintain the activity of ion pumps. Phloem unloading in the anaerobic root ceases and transport of metabolites and growth regulators between the root and shoot are therefore impeded.

4. SUMMARY AND CONCLUSIONS

A study was conducted to determine the potential ecological impact of flooding duration and N fertilization on yield, protein content and N uptake of bahiagrass under greenhouse conditions in 2008 and 2009. The overall results and observations in this study could be briefly summarized as follows:

1) Results disclosed that yield, crude protein content, and nitrogen uptake of bahiagrass varied significantly with levels of N fertilization;

2) Yield and nitrogen uptake of bahiagrass were not as significantly affected by flooding duration as N application. Bahiagrass can survive an extended periods of flood-

ing between 14 and 28 days;

3) Average NUP of bahiagrass was negatively affected by increasing duration of flooding which can be described by this relationship: $-10.6x + 130.2$; $R^2 = 0.82^{**}$;

4) The yield response, crude protein content, and nitrogen uptake of bahiagrass was linearly related to increasing levels of N application;

5) Negative impact of flooding on yield and nitrogen uptake of bahiagrass may have been mitigated by N fertilization.

REFERENCES

[1] Trought, M.C. and Drew, M.C. (1980) The development of waterlogging in wheat seedlings. I. Shoot and root growth in relation to changes in the concentration of dissolved gases and solutes in the soil solution. *Plant Soil*, **54**, 77-94.

[2] Jackson, M.B. (1985) Ethylene and the responses of plants to soil waterlogging and submergence. *Annual Review of Plant Physiology*, **36**, 145-174.

[3] Luttge, V. and Pittman, M.G. (1976) Transport and Energy. *Encyclopedia Plant Physiol*, **2**, 251-259.

[4] Chambliss, C.G. (1999) Bahiagrass. In: Chambliss, C.G., Ed., *Florida Forage Handbook*, University of Florida, Gainesville, 90.

[5] Naidoo, G. and Mundree, S.G. (1993) Relationship between morphological and physiological responses to waterlogging and salinity in *Sporobolus virginicus* (L.). *Oecologia*, **93**, 360-366.

[6] Rubio, G., Casasola, G. and Lavado, R.S. (1995) Adaptation and biomass production of two grasses in response to waterlogging and soil nutrient enrichment. *Oecologia*, **102**, 102-105.

[7] Crawford, R.M. (1993) Plant survival without oxygen. *Biologist*, **40**, 110-114.

[8] Drew, M.C., Sisworo, E.J. and Saker, L.R. (1979) Alleviation of waterlogging damage to young barley plants by application of nitrate and synthetic cytokinin. *New Phytologist*, **82**, 315-329.

[9] Hodgson, A.S. (1982) The effects of duration, timing and chemical amelioration of short-term waterlogging during furrow irrigation of cotton in cracking gray soil. *Australian Journal of Agricultural Research*, **33**, 1019-1028.

[10] Gallagher, R.N., Weldon, G.O. and Boswell, F.C. (1976) A semi-automated procedure for total nitrogen in plant and soil samples. *Soil Science Society of America Journal*, **40**, 887-889.

[11] SAS Institute (2000) SAS/STAT user's guide. Release 6.03. SAS Institute, Cary.

[12] Voesenek, L.A.C.J., Colmer, T.D., Pierik, R., Millenaar, F.F., Peeters, A.J. and Peeters, M. (2006) How plants cope with complete submergence. *New Phytologist*, **170**, 213- 226.

[13] Atkinson, D. (1973) Some general effects of phosphorus deficiency on growth and development. *New Phytologist*, **72**, 101-111.

[14] Aerts, R. and Vander, P.J.M. (1991) The relation between above- and belowground biomass allocation patterns and competitive ability. *Oecologia*, **87**, 551-559.

[15] Kozlowski, T.T. (1984) Plant responses to flooding. *BioScience*, **34**, 162-167.

[16] Heathcote, C.A., Davies, M.S. and Etherington, J.R. (1987) Phenotypic flexibility of *Carex flacca* (Shreb) tolerance of soil flooding by populations from contrasting habitats. *New Phytologist*, **105**, 381-392.

[17] Naidoo, G. and Naidoo, S. (1992) Waterlogging responses of *Sporobolus virginicus* (L.) Kunth. *Oecologia*, **90**, 445- 450.

[18] Loreti, J. and Oesterheld, M. (1996) Intraspecific variation in the resistance to flooding and drought in populations of *Paspalum dilitatum* from different topographic positions. *Oecologia*, **92**, 279-284.

[19] Crawford, R.M., Studer, C. and Studer, K. (1989) Deprivation indifference as a survival strategy in competition, advantages and disadvantages of anoxia tolerance in wetland vegetation. *Flora*, **182**, 189-201.

[20] Insausti, P., Chaneton, E.J. and Soriano, A. (1999) Flooding reverted grazing effects on plant community structure in mesocosms of lowland grassland. *Oikos*, **84**, 266-276.

[21] Laan, P., Tosserama, M., Blom, C.W.P.M. and Veen, B.W. (1990) Internal oxygen transport in *Rumex* species and its significance for respiration under hypoxic conditions. *Plant and Soil*, **122**, 39-46.

[22] Van der Samn, A.J.M., Voesenek, L.A.C.J., Blom, C.W.P.M., Harren, F.J.M. and Reuss, J. (1991) The role of ethylene in shoot elongation with respect to survival and seed output of flooded *Rumes maritimus* L. plants. *Functional Ecology*, **5**, 304-313.

[23] Jackson, M.B., Drew, M.C. (1984) Effects of flooding on growth and metabolism of herbaceous plants. In: Kozlowski, E.T., Ed., *Flooding and Plant Growth*, Academic Press, Orlando, 47-128.

[24] Hart, R.H., Burton, G.W. and Jackson, J.E. (1965) Seasonal variation in chemical composition and yield of coastal bermudagrass as affected by nitrogen fertilization schedule. *Agronomy Journal*, **57**, 381-385.

[25] Mathias, E.L., Bennet, O.L. and Lundberg, P.E. (1973) Effect of rates of nitrogen on yield, nitrogen use, and winter survival of Midland bermudagrass (*Cynodon dactylon*, L.) in Appalachia. *Agronomy Journal*, **65**, 65-67.

[26] Horn, F.P. and Taliaferro, C.M. (1974) Yield, composition, and IVDMD of four bermudagrass. *Journal of Animal Science*, **38**, 224.

[27] Barth, K.M., McLaren, J.B., Fribourg, H.A. and Carver, L.A. (1982) Crude protein content of forage consumed by steers grazing N-fertilized. Bermuda grass and Orchard

Grass-Ladino clover pastures. *Journal of Animal Science*, **55**, 1008- 1014.

[28] Chapin, F.S. (1980) The mineral nutrition of wild plants. *Annual Reviews of Ecology and Systematics*, **11**, 233-260.

[29] Struik, G. and Bary, J.R. (1970) Root-shoot ratios of native forest herbs and *Zea mays* at different soil-moisture levels.

Ecology, **51**, 892-893.

[30] Sigua, G.C. and Hudnall, W.H. (1992) Nitrogen and gypsum, management tools for revegetation and productivity improvement of brackish marsh in southwest Louisiana. *Communications in Soil Science and Plant Analysis*, **23**, 283-299.

3

Vermicompost, the story of organic gold: A review

Sujit Adhikary

Agriculture & Ecological Research Unit, Biological Sciences Division, Indian Statistical Institute, Kolkata, India

ABSTRACT

Earthworm has caught imagination of philosophers like Pascal and Thoreau. Yet its role in the nutrition of agricultural fields has attracted attention of researchers worldwide only in recent decades. Waste management is considered as an integral part of a sustainable society, thereby necessitating diversion of biodegradable fractions of the societal waste from landfill into alternative management processes such as vermicomposting. Earthworms excreta (vermicast) is a nutritive organic fertilizer rich in humus, NPK, micronutrients, beneficial soil microbes; nitrogen-fixing, phosphate solubilizing bacteria, actinomycets and growth hormones auxins, gibberlins & cytokinins. Both vermicompost & its body liquid (vermiwash) are proven as both growth promoters & protectors for crop plants. We discuss about the worms composting technology, its importance, use and some salient results obtained in the globe so far in this review update of vermicompost research.

Keywords: Vermicompost; Worms; Wastes; Nutrients; Worm Biology; Importance

1. INTRODUCTION

A revolution is unfolding in vermiculture studies for vermicomposting of diverse organic wastes by waste eater earthworms into a nutritive "organic fertilizer" and using them for production of chemical free safe food in both quantity & quality without recourse to agrochemicals. Heavy use of agrochemicals since the "green revolution" of the 1960s boosted food productivity at the cost of environment & society. It killed the beneficial soil organisms & destroyed their natural fertility, impaired the power of 'biological resistance' in crops making them more susceptible to pests & diseases. Chemically grown foods have adversely affected human health. The scientific community all over the world is desperately looking for an economically viable, socially safe & environmen-

tally sustainable alternative to the agrochemicals. Several farms in world especially in North America, Australia and Europe are going organic as the demand for "organic foods" are growing in society. In 1980, the US Board of Agriculture published a Report and Recommendations on Organic Farming based on case studies of 69 organic farmers in US and reported that over 90,000 to 100,000 farmers in US had already switched over to organic farming [1].

2. GENERAL CHARACTERISTICS OF VERMICOMPOST AND VERMICULTURE

Vermicompost is the excreta of earthworm, which are capable of improving soil health and nutrient status. Vermiculture is a process by which all types of biodegradable wastes such as farm wastes, kitchen wastes, market wastes, bio-wastes of agro based industries, livestock wastes etc. are converted while passing through the worm-gut to nutrient rich vermicompost. Vermi worms are used here act as biological agents to consume those wastes and to deposit excreta in the process called vermicompost.

3. VERMICOMPOSTING

Vermicomposting is a simple biotechnological process of composting, in which certain species of earthworms are used to enhance the process of waste conversion and produce a better product. Vermicomposting differs from composting in several ways [2]. It is a mesophilic process that utilizes microorganisms and earthworms that are active at 10°C to 32°C (not ambient temperature but temperature within the pile of moist organic material). The process is faster than composting; because the material passes through the earthworm gut, a significant but not fully understood transformation takes place, whereby the resulting earthworm castings (worm manure) are rich in microbial activity and plant growth regulators, and fortified with pest repellence attributes as well. In short, earthworms through a type of biological alchemy are capable of transforming garbage into "gold" [3,4].

4. NUTRIENTS IN VERMICOMPOST

Vermicompost is an excellent soil additive made up of digested compost. Worm castings are much higher in nutrients and microbial life and therefore, are considered as a higher value product (**Table 1**). Worm castings contain up to 5 times the plant available nutrients found in average potting soil mixes. Chemical analysis of the castings was conducted [5,6] and found that it contains 5 times the available nitrogen, 7 times the available potash and 1.5 times more calcium than that found in 15 cm of good top soil. In addition, the nutrient life is up to 6 times more in comparison to the other types of potting mixes. It is reported that phosphorous while passage through gut of worms is converted to the plant available form [7]. Phosphorous is usually considered as a limiting element for plant growth. Therefore, any process that significantly increases phosphorous availability through plants and organic matter will be very important for agriculture. The average potting soil mixes that is found in the market are usually sterile and do not have a microbial population. The combination of nutrients and microbial organisms are essential for growing healthy and productive plants. Vermicompost not only adds microbial organisms and nutrients that have long lasting residual effects, it also modulates structure to the existing soil, increases water retention capacity. Vermicompost may also have significant effects on the soil physical properties. It was observed that addition of vermicompost @ 20 t·ha^{-1} to an agricultural soil in two consecutive years significantly improved soil porosity and aggregate stability [8]. The number of large, elongated soil macro pores increased significantly after a single application of a dose of vermicompost equivalent to 200 kg·ha^{-1} of nitrogen to a cornfield [9]. Similarly, a significant decrease in soil bulk density and a significant increase in soil pH and total organic carbon after application of vermicompost in two consecutive growing seasons, at a rate equivalent to 60 kg·ha^{-1} of N. Together these changes in soil properties improve the availability of air and water, thus encouraging seedling emergence and root growth [10].

Vermicompost contains an average of 1.5% - 2.2% N, 1.8% - 2.2% P and 1.0% - 1.5% K. The organic carbon is ranging from 9.15 to 17.98 and contains micronutrients

Table 1. A comparison of the chemical, microbiological properties of soil, vermicompost and manure are given.

Parameters	Nutrient available from		
	Soil	Vermi-compost	Manure
pH	5.96 ± 0.11	8.09 ± 0.09	8.59 ± 0.14
Electrical conductivity (mS·cm^{-1})	0.33 ± 0.04	0.18 ± 0.02	3.05 ± 0.08
Moisture content (g·kg^{-1})	249 ± 4	535 ± 3	864 ± 5
Water holding capacity (g·kg^{-1})	361 ± 4	1103 ± 13	ND
DOC (mg·g^{-1} dry matter)	0.13 ± 0.03	0.60 ± 0.24	15.4 ± 7.91
DN (mg·g^{-1} dry matter)	0.04 ± 0.01	0.07 ± 0.03	1.89 ± 1.07
Total C (g·kg^{-1})	31 ± 1	181 ± 3	299 ± 6
Total N (g·kg^{-1})	3.0 ± 0.3	8.7 ± 0.7	14.2 ± 1.5
C-to-N ratio	10.2	20.9	21.1
NO$_3^-$ (mg·g^{-1} dry matter)	<0.1	<0.1	<0.1
NH$_4^+$ (mg·g^{-1} dry matter)	<0.1	<0.1	1.0 ± 0.7
P (mg·g^{-1} dry matter)	<0.1	<0.1	2.2 ± 1.6
K (mg·g^{-1} dry matter)	0.9 ± 0.2	1.3 ± 0.1	2.1 ± 0.1
Ca (mg·g^{-1} dry matter)	10.5 ± 3.4	26.3 ± 2.2	0.3 ± 0.1
Na (mg·g^{-1} dry matter)	0.05 ± 0.05	0.21 ± 0.04	0.42 ± 0.02
Background heterotrophic bacteria (log$_{10}$CFU·g^{-1})	7.85	8.41	8.93
Escherichia coli O157:H7 (log$_{10}$CFU·g^{-1})	0.00	0.00	0.00

ND not determined.

Values represent means ± SEM ($n = 3$).

like Sodium (Na), Calcium (Ca), Zinc (Zn), Sulphur (S), Magnesium (Mg) and Iron (Fe). Chemical, microbiological properties of soil, vermicompost and organic manure are given (**Table 1**) with the following details: Soil (Eutric cambisol of the "Denbigh" series, 0 - 15 cm) and earthworms (*L. terrestris*) were collected from a sheep-grazed pasture at Abergwyngregyn, North Wales, UK (53°13.9'N, 4°0.9'W). Earthworm bedding material (digested paper pulp and green waste and earthworms (*Dendrobaena veneta*) were collected from commercial composting beds at the same site. Aged (>1 month old) cattle manure was collected from a commercial farm in North Wales. After collection, all samples were stored in a climate-controlled room (Hemsec Ltd., Kirkby, UK) at 20°C, 70% relative humidity for the duration of the experimental period. This temperature was selected to reflect summertime soil and compost temperatures [11].

In another report [12] it is observed that the worm castings contain higher percentage (nearly two fold) of both macro and micronutrients than the garden compost (**Table 2**).

Earthworms consume various organic wastes and reduce the volume by 40% - 60%. Each earthworm weighs about 0.5 to 0.6 g, eats waste equivalent to its body weight and produces cast equivalent to about 50% of the waste it consumes in a day. The moisture content of castings ranges between 32% to 66% and the pH is around 7.0.

From various studies it is also, evident that vermincompost provides all nutrients in readily available form and enhances uptake of nutrients by plants. Soil available nitrogen increased significantly with increasing levels of vermicompost and highest nitrogen uptake was obtained

Table 2. Nutrient composition of vermicompost and garden compost are given.

Nutrient element	Vermicompost (%)	Garden compost (%)
Organic carbon	9.8 - 13.4	12.2
Nitrogen	0.51 - 1.61	0.8
Phosphorus	0.19 - 1.02	0.35
Potassium	0.15 - 0.73	0.48
Calcium	1.18 - 7.61	2.27
Magnesium	0.093 - 0.568	0.57
Sodium	0.058 - 0.158	<0.01
Zinc	0.0042 - 0.110	0.0012
Copper	0.0026 - 0.0048	0.0017
Iron	0.2050 - 1.3313	1.1690
Manganese	0.0105 - 0.2038	0.0414

at 50% of the recommended fertilizer rate plus 10 t·ha^{-1} vermicompost. Similarly, the uptake of nitrogen (N), phosphorus (P), potassium (K) and magnesium (Mg) by rice (*Oryza sativa*) plant was highest when fertilizer was applied in combination with vermicompost [13]. The production of potato (*Solanum tuberosum*) by application of vermicompost in a reclaimed sodic soil in India was studied and observed that with good potato growth the sodicity (ESP) of the soil was also reduced from initial 96.74 kg/ha to 73.68 kg/ha in just about 12 weeks. The average available nitrogen (N) content of the soil increased from initial 336.00 kg/ha to 829.33 kg/ha [14]. Vermicompost contains enzymes like amylase, lipase, cellulase and chitinase, which can break down the organic matter in the soil to release the nutrients and make it available to the plant roots [15].

5. WORMS AND THEIR BIOLOGICAL FEATURES

About 3000 species of earthworms are found worldwide. Out of which, approximately 384 species are reported to be found in India and their detail taxonomic studies have been done already [16]. Majority of earthworm species live in the soil, except some species like *Pontodrilus burmudensis*, which lives in estuarine water. Earthworms vary greatly in length [viz., *Microscotex phosphoreus* (Duges) is around 20 mm long while *Drawida grandus* (Bourus) may be one meter in length]. Earthworms are known to inhabit in diverse ecological niches. Besides, they are also found in organic materials like manures litter, compost, and hydrophilic environments near fresh and brackish water and also in snowy patches. Most of the earthworms are omnivorous; however, *Agastrodrilus* a carnivorous genus of earthworms from the Ivory Coast of Africa has been reported to feed upon other earthworms of the family Eudrilidae [17].

The most effective use of earthworms in organic waste management could be achieved when a detailed understanding of biology of all potentially useful species and their population dynamics, productivity and the life cycles of earthworms are known. Detail studies on Indian species [18] and tropical species [19] and knowledge about the reproductive strategies of earthworms have been done. Earthworms belong to the family Lumbricidae. Earthworms are hermaphrodites but self-fertilization is rarity. Cocoons or eggs are small varying according to earthworm species. Cocoon color changes with aging. At the age of 6 weeks, earthworm starts laying cocoons. In favorable food and weather conditions one pair of earthworms could produce approximately 100 cocoons in 6 weeks to 6 months [20]. Cocoons incubate roughly for about 3 - 5 weeks. Earthworms possess the ability to regenerate body segments, which are lost by accident or coercion. The doubling time *i.e.* the time taken by a given earthworm population to double in its number or biomass, specifically depends upon the earthworm species, type of food, climatic condition etc. For example, the mean doubling time of *Lampito mauritii* in different organic inputs ranges from 33.77 - 38.05 days while the value for Perionyx excavatus is 11.72 - 16.14 days [20]. The adult worm might live for about two years. Full-grown worms could be separated and dried in an oven to make "worm meal" which is a rich source of protein (70%), which are often used as animal, poultry and fish feed. *E. eugeniae* is a manure worm, which has been extensively used in North America and Europe for vermin composting because of its voracious appetite, high rate of growth, and reproductive ability [21]. A few years back it was brought to India and it has been progressively increasing application in the vermicomposting of animal manure and other forms of biomass [22]. The other epigeic species used in large-scale vermin culture is *E. foetida*, which has high potential for bio-converting organic waste into vermin casts [23].

6. MULTIPLICATION OF WORMS

Earth worms can be multiplied in 1:1 mixture of cow dung and decaying leaves kept in a cement tank or wooden box or plastic bucket with proper drainage facilities. The nucleus culture of the worms needs to be introduced into the above mixture at the rate of 50 worms per 10 kg of organic wastes properly mulched with dried grass or straw in a wet gunny bag. The unit should be kept in shade. Sufficient moisture level should be maintained by occasional sprinkling of water. Within 1 - 2 months, the worms multiply 300 times, which can be used for large-scale vermin composting. Suitability of dry olive cake, municipal biosolids and cattle manure as substrates for Vermicomposting was evaluated and reported that larger weights of newly hatched earthworms were obtained in substrate containing dry olive cake [23]. In another study, maize straw was found to be the most suitable feed material compared to soybean (*Glycine max*) straw, wheat straw, chickpea (*Cicer arientinum*) straw and city refuse for the tropical epigeic earthworm, *Perionyx excavatus* [24].

7. DIFFERENT SOURCES OF VERMICOMPOST

Worms are used to convert organic waste into dark brown nutrient-rich humus. Worms leave behind while reducing the household wastes turn into a good source of manure for plants the excreta. In specific cases, worms could degrade specific pollutants and might allow community formation of useful microorganisms. Due to low cost nature of inputs, the price of vermicompost in the market is usually low in South Asian countries like India. Earthworms bio engineering principles which could potentially act as a substitute to thermophilic composting is becoming increasingly common and numerous studies have shown that increased plant growth and yield could be achieved when plants grown in the presence of vermicompost [25-28]. Vermicompost prepared from paper mill waste, application also showed better growth of Rehu fish (*Labeo rohita*, Hamilton) when compared with other commercially available organic manures [29].

7.1. Vermicomposting from Household Wastes

Following method could be adopted for making vermicompost from household wastes. A wooden box of 45 × 30 × 45 cm or an earthen/plastic container with broad base and drainage holes should be used for this purpose. A plastic sheet with small holes should be placed at the bottom of the wooden box. 3 cm layer of soil and a 5 cm layer of coconut fiber for draining of excess moisture below it is kept inside the box. A thin layer of compost along with worms as inoculums was placed above it. About 250 worms are sufficient for the box. Vegetable wastes should be added in layers daily on top of the inoculums in daily basis. The top of the box should be covered with a piece of gunny bag to provide dim light inside the box. When the box is full, the box should be left undisturbed for a week. When the compost seems to be ready, the box should be kept in light for 2 - 3 hours so that the worms go down to the lowermost coconut fiber layer. The composted materials should be removed from the top of the box and gradually down and sieved for use in the urban or intensive horticultural and agricultural systems.

In Australia and New Zealand, vermiculture is being implemented from home worm bins to large scale composting of municipal biosolids and yard trimmings. A thriving industry is evolving to support these developments. Research continues in both the countries for further expand applications for earthworms and vermicomposting. At the household level, vermicomposting of food trimmings is becoming popular enough that a number of entrepreneurs have designed and are marketing home worm bins. Worm composting also is becoming more popular as an educational activity in schools [30].

7.2. Vermicomposting of Farm Wastes

Pits of sizes 2.5 m × 1 m × 0.3 m (length, breadth and depth) are taken in thatched sheds with sides left open. The bottom and sides of the pit are made hard by compacting with a wooden mallet. At the bottom of the pit a layer of coconut husk is spread with the concave side upward to ensure drainage of excess water and for proper aeration. The husk is moistened and above this, bio-waste mixed with cow dung in the ratio of 8:1 is spread up to a height of 30 cm above the ground level and water is sprinkled daily. After the partial decomposition of wastes for 7 to 10 days, the worms are introduced @ 500 to 1000 in numbers per pit. The pit is covered with jute bags. Moisture is maintained at 40 to 50 per cent population density and a temperature of 20°C - 30°C by sprinkling water over the bed. At higher temperature the worms is found to aestivate and at lower temperature, they will hibernate. When the compost is ready, it is removed from the pit along with the worms and heaped in shade with ample light. The worms will move to bottom of the heap. After one or two days, the compost from the top of the heap is removed. The undecomposed residues are put back to the pit with worms for further composting.

7.3. Harvesting of Vermicompost

Harvesting the compost means removing finished castings from the beds. The finished product is black or dark brown and is called crumbly worm compost. Harvesting the compost and adding fresh bedding, at least twice a year is necessary to keep the worms healthy. The compost can be harvested by spreading a sheet of plastic under a bright light or in the sun. The contents of the bed leaving the bedding materials are divided into a number of heaps on the sheet. The worms will crawl away from the light into the center of each heap and the worm compost can be brushed away on the outside by hand. The crawling worms will be collected for re-use.

7.4. Precautions during the Process

The following precautions should be taken during vermicomposting:
* The African species of earthworms, *Eisenia fetida* and *Eudrilus eugenae* are ideal for the preparation of vermicompost. Most Indian species are not suitable for the purpose.
* Only plant-based materials such as grass, leaves or vegetable peelings should be utilized in preparing vermicompost.
* Materials of animal origin such as eggshells, meat, bone, chicken droppings, etc. are not suitable for preparing vermicompost.
* *Gliricidia* loppings, tobacco leaves, onion, garlic, chilli etc. of kitchen wastes are not suitable for rearing earthworms.
* The earthworms should be protected against birds, termites, ants and rats.
* Adequate moisture should be maintained during the process. Either stagnant water or lack of moisture could kill the earthworms.

* After completion of the process, the vermicompost should be removed from the bed at regular intervals and replaced by fresh waste materials.

8. BENEFICIAL ROLES OF VERMICOMPOST

1) Red worm castings contain a high percentage of humus. Humus helps soil particles form into clusters, which create channels for the passage of air and improve its capacity to hold water. Presence of worms regenerate compacted soils and improves water penetration in such soils by over 50%. [31-33]. US study indicate that 10,000 worms in a farm plot provides the same benefit as three farmers working 8 hours in shift all year round with 10 tons of manure applied in the plot [34]. Humic acid present in humus provides binding sites for the plant nutrients, such as calcium, iron, potassium, sulfur and phosphorus. These nutrients are stored in the humic acid in a form readily available to plants, and are released when the plants require them. The "humic acid" in vermicompost stimulates plant growth even in small amount [35]. The humic acid in humus are essential to plants in four basic ways: a) Enables plant to extract nutrients from soil; b) Help to dissolve unresolved minerals to make organic matter ready for plants to use; c) Stimulates root growth; and d) Helps plant to overcome stress. Presence of humus in soil even helps chemical fertilizers to work better [36].

2) Humus is believed to aid in the prevention of harmful plant pathogens, fungi, nematodes and bacteria [37]. Vermicompost has an ability to fight soil-borne plant diseases such as root rot. Humus also increases water permeability and water retention capacity, contributing to better plant health and more efficiently use in soil moisture. It is found that nitrogen concentrations are higher in vermicompost than in aerobic compost piles. There are other agronomic benefits of composts application, such as high levels of soil-borne disease suppression and removal of soil salinity. One study reported that mean root disease was reduced from 82% to 18% in tomato and from 98% to 26% in capsicum in soils amended with compost [38].

3) A worm casting (also known as worm cast or vermicast) is a biologically active mound containing thousands of bacteria, enzymes, and remnants of plant materials that were not digested by the worms. In fact, the bacterial population of a cast is much greater than the bacterial population of either ingested soil, or the worm's gut. Microbial activity of beneficial microorganisms in worm castings is ten to twenty times higher than that of in the soil and other organic matter [39]. Among beneficial soil microbes stimulated by earth worms are "nitrogen-fixing & phosphate solubilizing bacteria", the "ac-

tinomycetes" & "mycorrhizal fungi". Studies found that the total bacterial count was more than 10/gm of vermincompost. It included Actinomycetes, Azotobacter, Rhizobium, Nitrobacter & Phosphate solubilizing Bacteria ranges from 102 - 106 per gm of vermicompost [40].

4) Castings contain slow released nutrients that are readily available to plants. Castings contain the plant nutrients that are encased in mucus membranes that are secreted by the earthworms. They dissolve slowly rather than allowing immediate nutrient leaching. The product has excellent soil structure, porosity, and aeration and water retention capabilities. Castings can hold 2 - 3 times more water than their weight in soil. Worm castings do not burn root systems. The product can insulate plant roots from extreme temperatures, reduce erosion and control weeds. It is odorless and consists of 100% recycled materials. Vermicompost also has very "high porosity", "aeration", "drainage" and "water holding capacity" than the conventional compost and this again due to humus contents [40].

5) The activity of the worm gut is like a miniature composting tube that mixes conditions and inoculates the residues [41]. Moisture, pH, and microbial populations in the gut are favorably maintained for a synergistic relationship, and then a terrific byproduct [42]. They swallow large amount of soil with organics (microbes, plant & animal debris) everyday, grind them in their gizzard and digest them in their intestine with aid of enzymes. Only 5 - 10 percent of the chemically digested and ingested material is absorbed into the body and the rest is excreted out in the form of fine mucus coated granular aggregates called "vermicastings" which are rich in NKP (nitrates, phosphates and potash), micronutrients and beneficial soil microbes [43].

6) Worm castings are the best imaginable potting soil for greenhouses or houseplants, as well as gardening and farming. It will not burn even the most delicate plants and all nutrients are water soluble, making it an immediate plant food. Earthworm castings, in addition to their use as a potting soil, can be used as a planting soil for trees, vegetables, shrubs, and flowers. They may be used as mulch so that the materials leach directly into the ground when watered.

7) Plant Growth Regulating Activity: Some studies speculated that the growth responses of plants from vermicompost appeared more like "hormone induced activity" associated with the high levels of nutrients, humic acids and humates in vermicompost [25,44]. Researches show that vermicompost use further stimulates plant growth even when plants are already receiving "optimal nutrition". It consistently improved seed germination, enhanced seedling growth and development, and increased plant productivity significantly much more than would be possible from the mere conversion of mineral nutrients into plant available forms. Some studies have also reported that vermicompost contained growth promoting hormone "auxins", "cytokinins" and flowering hormone "gibberellins" secreted by earth-worms [40,45, 46]. Growth promoting activity of vermicompost was tested [12] using a plant bioassay method. The plemule length of maize (*Zea mays*) seedling was measured 48 h after soaking in vermicompost water and in normal water. The marked difference in plemule length of maize seedlings indicated that plant growth promoting hormones are present in vermicompost (**Table 3**). Further, vermicompost makes plants grow fast and strong. Nematodes and diseases will not ruin gardens or plants if the soil is rich enough for them to grow fast. It is the weak plant in poor soil that is destroyed by nematodes and diseases [47].

Positive effects of vermicompost include stimulated seed germination in several plant species such as green gram [48], tomato plants [49,50], petunia [51] and pine trees [52]. Vermicompost also has a positive effect on vegetative growth, stimulating shoot and root development [53]. The effects include alterations in seedling morphology such as increased leaf area and root branching [54] and also has been shown to stimulate plant flowering, increasing the number and biomass of the flowers produced [51,55] as well as increasing fruit yield [26,27,49,56].

8) Ability to Develop Biological Resistance in Plants: Vermicompost contains some antibiotics and actinomycetes that help in increasing the "power of biological resistance" among the crop plants against pest and diseases. Spray of chemical pesticides was significantly reduced by over 75% where earthworms and vermicompost were used in agriculture [40,57].

9) Ability to Minimize Pests Attack: There seems to be strong evidence that worm castings sometimes repel hard-bodied pests [58,59]. Studies reported statistically significant decrease in arthropods (aphids, buds, mealy bug, and spider mite) populations, and subsequent reduction in plant damage, in tomato, pepper, and cabbage trials with 20% and 40% vermicompost additions [60]. Munroe doing commercial vermicomposting in California, US, claims that his product repels many different insect pests. His explanation is that this is due to production of enzymes "chitinase" by worms which breaks down the chitin in the insect's exoskeleton [61]. As regards the effects of vermicompost on insect pests and mites, field studies have shown that the addition of vermicompost to soil significantly reduces the incidence of

Table 3. Plemule length of maize seedlings.

Treatment	Initial length (cm)	Final length (cm)
Tank water	16.5	16.6
Vermi-wash	17.6	18.6

the psyllids *Heteropsylla cubana* [62] the sucking insect *Aproaerema modicella* [63], jassids, aphids, beetles, and spider mites [64]. Studies also reported considerable suppression of root knot nematode (*Meloidogyne incognita*) and drastic suppression of spotted spider mites (*Tetranychus* spp.) and aphid (*Myzus persicae*) in tomato plants after application of vermicompost teas (vermiwash liquid) [65].

10) Ability to Suppress Plant Disease: Studies reported that vermicompost application suppressed 20% - 40% infection of insect pests *i.e.* aphids (*Myzus persicae*), mearly bugs (*Pseudococcus* spp.) and cabbage white caterpillars (*Peiris brassicae*) on pepper (*Capiscum annuum*), cabbage (*Brassica oleracea*) and tomato (*Lycopersicum esculentum*) [66]. Studies have also found that use of vermicompost in crops inhibited the soil born fungal diseases. They also found significant suppression of plant-parasitic nematodes in field trials with pepper, tomatoes, strawberries and grapes [60]. The scientific explanation behind this concept is that high levels of agronomic beneficial microbial population in vermincompost protects plants by outcompeting plant pathogens for available food resources *i.e.* by starving them and also by blocking their excess to plant roots by occupying all the available sites. They also reported the disease suppressing effects by the applications of vermicompost, on attacks by fungus *Pythium* on cucumber, *Rhizoctonia* on radishes in the greenhouse, by *Verticillium* on strawberries and by *Phomposis* and *Sphaerotheca fulginae* on grapes in the field. In all these experiments vermicompost applications suppressed the incidence of the disease significantly. They also found that the ability of pathogen suppression disappeared when the vermicompost was sterilized, convincingly indicating that the biological mechanism of disease suppression involved was microbial antagonism. Vermicompost has also been found to have a wide range of indirect effects on plant growth such as the mitigation or suppression of plant diseases. Suppression of plant diseases has been extensively investigated in other organic amendments such as manure and compost [67-69]. Likewise, some studies have shown that vermicompost can suppress a wide range of microbial diseases, insect pests and plant parasitic nematodes. As regards the suppression of fungal diseases, [70] it was observed that the addition of vermicompost extracts to three ornamental plant species significantly reduced sporulation of the pathogen *Phytophthora cryptogea*. Similarly, aqueous extracts of vermicompost were capable of reducing the growth of pathogenic fungi such as *Botrytis cinerea*, *Sclerotinia sclerotiorum*, *Corticium rolfsii*, *Rhizoctonia solani* and *Fusarium oxysporum* [71]. The addition of solid vermicompost to tomato seeds significantly reduced infection caused by *Fusarium lycopersici* [72] and *Phytophthora nicotianae* [73]. Never-

theless, they did not find any significant suppressive effects of a sewage sludge vermicompost on *Phytophthora nicotianae*, in comparison with peat. Edward *et al.*, observed that the suppressive effect exerted by several types of vermicompost on several plant pathogens such as *Pythium*, *Rhizoctonia*, *Verticillium*, and *Plectosporium*, disappeared after sterilization of the vermicompost, and concluded that disease suppression may be related to the presence of biological suppressive agents in vermicompost [74].

11) Vermimeal Production: With the increasing demand for animal feed protein bolstered by the continuing growth in human population and food source, the production of vermimeal may be considered as the most economically feasible application of vermiculture. According to Kale, vermiculture has bright prospects in the animal feed industry [75]. Vermimeal or earthworm meal is a feed preparation consisting of processed earthworm biomass. It is a rich source of animal protein as well as essential amino acids, fats, vitamins, and minerals for livestock, birds and fish. About 5.5 kg of fresh ANC biomass (18% dry matter) is needed to produce 1 kg of vermimeal. It can be packed in plastic bags and stored in a cool dry place out of direct sun for up to 3 months. Proximate analysis of an ANC vermimeal in dry and pulverized form revealed the following composition; 68% crude protein, 9.57% fat, 11.05% nitrogen-free extract, and 9.07% ash [76]. Numerous studies on different livestock animals, birds and fishes have shown excellent results of feeding the animals with vermimeal or earthworm meal [77]. This is not surprising, considering that earthworms are a natural source of nutrition for birds and other animals in the wild.

9. APPLICATION IN CROP PLANTS

There have been several reports that earthworms and its vermicompost can induce excellent plant growth and enhance crop production.

9.1. Cereal Crops

Glasshouse studies made at CSIRO Australia found that the earthworms (*Aporrectodea trapezoids*) increased growth of wheat crops (*Triticum aestivum*) by 39%, grain yield by 35%, lifted protein value of the grain by 12% & resisted crop diseases as compared to the control. The plants were grown in a "red-brown earth" with poor nutritional status and 60% moisture. There were about 460 worms per m^{-2} [78]. They also reported that in Parana, Brazil invasion of earthworms significantly altered soil structure and water holding capacity. The grain yields of wheat and soybean was increased by 47% and 51%, respectively [79]. Some studies were made on the impact of vermicompost and garden soil in different proportion on wheat crops in India. It was found that when the garden soil and vermicompost were mixed in 1:2 proportions, the growth was about 72% - 76% while in pure vermicompost, the growth increased by 82% - 89% [80]. Another study reported that earthworms & its vermicast improve the growth and yield of wheat by more than 40% [81] (Palanisamy, 1996). Other studies also reported better yield and growth in wheat crops applied with vermicompost in soil [82-84]. Studies made on the agronomic impacts of vermicompost on rice crops (*Oryza sativa*) reported greater population of nitrogen fixers, actinomycetes and mycorrhizal fungi inducing better nutrient uptake by crops and better growth [85]. Another study was made on the impact of vermicompost on rice-legume cropping system in India. Integrated application of vermicompost, chemical fertilizer and biofertilizers (*Azospirillum* & phosphobacteria) increased rice yield by 15.9% over chemical fertilizer used alone. The integrated application of 50% vermicompost, 50% chemical fertilizer and biofertilizers recorded a grain yield of 6.25 and 0.51 ton/ha in the rice and legume respectively. These yields were 12.2% and 19.9% higher over those obtained with 100% chemical fertilizer when used alone [86]. Studies made in the Philippines also reported good response of upland rice crops grown on vermicompost [87].

9.2. Fruit Crops

Study found that worm waste (vermicompost) boosted grape yield by two-fold as compared to chemical fertilizers. Treated vines with vermicompost produced 23% more grapes due to 18% increase in bunch numbers. The yield in grapes was worth additional value [88]. Farmer in Sangli district of Maharashtra, India, grew grapes on "eroded wastelands" and applied vermicasting @ 5 tons/ha. The grape harvest was normal with improvement in quality, taste and shelf life. Soil analysis showed that within one year pH came down from 8.3 to 6.9 and the value of potash increased from 62.5 kg/ha to 800 kg/ha. There was also marked improvement in the nutritional quality of the grape fruits [89]. Study was made on the impacts of vermicompost and inorganic (chemical) fertilizers on strawberries (*Fragaria ananasa*) when applied separately and in combination. Vermicompost was applied @ 10 tons/ha while the inorganic fertilizers (nitrogen, phosphorus, potassium) @ 85 (N):155 (P):125 (K) kg/ha. Significantly, the yield of marketable strawberries and the weight of the largest fruit was 35% greater on plants grown on vermicompost as compared to inorganic fertilizers in 220 days after transplanting. Also there were 36% more "runners" and 40% more "flowers" on plants grown on vermicompost. Also, farm soils applied with vermicompost had significantly greater "microbial bio-

mass" than the one applied with inorganic fertilizers [32]. Studies also reported that vermicompost increased the yield of strawberries by 32.7% and drastically reduced the incidence of physiological disorders like albinism (16.1% → 4.5%), fruit malformations (11.5% → 4%), grey mould (10.4% → 2.1%) and diseases like botrytis rot. By suppressing the nutrient related disorders, vermincompost use increased the yield and quality of marketable strawberry fruits up to 58.6% [56]. Impact of vermicompost on cherries found that it increased yield of "cherries" for three (3) years after "single application" inferring that the use of vermicompost in soil builds up fertility and restore its vitality for long time and its further use can be reduced to a minimum after some years of application in farms. At the first harvest, trees with vermicompost yielded an additional $63.92 and $70.42 per tree and after three harvests profits per tree were $110.73 and $142.21, respectively [90].

9.3. Vegetable Crops

Studies on the production of important vegetable crops like tomato (Lycopersicum esculentus), eggplant (Solanum melangona) and okra (Abelmoschus esculentus) have yielded very good results [89,91-93]. Another study was made on the growth impact of earthworms (with feed materials), vermicompost, cow dung compost and chemical fertilizers on okra (*Abelmoschus esculentus*). Worms and vermicompost promoted excellent growth in the vegetable crop with more flowers and fruits development. But the most significant observation was drastically less incidence of "Yellow Vein Mosaic", "Color Rot" and "Powdery Mildew" diseases in worm and vermicompost applied plants [94]. Study was made on the production of potato (*Solanum tuberosum*) by application of vermicompost in a reclaimed sodic soil in India. The overall productivity of potato was significantly high (21.41 tons/ha) on vermicompost applied @ 6 tons/ha as compared to control which was 04.36 tons/ha. The sodicity of the soil was also reduced and nitrogen (N) contents increased significantly [14]. Study was made on the growth impacts of organic manure (containing earthworm vermicast) on garden pea (*Pisum sativum*) and compared with chemical fertilizers. Vermicast produced higher green pod plants, higher green grain weight per plant, higher percentage of protein content and carbohydrates and higher green pod yield (24.8% - 91%) as compared to chemical fertilizer [95]. Studies made on the effects of vermicompost & chemical fertilizer on the hyacinth beans (*Lablab purpureas*) it was found that all growth & yield parameters e.g. total chlorophyll contents in leaves, dry matter production, flower appearance, length of fruits and fruits per plant, dry weight of 100 seeds, yield per plot and yield per hectare were significantly higher in those plots which received vermicompost either alone or in combination with chemicals. The highest fruit yield of 109 ton/ha was recorded in plots which received vermicompost @ 2.5 tons/ha [96].

In addition to increasing plant growth and productivity, vermicompost may also increase the nutritional quality of some vegetable crops such as tomatoes [97], Chinese cabbage [98], spinach [99], strawberries [56] and lettuce [100].

10. TROUBLESHOOTING

There are two major problems in the process of making vermin compost.

Death of worms in large and small numbers
- Worms are dying for the following reasons:
- If they are not getting enough food, therefore food should be buried into the bedding.
- Food may be too dry, so moisture should be maintained until it is slightly damp.
- Food may be too wet, in which case bedding should be added.
- The worms may be too hot, so the bin should be put in the shade.

Bad smells from the vermicomposting grounds
- It is due to that there is not enough air circulation. In this case, add dry bedding under and over the worms. Turning of the food may give better result.
- There may be present some materials such as meat, pet feces, or greasy foods, which are harmful in the compost, pit. These should be removed.

Important practical points for vermiculture
- No smell if the right products or bedding and feed are used.
- No need to turn the compost as the worms act like little ploughs turning the bedding and food.
- Air is circulating on a continuous period.
- Composting time is short in comparison to other composts.
- Composting can be done year round.

11. HUNGRY WORMS & FUTURE EARTHWORM BIOTECHNOLOGY

Various approaches were employed in recent past to understand the mechanism of odorant and pheromone perception in diverse organisms. This has led to the identification of the pathways and a number of molecules involved in signal transduction. Intelligent use of behavioral genetic screens in *C. elegans*, close to earthworm in evolutionary scale has revealed a broader array of proteins that participate in chemosensation including pheromone perception. In mammals, odorants or pheromones bind to a seven trans-membrane G-protein coupled receptor. This results in the activation of adenyl cyclase via

G-protein homologs. cAMP production in turn activates Ca^{2+}-permeant CNG channels that produces an electric signal recognized and processed by the worm brain. In *C. elegans*, there are increasing evidences that two distinct pathways of odor perception operates, one utilizing cAMP or cGMP and a CNG-like channel that is similar to the mammalian pathway, and a second mechanism that uses an unidentified second messenger and a capsasin like cation channel. Genetic screens for worms that are unable to chemotax to particular odor(s) have also allowed identification of peripheral players, such as ODR-4, that contribute to the transduction process. The discoveries arising from combining different experimental approaches in organisms where chemosensation plays fundamentally distinct roles are likely to provide insight into evolution and elaboration of sensory systems. Recently, in an exciting discovery Kawano *et al.*, 2005 reported that a crude extract of worm relative *C. elegans* including dauer pheromone could enhance the lifespan of worms. The possible mechanism of action is through insulin pathway [101].

12. CONCLUSIONS

"Vermiculture Movement" is going on in India with multiple objectives of community waste management, highly economical way of crop production, which replaces the costly chemical fertilizers, and poverty eradication programs in villages. Vermicomposting to a non-professional simply means making of compost by worms by utilizing worm's innate behavior. Vermicomposting process improves soil aeration and thereby promotes the survival and dispersal of the useful bacterium within such systems, which is slowly becoming clear day by day. Vermicomposts could be prepared from the kitchen waste, farm waste, market waste, even from biodegradable city waste. The most effective uses of earthworms are organic waste management and supplement of readily available plant nutrients and vermicompost demands the credit as it maintains as well as improves soil health.

The chemical fertilizers are produced from "vanishing resources" of earth. Farmers urgently need a sustainable alternative, which is both economical and productive while also maintaining soil health & fertility. The new concept is "Ecological Agriculture", which is by definition different from "Organic Farming" that was focused mainly on production of chemical free foods. Ecological agriculture emphasizes on total protection of food, farm & human ecosystems while improving soil fertility & development of secondary source of income for the farmers. UN has also endorsed it. Vermiculture provides the best answer for ecological agriculture, which is synonymous with "sustainable agriculture". Thereby it may be concluded that during the present time the most beneficiary from the scheme is our environment. This article opens the scope for further several researches.

REFERENCES

[1] US Board of Agriculture (1980) Report and recommendations on organic farming—Case studies of 69 organic farmers in USA. Publication of US Board of Agriculture.

[2] Gandhi, M., Sangwan, V., Kapoor, K.K. and Dilbaghi, N. (1997) Composting of household wastes with and without earthworms. *Environment and Ecology*, **15**, 432-434.

[3] Vermi Co. (2001) Vermicomposting technology for waste management and agriculture: An executive summary. Vermi Co., Grants Pass.

[4] Tara Crescent (2003) Vermicomposting. Development Alternatives (DA) sustainable livelihoods.

[5] Ruz-Jerez, B.E., Ball, P.R. and Tillman, R.W. (1992) Laboratory assessment of nutrient release from a pasture soil receiving grass or clover residues, in the presence or absence of *Lumbricus rubellus* or *Eisenia fetida*. *Soil Biology and Biochemistry*, **24**, 1529-1534.

[6] Parkin, T.B. and Berry, E.C. (1994) Nitrogen transformations associated with earth worm casts. *Soil Biology and Biochemistry*, **26**, 1233-1238.

[7] Reinecke, A., Viljoen, S.V. and Saayman, R. (1992) The suitability of *Eudrilus eugenie, Perionyx excavatus* and *Eisenia fetida* (Oligochaeta) for vermicomposting in southern Africa in terms of their temperature requirements. *Soil Biology and Biochemistry*, **24**, 1295-1307

[8] Ferreras, L., Gomez, E., Toresani, S., Firpo, I. and Rotondo, R. (2006). Effect of organic amendments on some physical, chemical and biological properties in a horticultural soil. *Bioresource Technology*, **97**, 635-640.

[9] Marinari, S., Masciandaro, G., Ceccanti, B. and Grego, S. (2000). Influence of organic and mineral fertilizers on soil biological and physical properties. *Bioresource Technology*, **72**, 9-17.

[10] Gopinath, K.A., Supradip, S., Mina, B.L., Pande, H., Kundu, S. and Gupta, H.S. (2008) Influence of organic amendments on growth, yield and quality of wheat and on soil properties during transition to organic production. *Nutrient Cycling in Agroecosystems*, **82**, 51-60.

[11] Williams, A.P., Roberts, P. and Avery, L.M. (2006) Earth worms as vectors of *Escherichia coli* O 157:H7 in soil and vermicomposts. *FEMS Microbiology Ecology*, **58**, 54-64.

[12] Nagavallemma, K.P., Wani, S.P., Stephane, L., Padmaja, V.V., Vineela, C., Babu Rao, M. and Sahrawat, K.L. (2004) Vermicomposting: Recycling wastes into valuable organic fertilizer. Global Theme on Agrecosystems Report No. 8. Patancheru 502 324, International Crops Research Institute for the Semi-Arid Tropics, Andhra, 20 p.

[13] Jadhav, A.D., Talashilkar, S.C. and Pawar, A.G. (1997) Influence of the conjunctive use of FYM, vermicompost and urea on growth and nutrient uptake in rice. *Journal of Maharashtra Agricultural Universities*, **22**, 249-250.

[14] Ansari, A.A. (2008) Effect of Vermicompost on the Productivity of Potato (*Solanum tuberosum*) Spinach (*Spinacia oleracea*) and Turnip (*Brassica campestris*). *World Journal of Agricultural Sciences*, **4**, 333-336.

[15] Chaoui, H.I., Zibilske, L.M. and Ohno, T. (2003) Effects of earthworms casts and compost on soil microbial activity and plant nutrient availability. *Soil Biology and Bio-Chemistry*, **35**, 295-302.

[16] Julka, J.M. (1983) A new genus and species of earthworm (Octochaetidae:Oligochaeta) from South India. *Geobioscience New Reports*, **2**, 48-50.

[17] Lavelle, P. (1983) *Agastrodrilus omodeo* (Vaillaud), a genus of carnivorous earthworm from the Ivory coast. In : Satchell, J.E., Ed., *Earthworm Ecology from Darwin to Vermiculture*, Chapman and Hall, New York and London, 1983, 425-429.

[18] Julka, J.M. (2001) Earthworm diversity and it's role in agroecosystem. VII National symposium on soil biology and ecology. Bangalore University of Agricultural Sciences, Bangalore, 13-17.

[19] Dash, M.C. and Senapati, B.K. (1980) Cocoons morphology, hatching and emergence pattern in tropical earthworms. *Pedobiologia*, **20**, 317-324.

[20] Ismail, S.A. (1997) Vermicology: The biology of Earthworms. Orient Longman Limited, Chennai, 1997, 92.

[21] Gajalakshmi, S., Ramasamy, E.V. and Abbasi, S.A. (2001) Potential of two epigeic and two anecic earth worm species in vermicomposting of water hyacinth. *Bioresource Technology*, **76**, 177-181.

[22] Garg P., Gupta, A. and Satya, S. (2006) Vermicomposting of different types of waste using *Eisenia foetida*: A comparative study. *Bioresource Technology*, **97**, 391-395.

[23] Garg, V.K., Yadav, Y.K. and Sheoran, A. (2006) Livestock excreta management through vermicomposting using an epigeic earth worm *Eisenia foetida*. *The Environmentalist*, **26**, 269-276.

[24] Manna, M.C., Singh, M., Kundu, S., Tripathi, A.K. and Takkar, P.N. (1997) Growth and reproduction of the vermicomposting earthworm Perionyx excavatus as influenced by food materials. *Biology and Fertility of Soils*, **24**, 129-132.

[25] Atiyeh, R.M., Subler, S., Edwards, C.A., Bachman, G., Metzger, J.D. and Shuster, W. (2000) Effects of Vermicomposts and Composts on Plant Growth in Horticultural Container Media and Soil. *Pedobiologia*, **44**, 579-590.

[26] Arancon, N.Q., Edwards, C.A. and Atiyeh, R. (2004) Effects of vermicomposts produced from food waste on the growth and yields of greenhouse peppers. *Bioresource Technology*, **93**, 139-144.

[27] Arancon, N.Q., Edwards, C.A. and Bierman, P. (2004) Influnces of vermicomposts on field strawberries: Effects on growth and yields. *Bioresource Technology*, **93**, 145-153.

[28] Lee, J.J., Park, R.D. and Kim, Y.W. (2004) Effect of food waste compost on microbial population, soil enzyme activity and lettuce growth. *Bioresource Technology*, **93**, 21-28.

[29] Deolalikar, A.V. and Mitra, A. (1997) Application of paper mill solid waste vermicompost as organic manure in Rohu (*Labeo rohita* Hamilton) culture—A comparative study with other commercial organic manure. In: Azariah, J., *et al.*, Eds., *Proc. Int. Bioethics Workshop*: *Biomanagement of Biogeoresources*, University of Madras, Chennai.

[30] Applehof, M., Webster, K. and Buckerfield, J. (1996) Vermicomposting in Australia and New Zealand. *BioCycle*, **37**, 63-66.

[31] Ghabbour, S.I. (1973) Earthworm in agriculture: A modern evaluation. *Indian Review of Ecological and Biological Society*, **111**, 259-271.

[32] Bhat, J.V. and Khambata, P. (1996) Role of earthworms in agriculture. Indian Council of Agriculture Research, New Delhi, **22**, 36.

[33] Capowiez, Y., Cadoux S., Bouchand P., Roger-Estrade, J., Richard G. and Boizard, H. (2009) Experimental evidence for the role of earthworms in compacted soil regeneration based on field observations and results from a semi-field experiment. *Soil Biology & Biochemistry*, **41**, 711-717.

[34] Li, K.M. (2005) Vermiculture industry in circular economy. *Worm Digest*.

[35] Canellas. L.P., Olivares, F.L., Okorokova, A.L. and Facanha, R.A. (2002) Humic acids isolated from earthworm compost enhance root elongation, lateral root emergence, and plasma membrane H$^+$—ATPase activity in maize roots. *Journal of Plant Physiology*, **130**, 1951-1957.

[36] Li, K. and Li, P.Z. (2010) Earthworms helping economy, improving ecology and protecting health. In: Sinha, R.K. *et al.*, Eds., *Special Issue on "Vermiculture Technology"*, *International Journal of Environmental Engineering*, Inderscience Publishing, Olney.

[37] Nielson, R. (1965) Presence of plant growth substances in Earthworms demonstrated by Paper Chromatography and the Went Pea Test. *Nature*, **208**, 1113-1114.

[38] Ayres, M. (2007) Suppression of soilborn plant disease using compost. *3rd National Compost Research and Development Forum Organized by COMPOST Australia*, Murdoch University, Perth.

[39] Edwards, C.A. (1995) Historical overview of vermicomposting. *Biocycle*, **36**, 56-58.

[40] Suhane, R.K. (2007) Vermicompost. Rajendra Agriculture University, Pusa, 88.

[41] Abbot, I. and Parker, C.A. (1981) Interactions between earthworms and their soil environments. *Soil Biology and Biochemistry*, **13**, 191-197.

[42] Becker B. (1991) The benefits of earthworms. *Natural Food and Farming*, 12.

[43] Scheu, S. (1987) Microbial activity and nutrient dynamics in earthworms casts. *Journal of Biological Fertility Soils*, **5**, 230-234.

[44] Edwards, C.A. and Burrows, I. (1988) The potential of earthworms composts as plant growth media. In: Edward, C.A. and Neuhauser, E.F., Eds., *Earthworms in Waste and Environmental Management*, SPB Academic Publishing, The Hague, 21-32.

[45] Tomati, U., Grappelli, A. and Galli, E. (1987) The presence of growth regulators in earthworm worked wastes. *Proceeding of International Symposium on "Earthworms"*, Bologna-Carpi, 31 March-4 April 1985, 423-436.

[46] Tomati, V., Grappelli, A. and Galli, E. (1995) The Hormone like Effect of Earthworm Casts on Plant Growth. *Biology and Fertility of Soils*, **5**, 288-294.

[47] Gaddie, R.E. and Douglas, D.E. (1975) Earthworms for ecology and profit. *Scientific Earthworm Farming*, Bookworm Publishing Company, **1**, 175.

[48] Karmegam, N., Alagumalai, K. and Daniel, T. (1999) Effect of vermicompost on the growth and yield of green gram (*Phaseolus aureus* Roxb.). *Tropical Agriculture*, **76**, 143-146.

[49] Atiyeh, R.M., Arancon, N.Q., Edwards, C.A. and Metzger, J.D. (2000) Influence of earthworm-processed pig manure on the growth and yield of green house tomatoes. *Bioresource Technology*, **75**, 175-180.

[50] Zaller, J.G. (2007) Vermicompost as a substitute for peat in potting media: Effects on germination, biomass allocation, yields and fruit quality of three tomato varieties. *Scientia Horticulturae*, **112**, 191-199.

[51] Arancon, N.Q., Edwards, C.A., Babenko, A., Cannon, J., Galvis, P. and Metzger, J.D. (2008) Influences of vermicomposts, produced by earthworms and microorganisms from cattle manure, food waste and paper waste, on the germination, growth and flowering of petunias in the greenhouse, *Applied Soil Ecology*, **39**, 91-99.

[52] Lazcano, C., Sampedro, L., Zas, R. and Domínguez, J. (2010a) Vermicompost enhances germination of the maritime pine (*Pinus pinaster* Ait.). *New Forest*, **39**, 387-400.

[53] Edwards, C.A., Domínguez, J. and Arancon, N.Q. (2004) The influence of vermicomposts on plant growth and pest incidence. In: Shakir, S.H. and Mikhaïl, W.Z.A., Eds., *Soil Zoology for Sustainable Development in the 21st Century*, Cairo, 397-420.

[54] Lazcano, C., Arnold, J., Tato, A., Zaller, J.G. and Domínguez, J. (2009). Compost and vermicompost as nursery pot components: Effects on tomato plant growth and morphology. *Spanish Journal of Agricultural Research*, **7**, 944-951.

[55] Atiyeh, R.M., Arancon, N., Edwards, C.A. and Metzger, J.D. (2002) The influence of earthworm-processed pig manure on the growth and productivity of marigolds. *Bioresource Technology*, **81**, 103-108.

[56] Singh, R., Sharma, R.R., Kumar, S., Gupta, R.K. and Patil, R.T. (2008) Vermicompost substitution influences growth, physiological disorders, fruit yield and quality of strawberry (Fragaria xananassa Duch). *Bioresource Technology*, **99**, 8507-8511.

[57] Singh, R.D. (1992) Harnessing the earthworms for sustainable agriculture. Publication of Institute of National Organic Agriculture, Pune, 1-16.

[58] Arancon, N. (2004) An interview with Dr. Norman Arancon. *Casting Call*, **9**.

[59] Anonymous (2001) Vermicompost as Insect Repellent. *Biocycle*.

[60] Edwards, C.A. and Arancon, N. (2004) Vermicompost suppresses plant pests and disease attacks. *Rednova News*.

[61] Munroe, G. (2007) Manual of on-farm vermicomposting and vermiculture. Organic Agriculture Centre of Canada, Nova Scotia.

[62] Biradar, A.P., Sunita, N.D., Teggelli, R.G. and Devaranavadgi, S.B. (1998) Effect of vermicomposts on the incidence of subabul psyllid. *Insect Environment*, **4**, 55-56.

[63] Ramesh, P. (2000) Effects of vermicomposts and vermin-composting on damage by sucking pests to ground nut (*Arachis hypo*gea). *Indian Journal of Agricultural Sciences*, **70**, 334.

[64] Rao, K.R. (2002) Induced host plant resistance in the management of sucking insect pests of groundnut. *Annals of Plant Protection Science*, **10**, 45-50.

[65] Edwards, C.A., Arancon, N.Q., Emerson, E. and Pulliam, R. (2007) Suppressing plant parasitic nematodes and arthropod pests with vermicompost teas. *BioCycle*, **48**, 38-39.

[66] Arancon, N.Q., Edwards, C.A. and Lee, S. (2002) Management of plant parasitic nematode population by use of vermicomposts. *Proceedings of Brighton Crop Protection Conference-Pests and Diseases*, Brighton, 705-716.

[67] Noble, R. and Coventry, E. (2005) Suppression of soil-borne plant diseases with composts: A review. *Biocontrol Science and Technology*, **15**, 3-20.

[68] Termorshuizen, A.J., Van Rijn, E., Van der Gaag, D.J., Alabouvette, C., Chen, Y., Lagerlöf, J., Malandrakis, A.A., Paplomatas, E.J., Rämert, B., Ryckeboer, J., Steinberg, C. and Zmora-Nahum, S. (2006) Suppressiveness of 18 composts against 7 pathosystems: Variability in pathogen response. *Soil Biology and Biochemistry*, **38**, 2461-2477.

[69] Trillas M.I., Casanova, E., Cotxarrera, L., Ordovás, J., Borrero, C. and Avilés, M. (2006) Composts from agricultural waste and the *Trichoderma asperellum* strain T-34 suppress *Rhizoctonia solani* in cucumber seedlings.

Biological Control, **39**, 32-38.

[70] Orlikowski, L.B. (1999) Vermicompost extract in the control of some soil borne pathogens. *International Symposium on Crop Protection*, **64**, 405-410.

[71] Nakasone, A.K., Bettiol, W. and de Souza, R.M. (1999) The effect of water extracts of organic matter on plant pathogens. *Summa Phytopathologica*, **25**, 330-335.

[72] Szczech, M. (1999) Supressiveness of vermicompost against *Fusarium wilt* of tomato. *Journal of Phytopathology*, **147**, 155-161.

[73] Szczech, M., Smolinska, U. (2001) Comparison of suppressiveness of vermicompost produced from animal manures and sewage sludge against *Phytophthora nicotianae* Breda de Haar var. *nicotianae*. *Journal of Phytopathology*, **149**, 77-82.

[74] Edwards, C.A., Arancon, N.Q. and Greytak, S. (2006) Effects of vermicompost teas on plant growth and disease. *BioCycle*, **47**, 28-31.

[75] Kale, R.D. (2006) The role of earthworms and research on vermiculture in India. In: Guerrero III, R.D., Guerrero-del Castillo, M.R.A., Eds., *Vermi Technologies for Developing Countries. Proceedings of the International Symposium-Workshop on Vermi Technologies for Developing Countries*, Los Baños, 16-18 November 2005, 66-88.

[76] Guerrero, R.D. (2009) Vermicompost and vermimeal production. *MARID Agribusiness Technology Guide*, 22 p.

[77] Guerrero, R.D. (2009) Commercial vermimeal production: Is it feasible? In: Guerrero, R.D., Eds., *Vermi Technologies for Developing Countries. Proceedings of the International Symposium-Workshop on Vermi Technologies for Developing Countrie*, Los Baños, 16-18 November 2005, 112-120.

[78] Baker, G.H., Williams, P.M., Carter, P.J. and Long, N.R. (1997) Influence of lumbricid earthworms on yield and quality of wheat and clover in glasshouse trials. *Journal of Soil Biology and Biochemistry*, **29**, 599-602.

[79] Baker, G.H., Brown, G., Butt K., Curry, J.P. and Scullion, J. (2006) Introduced earthworms in agricultural and reclaimed land: Their ecology and influences on soil properties, plant production and other soil biota. *Biological Invasions*, **8**, 1301-1316.

[80] Krishnamoorthy, R.V. and Vajranabhaiah, S.N. (1986) Biological activity of earthworm casts: An assessment of plant growth promoter levels in the casts. *Proceedings of Indian Academy of Sciences (Animal Science)*, **95**, 341-351.

[81] Palanisamy, S. (1996) Earthworm and plant interactions. ICAR Training Program, Tamil Nadu Agricultural University, Coimbatore.

[82] Roberts, P., Jones, G.E. and Jones, D.L. (2007) Yield responses of wheat (*Triticum aestivum*) to vermicompost. *Journal of Compost Science and Utilization*, **15**, 6-15.

[83] Suthar, S. (2005) Effect of vermicompost and inorganic

fertilizer on wheat (*Triticum aestivum*) production. *Nature Environment Pollution Technology*, **5**, 197-201.

[84] Suthar, S. (2010) Vermicompost: An environmentally safe, economically viable and socially acceptable nutritive fertilizer for sustainable farming; In: Sinha, R.K., *et al.*, Eds., *Special Issue on Vermiculture Technology, Journal of Environmental Engineering*, Inderscience Publishing, Olney.

[85] Kale, R.D., Mallesh, B.C., Kubra, B. and Bagyaraj, D.J. (1992) Influence of vermicompost application on the available macronutrients and selected microbial populations in a paddy field. *Soil Biology and Biochemistry*, **24**, 1317-1320.

[86] Jeyabal, A. and Kuppuswamy, G. (2001) Recycling of organic wastes for the production of vermicompost and its response in rice legume cropping system and soil fertility. *European Journal of Agronomy*, **15**, 153-170.

[87] Guerrero, R.D. and Guerrero, L.A. (2008) Effect of vermicompost on the yield of upland rice in outdoor containers. *Asia Life Sciences*, **17**, 145-149.

[88] Buckerfield, J.C. and Webster, K.A. (1998) Worm worked waste boost grape yield: Prospects for vermicompost use in vineyards. *The Australian and New Zealand Wine Industry Journal*, **13**, 73-76.

[89] Sinha, R.K., Herat, S., Valani, D. and Chauhan, K. (2009) Vermiculture and sustainable agriculture. *American-Eurasian Journal of Agricultural and Environmental Sciences*, IDOSI Publication, 1-55.

[90] Webster, K.A. (2005) Vermicompost increases yield of cherries for three years after a single application. *EcoResearch*, South Australia.

[91] Atiyeh, R.M., Subler, S., Edwards, C.A. and Metzger, J.D. (1999) Growth of tomato plants in horticultural potting media amended with vermicompost. *Pedobiologia*, **43**, 1-5.

[92] Gupta, A.K., Pankaj, P.K. and Upadhyava, V. (2008) Effect of vermicompost, farm yard manure, biofertilizer and chemical fertilizers (N, P, K) on growth, yield and quality of lady's finger (*Abelmoschus esculentus*). *Pollution Research*, **27**, 65-68.

[93] Guerrero, R.D. and Guerrero, L.A. (2006) Response of eggplant (*Solanum melongena*) grown in plastic containers to vermicompost and chemical fertilizer. *Asia Life Sciences*, **15**, 199-204.

[94] Agarwal, S, Sinha, R.K. and Sharma, J. (2010) Vermiculture for sustainable horticulture: Agronomic impact studies of earthworms, cow dung compost and vermicompost vis-à-vis chemical fertilizers on growth and yield of lady's finger (*Abelmoschus esculentus*). In: Sinha, R.K. *et al.*, Eds., *Special Issue on Vermiculture Technology, International Journal of Environmental Engineering*, Inderscience Publishing, Olney.

[95] Meena, R.N., Singh, Y., Singh, S.P., Singh, J.P. and Singh, K. (2007) Effect of sources and level of organic manure on yield, quality and economics of garden pea (*Pisum sativam* L.) in eastern Uttar Pradesh. *Vegetable Science*, **34**, 60-63.

[96] Karmegam, N. and Daniel, T. (2008) Effect of vermincompost and chemical fertilizer on growth and yield of Hyacinth Bean (*Lablab purpureas*). *Dynamic Soil, Dynamic Plant, Global Science Books*, **2**, 77-81.

[97] Gutiérrez-Miceli, F.A., Santiago-Borraz, J., Montes Molina, J.A., Nafate, C.C., Abdud-Archila, M., Oliva Llaven, M.A., Rincón-Rosales, R. and Deendoven L. (2007) Vermicompost as a soil supplement to improve growth, yield and fruit quality of tomato (*Lycopersicum esculentum*). *Bioresource Technology*, **98**, 2781-2786.

[98] Wang, D., Shi, Q., Wang, X., Wei, M., Hu, J., Liu, J. and Yang, F. (2010) Influence of cow manure vermicompost on the growth, metabolite contents, and antioxidant activities of Chinese cabbage (*Brassica campestris* ssp. chinensis). *Biology and Fertility of Soils*, **46**, 689-696.

[99] Peyvast, G., Olfati, J.A., Madeni, S. and Forghani, A. (2008) Effect of vermicompost on the growth and yield of spinach (*Spinacia oleracea* L.). *Journal of Food Agriculture and Environment*, **6**, 110-113.

[100] Coria-Cayupán, Y.S., De Pinto, M.I.S. and Nazareno, M. A. (2009) Variations in bioactive substance contents and crop yields of lettuce (*Lactuca sativa* L.) cultivated in soils with different fertilization treatments. *Journal of Agricultural and Food Chemistry*, **57**, 10122-10129.

[101] Kawano, T., Kataoka, N. and Abe, S. (2005) Lifespan extending activity of substances secreted by the nematode *Caenorhabditis elegans* that include the dauer-inducing pheromone. *Bioscience, Biotechnology and Biochem*, **69**, 2479-2481.

4

Investigating poverty in rural Iran: The multidimensional poverty approach

Abdoulrasool Shirvanian, Mohammad Bakhshoodeh[*]

Agricultural Economics, Shiraz University, Shiraz, Iran; [*]Corresponding Author

ABSTRACT

In this study, rural poverty in Iran is investigated applying a multidimensional approach, association rules mining technique, and Levine, F and Tukey tests to household data of 2008. The results indicate that poverty in its multi-dimensions is an epidemic problem in rural Iran. The results also exhibit that there are 11 patterns of poverty in the rural areas including four main patterns with 99.62% coverage and seven sub-patterns with nearly 0.38% coverage. In these patterns, housing and household education are the most important dimensions of poverty and income poverty is the least important dimension. Government income support policy to households, in enforcement the law of targeting subsidies, cannot be regarded as pro poor policy but it follows other political aspects.

Keywords: Multidimensional Poverty Approach; Rural Poverty; Data Mining; Iran

1. INTRODUCTION

Until the early 1990s, poverty definitions and its measuring methods were largely based on income approach where poverty was recognized as lack of minimum income. Accordingly, this approach only considers the welfare aspects of human life that can be expressed in terms of revenue [1-4]. Therefore, the income poverty approach cannot explain much of people capabilities and so cannot be a base for fully explanation of poverty phenomenon in society. Moreover, using the income approach in classifying individuals as poor and non-poor follows the basic abnormality. It is possible that in practice a poor be classified as a non-poor based on income approach [5]. So, focusing on this approach in studying poverty phenomenon and developing strategies and policies to support poor is a big risk [4]. With respect to these matters, moving from the income poverty approach to multidimensional poverty approach is an important pro-gress in the poverty literature [1-3,6,7].

In the multidimensional approach, poverty concentration lies on the deprivation from resources and opportunities that entitle to each person in society, and poverty structure is expressed by reflecting the human failure in different dimensions of human welfare [3,8]. Human welfare has many dimensions such as housing, health, feeding, education, income, etc. Housing concept is not only constraint on the shelter as physical location but also involves the residential environment, all services and facilities that are necessary for better family life, and relatively right and safe occupation. Providing these services and facilities, facilitate inhabitants' activities, increases their efficiency and is a factor in establishing a stable life. Accordingly, efforts to achieve these quality criteria determine the ability of referring the term housing to buildings and structures [5,9-16]. Health poverty focuses on people who need health care. In absence of these cares, they suffer from health deprivation [17]. Someone who has low access to health services drop into disease trap and so disable to obtain suitable food, housing and job. Food poverty is the latest and the most unacceptable sign of frustration in people basic needs and is considered as the most important poverty dimension at the community and occurs when a person is unable to consume enough food according to acceptable society manner [18,19]. Education poverty causes to reduction in the individuals' human capital and so deprives them from suitable position of social opportunities [20-23] and ascending the training level trepans more reduction in the poverty rate [21-25].

To sum up, income alone is not a strong criterion to describe poverty phenomenon and to determine welfare, and therefore paying attention to the other dimensions such as housing, health, food and education are essential in examining the phenomenon of poverty in communities. In investigating these dimensions through multidimensional poverty approach, it is important to note that each welfare dimensions concentrate on the clear and separable matters [4,26-29]. So, in order to calculate each dimension, its criteria should be separately and independently considered from calculation of the criteria of other dimensions [1-3].

Another issue in concerning to poverty phenomenon is related to poverty distribution. According to the literature, poverty distribution in worldwide is such that developing countries suffer more than developed countries. In developing countries, large portion of the population live in rural areas and most of them are poor. So, the rural area in developing countries is considered as poor habitat [30] and poor in developing countries often do not have access to adequate housing and related services [31,32]. In these countries, health inadequacy made health poverty as a feature of rural poverty, notwithstanding optimistic thoughts about health in rural communities [33]. In nourishment dimension, the persons suffer from food poverty belong to the poorest people in developing countries and most of them concentrate in rural areas [11]. Education poverty in these countries is a common matter among many segments of society, especially the villagers [20-23].

Thus, in order to further success in fighting with multidimensional poverty on a global scale, focusing on rural communities in developing countries is essentially and substantially attempt with high emergency.

Iran is one of the developing countries that suffers from most of welfare dimensions. For instance, in housing dimension, despite the ideals aspirations in providing housing and making different strategies to achieve these ideals, the gap of classes between minority groups with the best housing and groups without adequate housing has become deeper [34]. Health system is also poor and imposing heavy costs on households is the most inadequacy and insufficiency of this system [35,36]. In nourishment dimension, in spite of the extensive legal and executive power in order to combat poverty in the country, households are faced with shortages in energy and micronutrients and imbalances in food consumption are intense. Geographic distribution of food poverty is also such that poor are more concentrated in rural areas [37,38].

These collections formed footstone of this investigation and made it essential. Therefore, study of multidimensional poverty phenomenon in Iranian rural society is targeted.

2. METHODS

There are many dimensions to be considered in the multidimensional poverty approach that are restricted to data accessibility [10,39]. Accordingly, five rural poverty dimensions including housing, health, nutrition, education and income were examined in this study.

Following Ravallion [40], food poverty index was calculated based on food usage in the normal range (best nutritional status) considering food pyramid adjusted for age and gender [41,42]. Determining a normal diet based

household food poverty not only provides the body needed energy, but indicates the nature of households' food poverty and can be considered as a practical guideline in the household food management to reduce and eradicate food poverty [11].

The most common indicators of adequate housing, including security, the sewer system, ownership, and density indexes were considered as the housing dimension. Efforts to achieve these quality criteria determine the ability of referring the term housing to buildings and structures [5,9-11,13-16].

Quality of remedy financial management was considered as the indicator of health poverty [43,44]. The household health expenditure as proportion of income was used to identify rural households that suffer from health poverty and to determine their health poverty gap [40,45,46]:

$$x_i = \frac{HC_i}{I_i} \tag{1}$$

where x_i is the health expenditure to income ratio for ith household, and HC_i and I_i are respectively the health expenditure and income of ith household. It should be noted that in the above relationship, household health expenditure is perfectly unexpected and household income, in comparing with this expenditure, is constant [47].

In order to examine education poverty, the information literacy indexes including information admission criterion and indicators of literacy skills were used [48]. The former index focuses on receiving information from various sources, including publications (newspaper, magazine and journal, and books), variety of media-aural visuals (fixed and mobile telephones, radio, television, computer, video and similar devices), and internet [48,49]. Indicators of literacy skills show the status of formal training in households and are introduced as a prerequisite for implementing information literacy skills. Despite the availability of information, lack of these skills can make the usage of these information impossible [48]. In this study, literacy skills were assessed by net enrolment rate [48,49] that shows the percentage of family members gaining education opportunities and calculated as [48,50]:

$$NER_i = \frac{NSL_i}{PN_i} * 100 \tag{2}$$

where NER_i is net enrolment rate at ith level of education, NSL_i is all students in household at ith education level and PN_i represents all household members that potentially lie in the ith education level.

In the multidimensional approach to poverty, income dimension must be calculated independently from other dimensions of poverty [1-3,7] whereas it is the cumulative measure of the monetary needs of individuals in the income poverty approach and so it is not independent of other dimensions of poverty. Therefore, the multidimensional approach to poverty cannot use the methods of calculation poverty line based on the income approach. Due to this, some studies have focused on the inability to earn appropriate income [16,51,52]. Combining information on household expenditure with income is an appropriate approach in order to complete the income criteria in the estimation of income poverty by use of household expenditure survey data [53-55]. In this study, the ratio of net expenditure (expenditure minus investment) to disposable income of household was used for this purpose [53,55,56] as expressed by [57]:

$$IP_i = \frac{TX_i}{TI_i} \qquad (3)$$

where IP_i is the income poverty criteria for the ith household, TX_i and TI_i are respectively total expenditure and total disposable income of the ith household.

Following Grootaert, $et\ al.$ [58] and Okunmadewa, $et\ al.$ [59], in order to aggregate indicators and indexes and then to express household poverty status in an overall index, the values of each indicator and index are normalized by **Eq.4**:

$$p_{ij} = \frac{z_j - x_{ij}}{z_j} \times 100 \qquad (4)$$

in which p_{ij} represents poverty status of the ith household taking the jth indicator or index, z_j is the acceptable value of jth index or indicator and x_{ij} is the amount of the ith household's owners from the jth indicator or index. Then, the overall index of poverty for each household (P) is expressed as [3,60]:

$$P = \frac{1}{n}\sum_{j}^{n} a_j p_j \qquad (5)$$

where n is the number of indicators or indices, a_j indicated the weight of jth indicator or index, and p_j is the poverty rate for each household in the jth indicator or index. It should be noted that the entropy weighting method was used to determine appropriate weights of indicators and indices [61-64].

Furthermore, determining the overall poverty situation in rural society needs to assess the level of the headcount ratio and the poverty gap indexes for each poverty indicator or index. In this study, the FGT indices are utilized to measure poverty rate ($\alpha = 0$) that shows the frequency distribution of poor households and poverty gap ($\alpha = 1$) that expresses the depth of poverty in rural Iran [65]:

$$FGT = P(\alpha) = \frac{1}{n}\sum_{i=1}^{q}\left(\frac{z - x_i}{z}\right)^{\alpha} \qquad (6)$$

where n and q are total and poor households respectively,

z is the acceptable poverty line and x_i is the owner level of ith household.

Moreover, the association rules mining technique, one of the most important non supervisory data mining techniques, was used for extracting poverty patterns in the society. This technique discovers and extracts patterns related to the nature of poverty without providing any previous hypothesis on the extraction of patterns in the society. The advantage of using the association rules mining technique, in comparison to pattern making based on specified hypothesizes, is that it allows the extraction of significant and unpredictable patterns without any information about them [66]. The mining association rules technique identifies those features that engage together. Accordingly, the general form of an association rule is as $X \Rightarrow Pov$ where X represents a set of characteristics of household and Pov represents the overall poverty situation of household and show antecedent and consequent of rule, respectively [66-70].

The discovery of association rules needs some criteria to express certainty degrees of discovered rules. These criteria allow for the rules with high certainty are selected and presented from the set of possible rules. These criteria are the most commonly and applicable criteria to evaluate and assess the accuracy and valuable of the discovered rules. The support criterion expresses as probability and shows the amount of protection of rule based on the individuals' communication level. Simply, this criterion represents the proportion of individuals with a set of features (X) occurring with the expected poverty (Pov), simultaneously. Mathematical expression of this criterion is as follows [67,68,70]:

$$\text{Support}(X \Rightarrow P) = P(X \cap Pov) \qquad (7)$$

in which $P(X \cap Pov)$ is the occurrence probability of the features sets X and Pov, simultaneously.

Confidence criterion expresses the occurrence probability of two or more features together. Thus, this criterion shows the degree of dependence between two features sets, X and Pov. This affiliation is calculated as follows [67,68,70]:

$$\text{Confidence}(X \Rightarrow Pov) = P(Pov/X) = \frac{P(X \cap Pov)}{P(X)} \qquad (8)$$

where $P(Pov/X)$ represents the occurrence probability of poverty with respect to occurrence attribute set X, and $P(X)$ represents the occurrence probability of features set X, regardless Pov. Other notations are defined previously. The more the confidence criterion, the higher the validation of pattern discovery would be.

Finally, lift rate criterion represents the ability level of pattern to provide the expected confidence. This criterion compares the pattern confidence with the expected confidence. The expected confidence is the confidence level that obtain when antecedent part (*X*) cannot increase the probability of occurrence poverty. Mathematical expression of this criterion is as follows [70]:

$$\text{Lift}\left(X \Rightarrow Pov\right) = \frac{\text{Confidence}\left(X \Rightarrow Pov\right)}{P\left(Pov\right)}$$
$$= \frac{P\left(X \cap Pov\right)}{P\left(X\right).P\left(Pov\right)} \qquad (9)$$

where $P\left(Pov\right)$ represents the occurrence probability of poverty regardless of the features set *X*. Other notations are defined previously.

In the extract patterns of rural poverty, one-way ANOVA test were used in order to assess dispersion of poverty dimensions. With respect to the fact that the ANOVA test is possible in two state including variance homogeneity and variance heterogeneity, it is needed to check homoscedasticity and heteroscedasticity in the patterns of rural poverty before applying this test. For this, several tests including the Fisher's test, Bartlett's test and Levine test are referred. Contrary to other tests, Levine test is less sensitive to the normal distribution of the population and so is used in this study [71]. The F test is also used to examine differences between the patterns of rural poverty in each of poverty dimensions. The test is overall test in examining differences between the patterns of rural poverty [71]. Based on F test, if average difference between each of poverty dimensions in the patterns of rural poverty is more than inter group differences, it inferred that these patterns are totality different in that poverty dimension.

Following by F statistic calculation and overall comparison of the patterns of rural poverty, Tukey test, that is the honestly significant test of differences, was used to assess the signification of average difference between pair patterns in each of rural poverty dimension [71].

In the conventional definitions of poverty and determining its level, planners are often inclined to use concept of the household [11,13,72]. In this regard, the household survey data published by the Iranian Statistics Center (2008) run at the national level and covering data in housing, education, food, health and income dimensions of Iranian households were used in this study.

3. RESULTS AND DISCUSSION

Table 1 provides information on the various dimensions of poverty in Iranian rural society. According to the table, all rural households have been dominated under education poverty. Based on the poverty gap, the depth and quality of the education poverty of households is such that rural households, on average, do not have access to nearly 44% of education facilities. Poverty rate also indicates that the vast majority of rural households

Table 1. Poverty rate and gap indexes in each poverty dimensions, whether or not prevail other dimensions, in the sample of rural households.

Dimensions of rural poverty	Poverty indexes (%)	
	Headcount ratio	Poverty gap
Education poverty	100.00	43.89
Housing poverty	99.98	38.46
Food poverty	99.64	41.85
Income poverty	57.04	1.84
Health poverty	36.96	0.35
Overall poverty	100.00	37.43

(nearly 100%) experience housing poverty. The depth and quality of the housing poverty suggests that rural households, on average, deprived from 38.46% of the standard of housing indicators. In the food dimension, headcount ratio shows 99.64% of rural households suffer from food poverty. This situation, similar to the state of headcount ratios in education and housing poverties, represents a broad range of food poverty in Iranian rural society. Based on the poverty gap, the quality of food poverty in Iranian rural community is such that on average, rural households use foods 41.85% below the recommended levels. As far as the income dimension is concerned, more than half (57.04%) of rural households are recognized to be poor. The income poverty gap among rural households is equal to 1.84% on average. Finally, 36.96% of rural households are faced with health poverty and the quality and depth of health poverty gap index in rural areas is equal to 0.35%.

In comparison, the largest proportions of poverty in these areas are attributed to education poverty as well as housing and food poverties. Minimum coverage of poverty in Iranian rural community is also related to health poverty. From the perspective of depth and quality of domination of poverty dimensions, poverty gap indicates that education poverty has the greatest and health poverty has the lowest depth. Based on this, not only the housing poverty lies in warning status, but also this warning is in the other dimensions of Iranian rural poverty, including education and food poverties. In the field of education poverty, the alert status that exist in both outer (headcount ratio) and inner (poverty gap) layers is more severe than housing poverty. In the field of food poverty, warning status merely in the perspective of the depth of poverty is more severe than the housing poverty. These situations present poverty in its multi-dimensions as an epidemic problem in Iranian rural society. The amount of headcount ratio (100%) in overall poverty index corroborates this phenomenon. On the other hand, overall overview of depth and quality of multidimensional poverty indicate that rural households deprive from 37.43% of welfare dimensions.

Table 2 presents patterns of rural poverty among rural households in the sample. As passed, five dimensions of poverty have been studied in this study. Accordingly, 32 rural poverty patterns could be derived independently, where each rural household merely lies in one of them. Table 2 suggests that, 11 poverty patterns are merely visible in the Iranian rural community. The values obtained for the lift and confidence criteria in these 11 poverty patterns indicate that each of these patterns is able to earn the highest confidence level (100%) with the highest lift (100%). In the perspective of support criterion, first to fourth poverty patterns allocate the highest values of this criterion to themselves. The fourth poverty pattern that reflects merely prevail education, housing and food poverties in rural households, with 34.30% of all households have the highest proportion of rural households. After that, the third poverty pattern lay, where rural households are faced with income poverty in addition to poverty dimensions mentioned in the previous poverty pattern. This poverty pattern allocates 28.51% of rural households to itself. In continue, the first pattern of rural poverty with a share equal to 28.22% of rural households is located. This poverty pattern includes all rural poverty dimensions, and so, it is the most complete pattern of

rural poverty. In the perspective of proportion of rural households, the second poverty pattern is located after these three patterns. In this poverty pattern, all poverty dimensions, in the absence of income dimension, are prevailed and it covers 8.59% of rural households. These four patterns, totality, cover 99.62% of rural households. Accordingly, first to fourth poverty patterns are considered as the main patterns of rural poverty. Seven other patterns of rural poverty, totally, have taken 0.38% of rural households. So, these patterns are regarded as subpatterns of rural poverty in Iranian rural society.

Table 3 shows mathematical structure of main patterns of poverty among the rural households. As can be seen, housing poverty is the most important dimension of rural poverty in the formation overall poverty in all main rural poverty patterns. So, by including the weights between 0.55 till 0.63 in rural poverty patterns, this dimension of rural poverty contributes over 50% in forming the overall poverty index. After the housing poverty, education poverty in the main rural poverty patterns with weights in the range of 0.37 until 0.42 is the most important dimension of rural poverty. Based on their importance, these dimensions are common in all main patterns of rural poverty to forming overall poverty. Other dimensions of rural poverty, including food, health and income poverties are devoted much lower weights than housing and education poverties weights in the rural poverty patterns,

Table 2. Poverty patterns of Iranian rural society and their evaluation criteria.

Patterns No.	Nature of rural poverty patterns	Support	Confidence	Lift	Observations	Aggregated frequency
1	Income Poverty = 1, Health Poverty = 1, Food Poverty = 1, Housing Poverty = 1, Education Poverty = 1 → Overall Poverty = 1	28.22	100	100	5561	28.22
2	Income Poverty = 0, Health Poverty = 1, Food Poverty = 1, Housing Poverty = 1, Education Poverty = 1 → Overall Poverty = 1	8.59	100	100	1692	36.81
3	Income Poverty = 1, Health Poverty = 0, Food Poverty = 1, Housing Poverty = 1, Education Poverty = 1 → Overall Poverty = 1	28.51	100	100	5619	65.32
4	Income Poverty = 0, Health Poverty = 0, Food Poverty = 1, Housing Poverty = 1, Education Poverty = 1 → Overall Poverty = 1	34.30	100	100	6759	99.62
5	Income Poverty = 1, Health Poverty = 0, Food Poverty = 0, Housing Poverty = 1, Education Poverty=1 → Overall Poverty=1	0.15	100	100	30	99.77
6	Income Poverty = 1, Health Poverty = 1, Food Poverty = 0, Housing Poverty = 1, Education Poverty = 1 → Overall Poverty = 1	0.14	100	100	28	99.91
7	Income Poverty = 0, Health Poverty = 0, Food Poverty = 0, Housing Poverty = 1, Education Poverty = 1 → Overall Poverty = 1	0.06	100	100	12	99.97
8	Income Poverty = 0, Health Poverty = 1, Food Poverty = 0, Housing Poverty = 1, Education Poverty = 1 → Overall Poverty = 1	0.01	100	100	2	99.98
9	Income Poverty = 1, Health Poverty = 0, Food Poverty = 1, Housing Poverty = 0, Education Poverty = 1 → Overall Poverty = 1	0.01	100	100	2	99.99
10	Income Poverty = 1, Health Poverty = 1, Food Poverty = 1, Housing Poverty = 0, Education Poverty = 1 → Overall Poverty = 1	5.07−E03	100	100	1	99.99
11	Income Poverty = 0, Health Poverty = 0, Food Poverty = 1, Housing Poverty = 0, Education Poverty = 1 → Overall Poverty = 1	5.07−E03	100	100	1	100
	Total patterns	100	-	-	19,707	-

Table 3. The mathematical structure of the main poverty patterns.

Main Patterns	Mathematical structure
First pattern	1.12−E04 * Income Poverty + 0.04 * Health Poverty + 8.68−E04 * Food Poverty + 0.59 * Housing Poverty + 0.38 * Education Poverty
Second pattern	0.03 * Health Poverty + 1.75−E03 * Food Poverty + 0.55 * Housing Poverty + 0.42 * Education Poverty
Third pattern	3.54−E04 * Income Poverty + 1.53−E03 * Food Poverty + 0.63 * Housing Poverty + 0.37 * Education Poverty
Fourth pattern	3.22−E03 * Food Poverty + 0.62 * Housing Poverty + 0.37 * Education Poverty

and thus they are at lower importance levels in the overall poverty. Among the recent three rural poverty dimensions, the food poverty is common among all main patterns of rural poverty. The heath dimension has the highest weight in the pattern that include food, health and income poverty dimensions. Finally, income dimension, with the lowest weight is considered as the least important among all rural poverty dimensions.

As shown in **Table 2**, the frequency distribution of poor rural households in four main patterns are 28.22%, 8.59%, 28.51% and 34.30% of rural households, respectively. Accordingly, the fourth pattern is the most important pattern of rural poverty from the perspective of households' coverage. The third, first and second patterns are lie after the first one. Reviewing this issue from the perspective of poverty gap in overall poverty index and in each of poverty dimensions require procedures such as Levine test, F test and Tukey's test. **Table 4** indicates the results of Levin and F statistics. The Levine test results for all poverty dimensions and overall poverty index in the main patterns of rural poverty show that the variances of all dimensions are equal in all main patterns. Thus, the main patterns in different dimensions of rural poverty are homoscedastic and so, we can use one-way ANOVA test with assuming the existence of homogeneity of variance between them in order to comparing the poverty gap in the different dimensions of poverty in these patterns.

F test results in all poverty dimensions and in overall poverty index of mentioned rural poverty patterns suggests that rural poverty in all configurations are distinct in all patterns. So that, in the mentioned rural poverty patterns the differences of average poverty gaps in each of poverty dimensions are statistically significant and this situation exists in the average of overall poverty index (**Table 4**).

Table 5 provides more detail information related to main patterns of rural poverty and shows significantly differences between the averages of poverty gap in each poverty dimensions in each pair of these patterns. Reviewing this issue suggests that the first pattern, by including 1.13% and 3.77% of poverty gap, respectively in the fields of health and income poverty is the most important pattern of poverty in rural society. The third and fourth patterns with respected 39.40% and 47.07% of

Table 4. Levine and F statistics for each of dimensions in the main rural patterns.

Dimensions of rural poverty	Levine statistics	F statistics
Education Poverty	9.54***	5.96***
Housing Poverty	9.59***	5.06***
Food Poverty	23.50***	283.67***
Health Poverty	684.30***	415.76***
Income Poverty	1491.60***	1416.63***
Overall Poverty	22.67***	29.80***

***Significant at 1%.

poverty gap are the most important poverty patterns on housing poverty and perspective food poverty in Iranian rural society in. In the field of education poverty, although the fourth poverty pattern has the biggest poverty gap, this value is not statistically significant from the poverty gap values in the first and third patterns. Therefore, these three patterns are commonly the most important patterns of rural poverty in this perspective. The overall poverty outcome, in form of overall poverty index, indicates that the fourth poverty pattern has the highest value of poverty gap.

Also, according to **Table 5**, the first pattern, with 37.77% of poverty gap in field of food poverty, has the lowest poverty gap, whilst the second and fourth poverty patterns exclude health poverty the third and fourth patterns do income poverty dimension. The second poverty pattern has the lowest poverty gap in housing, not statistically significant different from the corresponding values for the first and fourth patterns and therefore the lowest rate of poverty gap is commonly devoted to these three patterns. Similarly, the second pattern has the lowest education poverty gap not statistically significant from that of the first pattern and so these two patterns are commonly categorized similar in this context. The overall poverty outcome, in form of overall poverty index, indicates that the first poverty pattern has the lowest value of poverty gap.

Important note with regard to **Table 5** is that rural households are close to each other in term of the overall poverty index. Accordingly, it seems that the same level

Table 5. Average poverty gap in dimensions of main rural poverty patterns and its comparisons.

The main patterns of rural poverty	Dimensions of rural poverty					
	Education poverty	Housing poverty	Foods poverty	Health poverty	Income poverty	Overall poverty
First pattern	43.91ab	38.46a	37.77a	1.13a	3.77a	39.18a
Second pattern	43.69a	38.02a	38.95b	0b	2.64b	40.10b
Third pattern	43.95b	39.40b	45.83c	0.36c	0c	40.03b
Fourth pattern	44.03b	38.65a	47.07d	0b	0c	40.69c

Note: In each column, common letters indicate no significant difference and non-shared letters indicate significant differences in the level of 10%.

of facilities and resources are needed and the same programs should be developed to combat poverty. But what lies behind this similarity suggests existence of different structures of poverty in the rural society, despite the similarity in the overall index of poverty. So, combating rural poverty requires different plans and different facilities and resources that cannot be provided merely by government income support.

4. CONSEQUENCE OF USING MERELY INCOME POVERTY IN IDENTIFICATION OF POOR HOUSEHOLDS

As previously revealed in **Table 1**, nearly 57% of rural households suffer from income poverty and the rest of them (43%) are free of it. According to enforcement process of targeting subsidies law in Iran, determining the poor and vulnerable households who need government support, is based on household per capita income. Thus, 43% of rural households who do not suffer from income poverty cannot receive the government support program. **Table 6** provides information regarding the number and frequency of rural households who do not suffer from income poverty, but suffer from poverty in other dimensions. According to this table, all households that are free of income poverty suffer from education poverty. The vast majority of these households also suffer from food and housing poverties. In addition, about 20% of such households suffer from health poverty. Reviewing these cases at all households in the sample are also noteworthy. According to the third column of **Table 6**, despite the lack of income poverty, 42.96%, 42.95%, 42.89% and 8.60% of all rural households suffer from education, housing, food and health poverties, respectively. So, it can be deduced that in Iranian rural society, not only households with income poverty need to be supported but also the vast majority of households without income poverty, need assistance and support to deal with education, housing, food and health poverties. If the support in the targeting subsidy scheme confine to households with income poverty, the mentioned groups of rural households will be ignored. Thus, income sup-

port in targeting subsidies program is not in favor of these groups of poor rural households and does not lead them to exit from poverty.

Reviewing this issue in the patterns of rural poverty is also considerable. According to **Table 2**, among the 11 patterns obtained for Iranian rural poverty, income poverty along with other poverty dimensions govern in six patterns. The rest of patterns are free of income poverty but prevail the other dimensions of poverty. With respect to that in enforcement the law of targeting subsidies, support of families developed based on their income level and in the early years of its implementation, support packages of targeting subsidies program is merely income. Therefore, enforcing the law of targeting subsidies will be last different effects on the mentioned patterns. Thus, Income support to poor households does not effect on income poverty in five poverty patterns that cover 42.96% of rural poor households, and merely affect on this dimension in six patterns that cover 57.04% of them (**Table 2**).

5. CONCLUSIONS AND RECOMMENDATIONS

The finding showed that education poverty in perspective headcount ratio, among the various dimensions of poverty in Iranian rural society, is the vastest and then with small differences housing and food poverties are located. Minimum coverage of poverty in Iranian rural society is also related to health poverty. From perspective of depth and quality of different poverty dimensions those dominated on rural society, the greatest and least depth of poverty are devoted to education and health poverties, respectively. Accordingly, not only the condition of housing poverty in Iranian rural society, similar to the situation of housing poverty in developing countries [31,32], is on alert status, but also this warning status are in the other dimensions of rural poverty, including education and food poverties. In the field of education poverty, alert state in the term of level and depth are much severer than housing poverty. In the food poverty field, alert status merely in the perspective of the depth of

Table 6. Frequency distribution of poverty dimensions amongst non income poor households in rural Iran.

Dimensions of multidimensional poverty	Suffered households	Frequency with respect to households that are not income poor	Frequency with respect to all households
Education poverty	8466	100.00	42.96
Housing poverty	8465	99.99	42.95
Foods poverty	8452	99.83	42.89
Health poverty	1694	20.01	8.60

poverty is much severer than housing poverty. These situations present poverty in its multi-dimensions as an epidemic problem in Iranian rural society. The high headcount ratio (100%) in overall poverty index corroborates this phenomenon. Important note in estimating poverty using the multidimensional approach in Iranian rural society is that the estimates indicated that whole Iranian rural society is suffering from poverty. This is confirmed in the literature of poverty. Bossert, *et al.* [73] expressed that the non deprivation of people in the real world so much rarely happens that it can be ignored. Therefore, all individuals in the society typically suffer from poverty. But in the Iranian rural society, the depth and quality of multidimensional poverty is such that, in total, the poor households are suffering from deprivation of 37.43% of welfare indexes.

With respect to investigated dimensions, merely 11 poverty patterns are visible in the Iranian rural community. These poverty patterns emphases that prevailed poverty on rural society in Iran is not merely income poverty. Rural households with respect to their situations are depriving from one or more dimensions which income poverty may be one of their poverty dimensions. From these 11 patterns, four patterns cover 99.62% of the rural households. Accordingly, these poverty patterns consider as main patterns of rural poverty. Seven other patterns of rural poverty, totally, have taken 0.38% of rural households. So, these patterns are regarded as subpatterns of rural poverty in Iranian rural society.

Important note related to the Iranian rural poverty patterns is that the overall poverty index of these patterns is close to each other. Accordingly, it seems that the same level of facilities and resources are needed and the same program should be developed to combat poverty. But what lies behind this similarity suggests existence of different structures of poverty in the rural society, despite the similarity in the overall index of poverty. Therefore, combating rural poverty requires different plans and different facilities and resources that cannot be provided merely by government income support.

In this regard, the results showed that the inliers of poverty dimensions (quality and depth of poverty) in Iranian rural society made different orders in rural poverty patterns. Thus, in the perspective of health and income

poverties the first pattern, in the perspective of housing poverty the third pattern, in the perspective of food poverty the fourth pattern, in the perspective of education poverty, commonly, the first, third and fourth patterns, and in the perspective of overall poverty index the fourth pattern are the most important patterns in rural society, respectively. Similarly, in the perspective of food poverty the first pattern, in the perspective of health poverty, commonly, the second and fourth patterns, in the perspective of income poverty, commonly, the third and fourth patterns, in the perspective of housing poverty, commonly, the first, second and fourth patterns, in the perspective of education poverty, commonly, the first and second patterns, and in the perspective of overall poverty index the first pattern are the least important patterns in rural society, respectively.

In addition, study of structure of overall poverty in main patterns of Iranian rural poverty indicated that housing poverty is the most important dimension in the formation overall poverty in all poverty patterns. Moreover, educational poverty, after the housing poverty, is the most important dimension of rural poverty. Degree of importance of these rural poverty dimensions is such that these dimensions are common in all main patterns of rural poverty to forming overall poverty. Other dimensions of rural poverty, including food, health and income poverty have much lower importance than housing and educational poverties in the rural poverty patterns. Among the recent three rural poverty dimensions, the food poverty is common among all the main patterns of rural poverty. Finally, income poverty among all rural poverty dimensions is considered as the least important.

With respect to enforcement process of targeting subsidies law in Iran, determining the poor and vulnerable households those need government support, is based on household per capita income. Thus, 42.96% of rural households, those do not suffer from income poverty, cannot receive the government support program. The results showed that all households who are free of income poverty suffer from education poverty. The vast majority of these households also suffer from food and housing poverties. In addition, about 20% of such households suffer from health poverty. Accordingly, it can be deduced that in Iranian rural society, not only households with income poverty need to be supported but also, the vast majority of households without income poverty, need assistance and support to deal with education, housing, food and health poverties. If the support in the targeting subsidy scheme confine to households with income poverty, the mentioned groups of rural households will be ignored. Thus, income support in targeting subsidies program is not in favor of these groups of poor rural households and does not lead them to exit from poverty.

Reviewing this issue in the patterns of rural poverty is also considerable. Among the 11 patterns obtained for Iranian rural poverty, in six patterns, income poverty along with other poverty dimensions govern on the rural households. The rest of patterns are free of income poverty but prevail the other dimensions of poverty. With respect to that in enforcement the law of targeting subsidies, support of families developed based on their income level and in the early years of its implementation, support packages of targeting subsidies program is merely income. Therefore, enforcement of the law of targeting subsidies may have different effects on the mentioned patterns. Thus, Income support to poor households does not influence income poverty in five poverty patterns that cover 42.96% of rural poor households, and merely affect on this dimension in six patterns that cover 57.04% of them. The government successful or unsuccessful in social support policy depends on ability to identifying deprived households in welfare dimensions [52], so government social support policy to households is inefficient and it is not pro poor policy but follows other political aspects.

REFERENCES

[1] Alkire, S. and Foster, J. (2008) Counting and multidimensional poverty measurement. Oxford Poverty and Human Development Initiative, University of Oxford, Oxford.

[2] Bossert, W., Chakravarty, S.R. and Ambrosio, C.D. (2009) Measuring multidimensional poverty: The generalized counting approach. The Social Sciences and Humanities Research Council of Canada, Ottawa.

[3] Bourguignon, F. and Chakravarty, S. (2003) The measurement of multidimensional poverty. *Journal of Economic Inequality*, **1**, 25-49.

[4] Rojas, M. (2008) Experienced poverty and income poverty in Mexico: A subjective well-being approach. *World Development*, **36**, 1078-1093.

[5] Whitener, L.A. (2000) Housing poverty in rural areas greater for racial and ethnic minorities. *Rural America*, **15**, 1-8.

[6] Moisio, P. (2004) A latent class application to the multi-

dimensional measurement of poverty. *Quality and Quantity*, **38**, 703-717.

[7] Tsui, K.Y. (1995) Multidimensional generalizations of the relative and absolute indices: The Atkinson-Kolm-Sen approach. *Journal of Economics Theory*, **67**, 251-265.

[8] Sen, A. (1976) Poverty: An ordinal approach to measurement. *Econometrica*, **44**, 219-231.

[9] Dewilde, C. (2004) The multidimensional measurement of poverty in Belgium and Britain: A categorical approach. *Social Indicators Research*, **68**, 331-369.

[10] Krishnakumar, J. and Ballone, P. (2008) Estimating basic capabilities: A structural equation model applied to Bolivia. *World Development*, **36**, 992-1010.

[11] Park, K. (2005) Text book of preventive and social medicine. 18th Edition, Banarsidas Bhanot Publisher, Jabalpur.

[12] Reckford, J. (2010) Housing and health: Partners against poverty, Shelter Report. Habitat for Humanity International, Atlanta.

[13] Sato, H. (2006) Housing inequality and housing poverty in urban China in the late 1990s. *China Economic Review*, **17**, 37-50.

[14] Shinns, L.H. and Lyne, M.C. (2003) Symptoms of poverty within a group of land reform beneficiaries in the Midlands of KwaZulu-Natal: Analysis and policy recommendations. US Agency for International Development (USAID), Washington DC.

[15] Thalmann, P. (2003) House poor or simply poor? *Journal of Housing Economics*, **12**, 291-317.

[16] Zeller, M., Sharma, M., Henry, C. and Lapenu, C.C. (2006) An operational method for assessing the poverty outreach performance of development policies and projects: Results of case studies in Africa, Asia, and Latin America. *World Development*, **34**, 446-464.

[17] Folland, S., Goodman, A. and Stano, M. (2010) Economics of health and health care. 6th Edition, Prentice Hall Inc., Upper Saddle River.

[18] Balanda, K.P., Hochart, A., Barron, S. and Fahy, L. (2008) Tackling food poverty: Lessons from the decent food for all (DFfA) intervention. Institute of Public Health in Ireland, Dublin.

[19] Riches, G. (2002) Food banks and food security: Welfare reform, human rights and social policy, lessons from Canada? *Social Policy and Administration*, **36**, 648-663.

[20] Anand, S. and Sen, A. (2000) Human development and economic sustainability. *World Development*, **28**, 2029-2049.

[21] Behr, T., Christofides, C. and Neelakantan, P. (2004) The effects of state public k-12 education expenditures on income distribution. National Education Association (NEA) Research, Washington DC.

[22] Galbraith, K.J. (1991) Economics in the century ahead.

The Economic Journal, **101**, 41-46.

[23] Psacharopoulos, G. and Woodhall, M. (1985) Education for development: An analysis of investment choices. Oxford University Press, Oxford.

[24] Nordtveit, B.H. (2008) Poverty alleviation and integrated service delivery: Literacy, early child development and health. *International Journal of Educational Development*, **28**, 405-418.

[25] Tilak, J.B.G. (2007) Post-elementary education, poverty and development in India. *International Journal of Educational Development*, **27**, 435-445.

[26] Rojas, M. (2006) Life satisfaction and satisfaction in domains of life: Is it a simple relationship? *Journal of Happiness Studies*, **7**, 467-497.

[27] Rojas, M. (2006) Well-being and the complexity of poverty: A subjective well-being approach. In: McGillivray, M. and Clarke, M., Eds., *Understanding Human Well-Being*, United Nations University Press, Tokyo, 182-206.

[28] Rojas, M. (2007) The complexity of well-being: A life satisfaction conception and a domains-of-life approach. In: Gough, I. and McGregor, A., Eds., *Wellbeing in Developing Countries: From Theory to Research*, Cambridge University Press, Cambridge, 242-258.

[29] Van-Praag, B.M.S., Frijters, P. and Ferrer-i-Carbonell, A. (2003) The anatomy of subjective well-being. *Journal of Economic Behavior and Organization*, **51**, 29-49.

[30] Dixon, C. (1990) Rural development in the third world. Routledge Chapman and Hall Inc., Cambridge.

[31] Miltin, D. (2001) Housing and urban poverty: A consideration of the criteria of affordability, diversity and inclusion. *Housing Studies*, **16**, 509-522.

[32] Sengupta, U. (2010) The hindered self-help: Housing policies, politics and poverty in Kolkata, India. *Habitat International*, **34**, 323-331.

[33] Antony, G.M. and Rao, K.V. (2007) A composite index to explain variations in poverty, health, nutritional status and standard of living: Use of multivariate statistical methods. *Public Health*, **121**, 578-587.

[34] Nassiri, M. (2007) Geographic distribution of housing poverty and scattering householder divorced women in 22 regions of Tehran. *Journal of Social Welfare*, **24**, 240-223.

[35] Babai, N. (2003) Social policy and health. *Journal of Social Welfare*, **10**, 201-232.

[36] Ministry of Health and Medical Education (2004) National document of health sector development on the forth economic, social and cultural program. Deputy of harmony and community, Ministry of Health and Medical Education, Tehran.

[37] Endocrine and Metabolism Research Center (2001) Preliminary results in goiter prevalence in Iran provinces. Martyr Beheshti University of Medical Sciences, Tehran.

[38] Kimiagar, M. and Badjan, M. (2005) Poverty and malnutrition in Iran. *Journal of Social Welfare*, **18**, 112-191.

[39] Alkire, S. (2007) Choosing dimensions: The capability approach and multidimensional poverty. Department of International Development, Chronic Poverty Research Centre, University of Oxford, Oxford.

[40] Ravallion, M. (1992) Poverty comparisons: A guide to concepts and methods. The World Bank, Washington DC.

[41] Alexopoulos, Y., Hebberd, K. and Bays, H. (2008) Krause's food and nutrition therapy. 12th Edition, Elsevier Inc., Amsterdam.

[42] The Health Canada Web Site (2010) Canada's food guide, Farsi Version.

[43] Fasco, A. (2003) On the definition and measurement of poverty: The contribution of multidimensional analysis. *Proceedings of the 3rd Conference on the Capability Approach: From Sustainable Development to Sustainable Freedom*, Pavia, 7-9 September 2003, 1-39.

[44] Saisana, M. and Saltelli, A. (2010) The multidimensional poverty assessment tool (MPAT): Robustness issues and critical assessment. European Commission and Institute for the Protection and Security of the Citizen, Ispra.

[45] Veenstra, N. (2006) Social protection in a context of HIV/AIDS: A closer look at South Africa. *Social Dynamics*, **32**, 111-135.

[46] Doorslaer, E.V., O'Donnell, O., Rannan-Eliya, R.P., Somanathan, A., Adhikari, S.R., Garg, C.C., Harbianto, D., Herrin, A.N., Nazmul-Huq, M., Ibragimova, S., Karan, A., Wan-Ng, C., Pande, B.R., Racelis, R., Tao, S., Tin, K., Tisayaticom, K., Trisnantoro, L., Vasavid, C. and Zhao, Y. (2006) Effect of payments for health care on poverty estimates in 11 countries in Asia: An analysis of household survey data. *Lancet*, **368**, 1357-1364.

[47] Russell, S. (1996) Ability to pay for health care: Concepts and evidence. *Health Policy and Planning*, **11**, 219-237.

[48] UNESCO Institute for Statistics (2008) List of potential international indicators for information supply, access and supporting skills. UNESCO, Paris.

[49] UNESCO (2005) Education for all, literacy for life. United Nations Educational, Scientific and Cultural Organization (UNESCO), Paris.

[50] UNESCO (2010) Glossary.

[51] Henry, C., Sharma, M., Lapenu, C. and Zeller, M. (2003) Microfinance poverty assessment tool. Consultative Group to Assist the Poor (CGAP) and the World Bank, Washington DC.

[52] Naveed, A. and Ul-Islam, T. (2010) Estimating multidimensional poverty and identifying the poor in Pakistan: A alternative approach. Research Consortium on Educational Outcomes and Poverty (RECOUP), Cambridge.

[53] Saunders, P. (1997) Living standards, choice and poverty. *Australian Journal of Labour Economics*, **1**, 49-70.

[54] Saunders, P. and Hill, T. (2008) A consistent poverty approach to assessing the sensitivity of income poverty measures and trends. *The Australian Economic Review*, **41**, 371-388.

[55] Saunders, P., Hill, T. and Bradbury, B. (2007) Poverty in Australia sensitivity analysis and recent trends. Social Policy Research Centre, University of New South Wales, Kensington.

[56] Schubert, R. (1994) Poverty in developing countries: Its definition, extent and implications. *Economics*, **49-50**, 17-40.

[57] Smith, P. (1996) Measuring outcome in the public sector. Taylor and Francis Ltd., London.

[58] Grootaert, C. and Narayan, D. (2004) Local institutions, poverty and household welfare in Bolivia. *World Development*, **32**, 1179-1198.

[59] Okunmadewa, F.Y., Yusuf, S.A. and Omonona, B.T. (2007) Effects of social capital on rural poverty in Nigeria. *Pakistan Journal of Social Sciences*, **4**, 331-339.

[60] Muro, P.D., Mazziotta, M. and Pareto, A. (2009) Composite indices for multidimensional development and poverty: An application to MDG indicators. University of Roma Tra, Rome.

[61] Deutsch, J. and Silber, J. (2005) Measuring multidimensional poverty: An empirical comparison of various approaches. *Review of Income and Wealth*, **51**, 145-174.

[62] Shannon, C.E. (1948) The mathematical theory of communication. *The Bell System Technical Journal*, **27**, 379-423, 623-656.

[63] Tsui, K.Y. (1999) Multidimensional inequality and multidimensional generalized entropy measures: An axiomatic derivation. *Social Choice and Welfare*, **16**, 145-157.

[64] Zhi-Hong, Z., Yi, Y. and Jing-Nan, S. (2006) Entropy method for determination of weight indicators in fuzzy synthetic evaluation for water quality assessment. *Journal of Environmental Science*, **18**, 1020-1023.

[65] Foster, J., Greer, J. and Thorbecke, E. (2010) The Foster-Greer-Thorbecke (FGT) poverty measures: 25 years later. *Journal of Economic Inequality*, **8**, 491-524.

[66] Han, J. and Kamber, M. (2006) Data mining: Concepts and techniques. 2nd Edition, Morgan Kaufmann Publishers, San Francisco.

[67] Margahny, M.H. and Mitwaly, A.A. (2005) Fast algorithm for mining association rules. *Proceedings of the Artificial Intelligence and Machine Learning 05 Conference*, Cairo, 19-21 December 2005, 19-21.

[68] Olson, D.L. and Delen, D. (2008) Advanced data mining techniques. Springer, New York.

[69] Russell, G.J. and Petersen, A. (2000) Analysis of cross category dependence in market basket selection. *Journal of Retailing*, **78**, 367-392.

[70] Thakur, M., Olafsson, S., Lee, J.S. and Hurburgh, C.R. (2010) Data mining for recognizing patterns in foodborne disease outbreaks. *Journal of Food Engineering*, **97**, 213-227.

[71] Freund, R.J., Mohr, D.L. and Wilson, W.J. (2010) Statistical methods. 3th Edition, Elsevier Inc., Amsterdam.

[72] Silber, J. (2007) Measuring poverty: Taking a multidimensional perspective. *Hacienda Publica Espanola/Revista de Economia Publica*, **182**, 29-73.

[73] Bossert, W., Ambrosio, C.D. and Peragine, V. (2007) Deprivation and social exclusion. *Econonica*, **74**, 777-803.

5

Improving crop yield, N uptake and economic returns by intercropping barley or canola with pea

Sukhdev S. Malhi

Agriculture and Agri-Food Canada, Melfort, Canada

ABSTRACT

Two field experiments were conducted from 2009 to 2011 on a Gray Luvisol (Typic Haplocryalf) loam at Star City, Saskatchewan, Canada, to determine the effectiveness of intercropping barley or canola with pea in improving crop yield, total N uptake, seed quality, Land Equivalency Ratio (LER) and economic returns compared to barley, canola or pea grown as monocultures. Average seed yields of barley-pea or canola-pea intercrops were usually greater than those of barley, canola or pea as sole crops. In intercrops, application of N fertilizer increased seed yield of barley or canola but had only slight beneficial effect on the combined seed yield of both crops together. The LER values for intercrops were usually much greater than 1, suggesting less land requirements of intercropping systems than monoculture for the same seed yield. Net returns were lowest for barley as sole crop. Without applied N, net returns were slightly lower for barley-pea intercrop and slightly greater for canola-pea intercrop than pea as a sole crop. Generally, protein concentration in canola or barley seed was higher and oil concentration in canola seed was lower in intercrop combinations compared to sole crops. Response trends of total N uptake in seed or straw were usually similar to that of seed or straw yield. In conclusion, intercropping barley or canola with pea improved yield, N uptake and net returns, suggesting the potential of barley-pea or canola-pea intercrops and pea for organic farming systems.

Keywords: Barley; Canola; Crop yield; Economic returns; Intercrop; N Fertilizer; N Uptake; Oil; Pea; Protein

1. INTRODUCTION

Intercropping refers to growing two or more distinct crops in the same field at the same time. Intercropping, especially a mix of non-legume and legume crops, can have many benefits. Intercropping adds diversity to the cropping system, resulting in the stability of production by lowering risk of crop failure in barley-pea intercropping [1,2]. Intercropping may reduce input costs by lowering fertilizer and pesticide requirements, and thus increase economic returns for mustard-pea or barley-pea intercrops [3,4], and also improve harvest ability of crops in cereals-pea or cereal-lentil intercrops [5,6]. Intercropping can also lead to extra yield [also called out-yielding (*i.e.*, when the yield produced by an intercrop is greater than the yield produced by the component crops grown in monoculture on the same total land area)] in cereal-legume [2,3,7-10] or oilseed-legume [3,10] intercropping systems, and grain quality [5,11]. Out-yielding can be calculated by measuring productivity efficiency of intercrops relative to sole crops by using various methods, such as Area x Time Equivalency Ratio (ATER) [12], Relative Yield Total (RYT) [13] or Land Equivalency Ratio (LER) [14]. However, the LER is most commonly used to make intercrop versus sole crop comparisons, and is defined as the relative land area under sole crops that is required to produce yields equivalent to intercrops.

There are many reasons for the occurrence of out-yielding in intercropping systems. Weed suppression [6,15] and lower susceptibility to insects and diseases may increase yields of intercrops [16-18]. This is also possible that a mixture of different crop plants will use resources (e.g., nutrients, water and light) more efficiently than crop plants of the same type, and may also provide mutual benefits to each other, such as fixed nitrogen (N) from legumes, root length [2,19-21]. The magnitude of beneficial effects of intercropping non-legumes with legumes on yield, produce quality, nutrient uptake and economic returns vary with crop species, soil type and climatic conditions (agro-ecological regions). In Canada, especially in the Parkland region, research information is lacking on the effects of intercropping non-legume and legume annual crops on crop yield, seed quality and economic returns. The purpose of this study was to determine the feasibility of intercropping annual non-legume (oilseeds or cereals) and legume (pea) crops for optimum yield, produce quality and economic returns.

2. MATERIALS AND METHODS

2.1. Field Experimentation

Two field experiments (Experiment 1 with barley-pea intercrop; Experiment 2 with canola-pea intercrop) were established in 2009 on a Gray Luvisol (Typic Haplocryalf) loam at Star City, Saskatchewan, Canada. Some characteristics of the soil used in these experiments are presented in **Table 1**. Soil was low in available N. Precipitation in the growing season (May, June, July and August) from 2009 to 2011, and long-term (30-year) average of precipitation and air temperature in May to August at the nearest Environment Canada Meteorological Station (AAFC Melfort Research Farm) are presented in **Table 2**. The amount of precipitation in the growing season over four months from May to August was 226.0 mm, 403.0 mm and 198.0 mm, in 2009, 2010, and 2011, respectively. The precipitation for the 30-year average at the nearest meteorological station (Melfort) was 243.7 mm. In 2009, the growing season precipitation (GSP) was near long-term average, with slightly lower than average precipitation in May and slightly higher than average precipitation in August. In 2010, the GSP was much higher than average (especially in June, and also in April prior to spring), and relatively cooler air temperatures in most

summer. In 2011, the GSP was below average (especially in May during seeding season and in August during seed formation/filling), with relatively cooler air temperatures and wet conditions in June, and relatively warmer/hotter air temperatures and dry moisture conditions in late July and August.

In both experiments, there were 10 treatments, where barley and pea (Experiment 1), or canola and pea (Experiment 2) were grown as monocrops and in combinations, and N fertilizer at 0, 40 and 80 $kg \cdot N \cdot ha^{-1}$ was applied to monocrops barley and canola and their combination with pea. In Experiment 1, the treatments were: 1) barley, 0 $kg \cdot N \cdot ha^{-1}$, 2) barley, 40 $kg \cdot N \cdot ha^{-1}$, 3) barley, 80 $kg \cdot N \cdot ha^{-1}$, 4) pea, 0 $kg \cdot N \cdot ha^{-1}$, 5) barley-pea in alternate rows, 0 $kg \cdot N \cdot ha^{-1}$, 6) barley-pea in alternate rows, 20 $kg \cdot N \cdot ha^{-1}$ to only barley, 7) barley-pea in alternate rows, 40 $kg \cdot N \cdot ha^{-1}$ to only barley, 8) barley-pea in same row, 0 $kg \cdot N \cdot ha^{-1}$, 9) barley-pea in same row, 20 $kg \cdot N \cdot ha^{-1}$, and 10. barley-pea in same row, 40 $kg \cdot N \cdot ha^{-1}$. In Experiment 2, the treatments were: 1) canola, 0 $kg \cdot N \cdot ha^{-1}$, 2) canola, 40 $kg \cdot N \cdot ha^{-1}$, 3) canola, 80 $kg \cdot N \cdot ha^{-1}$, 4) pea, 0 $kg \cdot N \cdot ha^{-1}$, 5) canola-pea in alternate rows, 0 $kg \cdot N \cdot ha^{-1}$, 6) canola-pea in alternate rows, 20 $kg \cdot N \cdot ha^{-1}$ to only canola, 7) canola-pea in alternate rows, 40 $kg \cdot N \cdot ha^{-1}$ to only canola, 8) canola-pea in same row, 0 $kg \cdot N \cdot ha^{-1}$, 9) canola-pea in same row, 20 $kg \cdot N \cdot ha^{-1}$, and 10) canola-pea in same row, 40 $kg \cdot N \cdot ha^{-1}$. In the two experiments, canola and barley were rotated in alternate years. Each treatment was

Table 1. Some characteristics of soil in spring 2009 at initiation of field experiments at Star City in northeastern Saskatchewan.

Site	Soil Great Group[z]	Depth (cm)	Texture	Organic Matter (%)	pH (1:2 water)	Nitrate-N ($mg \cdot kg^{-1}$)	Extractable P ($mg \cdot kg^{-1}$)	Sulphate-S ($mg \cdot kg^{-1}$)	Extractable K ($mg \cdot kg^{-1}$)
Star City	Gray Luvisol	0 - 15	Loam	3.1	6.6	7.5	13.9	4.5	202
		15 - 30				2.4	9.6	2.3	146
		30 - 60				3.0	7.8	1.6	180

[z]Based on Canadian Soil Classification System.

Table 2. Growing season monthly and total precipitation in 2009 to 2011 growing seasons, and average 30-yr average precipitation and temperature at Star City in northeastern Saskatchewan.

Month	Precipitation in the growing season (mm)[z] Star City			30-yr average (Melfort Research Farm)	
	2009	2010	2011	Precipitation (mm)	Temperature (°C)
May	21.2	66.6	10.5	45.6	9.1
June	46.6	113.2	103.5	65.8	16.9
July	75.6	63.6	73.3	75.5	18.3
August	81.6	56.8	10.7	56.8	19.6
Total	225.0	300.2	198.0	243.7	

[z]At the nearest Environment Canada Meteorological Station (Melfort).

replicated four times in a randomized complete block design. Each experimental unit (plot) was 1.8 m × 7.5 m. All plots received blanket applications of P, K and S fertilizers. The crops were seeded in rows 17.8 cm apart, with 10 rows per plot. In the barley-pea intercrop Experiment 1, no herbicide was applied and weeds were controlled manually by hoeing whenever needed. In the canola-pea intercrop Experiment 2, herbicide Solo (a.i., imazamox) was applied to control annual weeds in both canola and pea. We did not observe any visual symptoms of disease in any crop, so no fungicide was applied. At maturity, the crop was harvested with a combine for seed yield. Plant samples were also harvested for biomass and straw yield. Seed and straw samples were analysed for total N concentration [22] to calculate N uptake.

2.2. Land Equivalency Ratio

To compare crop growth/yield in intercrops relative to the respective sole crops, LER was calculated by using formula [LER = (Intercrop1/Sole Crop1) + Intercrop2/ Sole Crop2)] of Szumigalski and Van Acker [14]. The LER value greater than 1 indicated out-yielding with intercropping (*i.e.*, greater production and lower land requirement with intercropping compared to sole crops). On the other hand, the LER value less than 1 suggested under-yielding with intercropping (*i.e.*, lower productivity and greater land requirement with intercropping than the sole crops).

2.3. Statistical Analysis

The various treatments were applied to the same plots every year over three years to include any residual cumulative effects of the treatments over time. In addition, the response trends of crops to various treatments were generally similar in all three years. Because of these reasons, the data for each parameter in each experimental unit were averaged over three years [23], and then these calculated means were subjected to analysis of variance (ANOVA) using procedures as outlined in SAS [24]. Significant ($p \leq 0.05$) differences between treatments were determined using least significant difference ($LSD_{0.05}$). The LSD and standard error of the mean (SEM), along with significance for different parameters are presented in the various tables.

3. RESULTS

3.1. Barley-Pea Intercrop (Experiment 1)

3.1.1. Yield, Land Equivalency Ratio and N Uptake

Seed yield of barley as sole crop increased substantially with application of N, and also increased significantly in both intercrop combinations (**Table 3**). Seed yield of pea decreased with increasing N fertilizer application in both intercrop combinations, and more so seeding in the same row intercrop. The LER values for the corresponding N rate treatments in barley-pea intercrop combinations were greater than 1 in all cases. The LER values were highest when no N was applied to intercrops, but decreased with application of N fertilizer at 20 or 40 $kg \cdot N \cdot ha^{-1}$. Total N uptake in seed for barley and pea as sole crops as well as when sown as intercrop in both combinations showed trends similar to seed yield (**Table 3**). The LER values for total N uptake in seed also showed patterns similar to LER for seed yield, but the actual LER values were usually greater for total N uptake in seed than for seed yield.

Table 3. Seed yield, N uptake of seed and land equivalency ratio (LER) for barley and pea, grown as sole crops or in various intercrop combinations at Star City, Saskatchewan (average of 2009 to 2011, Experiment 1).

Treatment	Seed Yield ($kg \cdot ha^{-1}$)			N Uptake in Seed ($kg \cdot N \cdot ha^{-1}$)		
	Barley	Pea	LER	Barley	Pea	LER
1. Barley, 0 $kg \cdot N \cdot ha^{-1}$	2062			30.9		
2. Barley, 40 $kg \cdot N \cdot ha^{-1}$	3065			46.1		
3. Barley, 80 $kg \cdot N \cdot ha^{-1}$	3975			62.0		
4. Pea, 0 $kg \cdot N \cdot ha^{-1}$		3097			100.8	
5. Barley-Pea in Alternate Rows, 0 $kg \cdot N \cdot ha^{-1}$	1927	1790	1.50	36.3	59.9	1.75
6. Barley-Pea in Alternate Rows, 20 $kg \cdot N \cdot ha^{-1}$ to only Barley	2420	1395	1.24	41.3	47.1	1.39
7. Barley-Pea in Alternate Rows, 40 $kg \cdot N \cdot ha^{-1}$ to only Barley	2696	1371	1.15	47.5	46.3	1.25
8. Barley-Pea in Same Row, 0 $kg \cdot N \cdot ha^{-1}$	2034	1730	1.54	34.4	57.0	1.67
9. Barley-Pea in Same Row, 20 $kg \cdot N \cdot ha^{-1}$	2376	1365	1.22	40.7	46.6	1.37
10. Barley-Pea in Same Row, 40 $kg \cdot N \cdot ha^{-1}$	2707	1120	1.07	47.9	38.5	1.19
$LSD_{0.05}$	394	364	0.17	6.5	11.9	0.26
SEM and Significance Level	134.9***z	122.6***	0.057***	2.22***	4.02***	0.085**

z, ** and *** refer to significant treatment effects in ANOVA at $P \leq 0.01$ and $P \leq 0.001$, respectively.

The LER values in both barley-pea intercrop combinations were greater when sole crop of barley was grown at zero-N rate compared with application of N fertilizer, with the least at 80 kg·N·ha^{-1} rate (**Table 4**). This indicated crop yield would be highest and land requirement would be lowest for barley-pea intercropping, when sole crop is grown without any applied N fertilizer. The LER values were only slightly greater than 1 in many cases when barley received 80 kg·N·ha^{-1}, suggesting decrease in seed yield and increase in land requirement for barley sole crop receiving increasing rate of applied N. For barley as sole crop particularly at 0 and also at 40 kg·N·ha^{-1} rates, the LER values increased with increasing N rate from 0 to 40 kg·N·ha^{-1} in both barley-pea intercrop combinations, with the highest LER values when the bar-ley-pea intercrop combinations received N fertilizer at 40 kg·N·ha^{-1}. This suggested that seed yield increased and land requirement decreased for barley-pea intercrop with increasing rate of applied N. The LER values for total N uptake in seed also showed response trends similar to LER for seed yield (**Table 4**).

Total of seed and straw yield of barley + pea together in both intercrop combinations increased with applied N, but the increases were much smaller for seed yield than straw yield (**Table 5**). Seed and straw yield of barley as sole crop increased considerably with application of N. Seed and straw yield of pea as sole crop was greater than barley as sole crop receiving no N fertilizer for seed yield and 0 and 40 kg·N·ha^{-1} rates for straw yield, but were lower than both intercrop combinations in all cases.

Table 4. Land equivalency ratio (LER) for barley with or without applied N and pea without applied N grown as sole crops or in various intercrop combinations at Star City, Saskatchewan (average of 2009 to 2011, Experiment 1).

| Treatment | LER compared to barley at N rates (kg·N·ha^{-1}) | | | | | |
| | For seed yield | | | For N uptake in seed | | |
	Barley at 0 N	Barley at 40 N	Barley at 80 N	Barley at 0 N	Barley at 40 N	Barley at 80 N
5. Barley-Pea in Alternate Rows, 0 kg·N·ha^{-1}	1.50	1.21	1.06	1.75	1.39	1.17
6. Barley-Pea in Alternate Rows, 20 kg·N·ha^{-1} to only Barley	1.62	1.24	1.05	1.81	1.39	1.12
7. Barley-Pea in Alternate Rows, 40 kg·N·ha^{-1} to only Barley	1.85	1.38	1.15	2.09	1.58	1.25
8. Barley-Pea in Same Row, 0 kg·N·ha^{-1}	1.54	1.22	1.06	1.67	1.33	1.10
9. Barley-Pea in Same Row, 20 kg·N·ha^{-1}	1.61	1.22	1.04	1.81	1.37	1.12
10. Barley-Pea in Same Row, 40 kg·N·ha^{-1}	1.80	1.29	1.07	2.03	1.50	1.19
LSD$_{0.05}$	0.17	0.13	ns	0.17	0.14	0.11
SEM and Significance Level	0.055**^{z}	0.042*	0.037^{ns}	0.056***	0.047*	0.037*

^{z},*,**,*** and ^{ns} refer to significant treatment effects in ANOVA at $P \leq 0.1$, $P \leq 0.05$, $P \leq 0.01$, $P \leq 0.001$ and not significant, respectively.

Table 5. Yield and N uptake of seed, straw and seed + straw for both crops of barley and/or pea together, grown as sole crops or in various intercrop combinations at Star City, Saskatchewan (average of 2009 to 2011, Experiment 1).

| Treatment | Yield (kg·ha^{-1}) | | | N uptake (kg·N·ha^{-1}) | | |
	Seed	Straw	Seed + Straw	Seed	Straw	Seed + Straw
1. Barley, 0 kg·N·ha^{-1}	2062	1939	4001	30.9	12.5	43.4
2. Barley, 40 kg·N·ha^{-1}	3065	3123	6188	46.1	15.6	61.7
3. Barley, 80 kg·N·ha^{-1}	3975	4126	8101	62.0	22.9	84.9
4. Pea, 0 kg·N·ha^{-1}	3097	3641	6738	100.8	40.5	141.3
5. Barley-Pea in Alternate Rows, 0 kg·N·ha^{-1}	3717	4417	8134	96.1	38.2	134.3
6. Barley-Pea in Alternate Rows, 20 kg·N·ha^{-1} to only Barley	3815	4248	8063	88.4	33.8	122.2
7. Barley-Pea in Alternate Rows, 40 kg·N·ha^{-1} to only Barley	4067	4889	8956	93.8	37.0	125.4
8. Barley-Pea in Same Row, 0 kg·N·ha^{-1}	3764	4049	7813	91.4	34.5	125.9
9. Barley-Pea in Same Row, 20 kg·N·ha^{-1}	3741	4531	8272	87.3	36.2	123.5
10. Barley-Pea in Same Row, 40 kg·N·ha^{-1}	3827	4770	8597	86.5	39.2	125.7
LSD$_{0.05}$	469	554	932	11.8	6.0	16.3
SEM and Significance Level	161.7***^{z}	191.0***	321.2***	4.07***	2.05***	5.62***

^{z},*** refers to significant treatment effects in ANOVA at $P \leq 0.001$.

Seed + straw yields usually followed trends similar to straw yields. Response trends of total N uptake in straw or seed + straw to various treatments were also similar to corresponding straw yield or seed + straw yield (**Table 5**).

no effect on protein concentration in seed when both crops were seeded in the same row. There was no consistent effect of any treatment on protein concentration in pea seed.

3.1.2. Seed Quality

Protein concentration in barley seed was highest when barley was intercropped with pea in alternate rows without any applied N (**Table 6**). Protein concentration in barley seed tended to decrease with N fertilization in barley-pea intercrop combination in alternate rows, but

3.1.3. Economic Returns

Net returns of barley as a sole crop increased substantially with application of N, but still were much lower than both barley-pea intercrop combinations in all three crop price scenarios (**Table 7**). Net returns for barley-pea intercrop without applied N were slightly lower than pea

Table 6. Protein concentration in seed of barley and pea grown as sole crops compared to various combinations of barley and pea intercrop treatments for similar N rates at Star City, Saskatchewan (average of 2009 to 2011, Experiment 1).

Treatment	Protein concentration in seed ($g \cdot N \cdot kg^{-1}$)	
	Barley	Pea
1. Barley, $0 \ kg \cdot N \cdot ha^{-1}$	100	
2. Barley, $40 \ kg \cdot N \cdot ha^{-1}$	98	
3. Barley, $80 \ kg \cdot N \cdot ha^{-1}$	102	
4. Pea, $0 \ kg \cdot N \cdot ha^{-1}$		206
5. Barley-Pea in Alternate Rows, $0 \ kg \cdot N \cdot ha^{-1}$	127	206
6. Barley-Pea in Alternate Rows, $20 \ kg \cdot N \cdot ha^{-1}$ to only Barley	119	206
7. Barley-Pea in Alternate Rows, $40 \ kg \cdot N \cdot ha^{-1}$ to only Barley	117	206
8. Barley-Pea in Same Row, $0 \ kg \cdot N \cdot ha^{-1}$	116	204
9. Barley-Pea in Same Row, $20 \ kg \cdot N \cdot ha^{-1}$	118	210
10. Barley-Pea in Same Row, $40 \ kg \cdot N \cdot ha^{-1}$	117	211
$LSD_{0.05}$	4	5
SEM and Significance Level	$1.4^{***\,z}$	$1.68^{\bullet}$

^{z}, $^{\bullet}$ and *** and ^{ns} refer to significant treatment effects in ANOVA at $P \le 0.1$, and $P \le 0.001$, respectively.

Table 7. Economic returns for barley and pea grown as sole crops, and in various combinations as intercrop in alternate rows and in same row at low ($100 Mg^{-1} for barley and $200 ha^{-1} for pea), medium ($150 Mg^{-1} for barley and $300 ha^{-1} for pea) and high ($200 Mg^{-1} for barley and $400 ha^{-1} for pea) prices at Star City, Saskatchewan (average of 2009 to 2011, Experiment 1).

Treatment	Gross returns ($ ha^{-1})			Net returns above N fertilizer costs^{z} ($ ha^{-1})		
	Low	Med	High	Low	Med	High
1. Barley, $0 \ kg \cdot N \cdot ha^{-1}$	206	309	412	206	309	412
2. Barley, $40 \ kg \cdot N \cdot ha^{-1}$	307	460	613	247	400	553
3. Barley, $80 \ kg \cdot N \cdot ha^{-1}$	397	596	795	277	476	675
4. Pea, $0 \ kg \cdot N \cdot ha^{-1}$	619	929	1239	619	929	1239
5. Barley -Pea in Alternate Rows, $0 \ kg \cdot N \cdot ha^{-1}$	551	826	1101	551	826	1101
6. Barley -Pea in Alternate Rows, $20 \ kg \cdot N \cdot ha^{-1}$ to only Barley	521	782	1042	491	752	1012
7. Barley -Pea in Alternate Rows, $40 \ kg \cdot N \cdot ha^{-1}$ to only Barley	544	816	1088	484	756	1028
8. Barley -Pea in Same Row, $0 \ kg \cdot N \cdot ha^{-1}$	549	824	1099	549	824	1099
9. Barley -Pea in Same Row, $20 \ kg \cdot N \cdot ha^{-1}$	511	766	1021	481	736	991
10. Barley -Pea in Same Row, $40 \ kg \cdot N \cdot ha^{-1}$	495	742	990	435	682	930
$LSD_{0.05}$	73	110	147			
SEM and Significance Level	$25.3^{***\,y}$	37.9^{***}	50.6^{***}	25.3^{***}	37.9^{***}	50.6^{***}

^{z}The cost of N fertilizer was $1500 Mg^{-1} of N; $^{y\,***}$refers to significant treatment effects in ANOVA at $P \le 0.001$.

as a sole crop. Net returns decreased with N fertilization when the intercrops were seeded in alternate rows and more so in the same row, suggesting that N fertilization to barley-pea intercrops has no net economic benefit, in spite of slight increase in seed yield.

3.2. Canola-Pea Intercrop (Experiment 2)

3.2.1. Yield, Land Equivalency Ratio and N Uptake

Like barley, seed yield of canola as sole crop also increased considerably with application of N (**Table 8**). Seed yield of canola in intercrop increased with N fertilization, but the increases were small and significant only when both crops were seeded in the same row. There was also small increase (but not significant) in seed yield of pea with application of 20 kg·N·ha^{-1} in both intercrop combinations. The LER values for the corresponding N rate treatments were much greater than 1 for both intercropping combinations, but tended to decrease with increasing N rate. Total N uptake in seed for canola and pea as sole crops as well as when sown as intercrop in both combinations showed trends similar to seed yield (**Table 8**). The LER values for total N uptake in seed also showed patters similar to LER for seed yield.

The LER values in the canola-pea intercrop in both same row and alternative rows combinations were greater when sole crop of canola was grown at zero-N compared to application of N fertilizer, with the lowest LER values at 80 kg·N·ha^{-1} rate (**Table 9**). This indicated the highest crop yield and lowest land requirement for canola-pea intercrops, when sole crop of canola was grown

without any applied N fertilizer. For all canola as sole crop treatments, the LER values increased with increasing N rate in the canola-pea intercrop, especially when seeded in the same row. This suggests the increase in seed yield and decrease in land requirement for canola-pea intercrop with increasing rate of applied N.

Total of seed or straw yield of canola + pea together tended to increase with applied N in both intercrop combinations and more so for straw yield (**Table 10**). Seed and straw yield of canola as sole crop increased substantially with application of N. Seed yield of pea as sole crop (with no N applied) was much greater than canola as sole crop receiving 0 to 80 kg·N·ha^{-1} rates, but it was similar to canola-pea intercrop in alternate rows and slightly greater than when both were seeded in same row without applied N. Straw yield of pea as sole crop (with no N applied) was much greater than canola as sole crop at only 0 kg·N·ha^{-1} rate, but it was lower than that of in both intercrop combinations regardless of N application rate. Seed + straw yields usually followed trends similar to straw yields. Response trends of total N uptake in seed, straw or seed + straw to various treatments were similar to corresponding yield of seed, straw or seed + straw (**Table 10**).

3.2.2. Seed Quality

In treatments without any applied N, protein concentration in canola seed was highest when canola was intercropped with pea (**Table 11**). In the applied N treatments, protein concentration in canola seed tended to decrease with N rate in both canola-pea intercrop com-

Table 8. Seed yield, N uptake in seed and land equivalency ratio (LER) for canola or pea, grown as sole crops or in various intercrop combinations at Star City, Saskatchewan (average of 2009 to 2011, Experiment 2).

Treatment	Seed yield (kg·ha^{-1})			N uptake in seed (kg·N·ha^{-1})		
	Canola	Pea	LER	Canola	Pea	LER
1. Canola, 0 kg·N·ha^{-1}	834			26.5		
2. Canola, 40 kg·N·ha^{-1}	1167			36.5		
3. Canola, 80 kg·N·ha^{-1}	1596			52.1		
4. Pea, 0 kg·N·ha^{-1}		2742			92.0	
5. Canola-Pea in Alternate Rows, 0 kg·N·ha^{-1}	468	2234	1.45	17.3	74.3	1.55
6. Canola-Pea in Alternate Rows, 20 kg·N·ha^{-1} to only Canola	583	2464	1.50	20.6	79.7	1.53
7. Canola-Pea in Alternate Rows, 40 kg·N·ha^{-1} to only Canola	503	2462	1.31	17.6	82.8	1.33
8. Canola-Pea in Same Row, 0 kg·N·ha^{-1}	577	2310	1.56	20.9	75.1	1.65
9. Canola-Pea in Same Row, 20 kg·N·ha^{-1}	679	2435	1.51	23.9	79.3	1.56
10. Canola-Pea in Same Row, 40 kg·N·ha^{-1}	763	2260	1.40	26.9	73.9	1.41
LSD$_{0.05}$	126	528	0.22	4.4	17.4	0.23
SEM and Significance Level	43.0[***] [z]	177.8[ns]	0.073[ns]	1.52[***]	5.86[ns]	0.075[*]

[z], [*], [***] and [ns] refer to significant treatment effects in ANOVA at P ≤ 0.1, P ≤ 0.001 and not significant, respectively.

Table 9. Land equivalency ratio (LER) for canola with or without applied N and pea without applied N grown as sole crops or in various intercrop combinations at Star City, Saskatchewan (average of 2009 to 2011, Experiment 2).

	LER compared to canola at N rates (kg·N·ha⁻¹)					
	For seed yield			For N uptake in seed		
Treatment	Canola at 0 N	Canola at 40 N	Canola at 80 N	Canola at 0 N	Canola at 40 N	Canola at 80 N
5. Canola-Pea in Alternate Rows, 0 kg·N·ha⁻¹	1.45	1.29	1.19	1.55	1.36	1.21
6. Canola-Pea in Alternate Rows, 20 kg·N·ha⁻¹ to only Canola	1.71	1.50	1.36	1.76	1.53	1.35
7. Canola-Pea in Alternate Rows, 40 kg·N·ha⁻¹ to only Canola	1.62	1.44	1.31	1.67	1.51	1.33
8. Canola-Pea in Same Row, 0 kg·N·ha⁻¹	1.56	1.37	1.24	1.65	1.43	1.25
9. Canola-Pea in Same Row, 20 kg·N·ha⁻¹	1.75	1.51	1.38	1.83	1.56	1.38
10. Canola-Pea in Same Row, 40 kg·N·ha⁻¹	1.84	1.58	1.40	1.92	1.64	1.41
LSD$_{0.05}$	0.21	0.20	0.21	0.22	0.21	0.21
SEM and Significance Level	0.070^{*z}	0.067*	0.071^{ns}	0.073*	0.069*	0.071^{ns}

^{z*}, *, and ns refer to significant treatment effects in ANOVA at $P \leq 0.1$, $P \leq 0.05$ and not significant, respectively.

Table 10. Yield and N uptake of seed, straw and seed + straw for both crops of canola and/or pea together, grown as sole crops or in various intercrop combinations at Star City, Saskatchewan (average of 2009 to 2011, Experiment 2).

	Yield (kg·ha⁻¹)			N uptake (kg·N·ha⁻¹)		
Treatment	Seed	Straw	Seed + Straw	Seed	Straw	Seed + Straw
1. Canola, 0 kg·N·ha⁻¹	834	2591	3425	26.5	18.9	45.4
2. Canola, 40 kg·N·ha⁻¹	1167	3806	4973	36.5	25.1	61.6
3. Canola, 80 kg·N·ha⁻¹	1596	4815	6411	52.1	33.6	85.7
4. Pea, 0 kg·N·ha⁻¹	2742	3948	6690	92.0	53.8	145.8
5. Canola-Pea in Alternate Rows, 0 kg·N·ha⁻¹	2702	4634	7336	91.7	56.3	148.0
6. Canola-Pea in Alternate Rows, 20 kg·N·ha⁻¹ to only Canola	3047	4667	7714	100.3	53.2	153.5
7. Canola-Pea in Alternate Rows, 40 kg·N·ha⁻¹ to only Canola	2965	4892	7857	100.3	56.1	156.4
8. Canola-Pea in Same Row, 0 kg·N·ha⁻¹	2887	4292	7179	95.9	47.6	143.6
9. Canola-Pea in Same Row, 20 kg·N·ha⁻¹	3114	4633	7747	103.2	48.6	151.8
10. Canola-Pea in Same Row, 40 kg·N·ha⁻¹	3023	4544	7567	100.7	45.1	145.9
LSD$_{0.05}$	431	636	872	14.4	9.3	18.2
SEM and Significance Level	148.7^{***z}	219.3***	300.6***	4.96***	3.20***	6.28***

^{z***} refers to significant treatment effects in ANOVA at $P \leq 0.001$.

binations. Protein concentration in pea seed was greater when pea was grown in intercrop compared to pea as sole crop, but there was no additional benefit of N fertilization on protein concentration in pea seed in both intercrop combinations. Oil concentration in canola seed was lower in both canola-pea intercrop combinations compared to canola as sole crop, especially when no N was applied to intercrops (**Table 11**). Oil concentration in canola seed increased with N fertilization in both canola-pea intercrop combinations.

3.2.3. Economic Returns

Net returns of canola as sole crop increased dramatically with application of N, but these net returns were much lower than that in canola-pea intercrop combinations or pea as sole crop (**Table 12**). Net returns in canola-pea intercrops without applied N were highest when both crops were seeded in the same row, and this was followed closely by intercrops seeded in alternate rows, which in turn was slightly greater than pea as sole crop. In both intercrop combinations, the net returns first in-

Table 11. Protein and oil concentration in seed of canola and pea grown as sole crops compared to various combinations of canola and pea intercrop treatments for similar N rates at Star City, Saskatchewan (average of 2009 to 2011, Experiment 2).

Treatment	Protein ($g \cdot N \cdot kg^{-1}$)		Canola
	Canola	Pea	Oil ($g \cdot N \cdot kg^{-1}$)
1. Canola, $0 \ kg \cdot N \cdot ha^{-1}$	199		488
2. Canola, $40 \ kg \cdot N \cdot ha^{-1}$	195		489
3. Canola, $80 \ kg \cdot N \cdot ha^{-1}$	205		484
4. Pea, $0 \ kg \cdot N \cdot ha^{-1}$		172	
5. Canola-Pea in Alternate Rows, $0 \ kg \cdot N \cdot ha^{-1}$	234	207	436
6. Canola-Pea in Alternate Rows, $20 \ kg \cdot N \cdot ha^{-1}$ to only Canola	225	201	453
7. Canola-Pea in Alternate Rows, $40 \ kg \cdot N \cdot ha^{-1}$ to only Canola	223	210	455
8. Canola-Pea in Same Row, $0 \ kg \cdot N \cdot ha^{-1}$	232	204	433
9. Canola-Pea in Same Row, $20 \ kg \cdot N \cdot ha^{-1}$	224	203	436
10. Canola-Pea in Same Row, $40 \ kg \cdot N \cdot ha^{-1}$	224	202	451
$LSD_{0.05}$	8	37	11
SEM and Significance Level	2.8^{***z}	12.5^{ns}	3.6^{***}

[z],[***] and [ns] refer to significant treatment effects in ANOVA at $P \leq 0.001$ and not significant, respectively.

Table 12. Economic returns for canola and pea grown as sole crops, and in various combinations as intercrop in alternate rows and in same row at low ($300 Mg^{-1} for canola and $200 ha^{-1} for pea), medium ($450 Mg^{-1} for canola and $300 ha^{-1} for pea) and high ($600 Mg^{-1} for canola and $400 ha^{-1} for pea) prices at Star City, Saskatchewan (average of 2009 to 2011, Experiment 2).

Treatment	Gross returns ($ ha^{-1})			Net returns above N fertilizer costs[z] ($ ha^{-1})		
	Low	Med	High	Low	Med	High
1. Canola, $0 \ kg \cdot N \cdot ha^{-1}$	250	375	501	250	375	501
2. Canola, $40 \ kg \cdot N \cdot ha^{-1}$	350	525	700	290	465	640
3. Canola, $80 \ kg \cdot N \cdot ha^{-1}$	479	718	958	359	598	838
4. Pea, $0 \ kg \ N \ ha^{-1}$	548	823	1097	548	823	1097
5. Canola-Pea in Alternate Rows, $0 \ kg \cdot N \cdot ha^{-1}$	587	881	1174	587	881	1174
6. Canola-Pea in Alternate Rows, $20 \ kg \cdot N \cdot ha^{-1}$ to only Canola	668	1001	1335	638	971	1305
7. Canola-Pea in Alternate Rows, $40 \ kg \cdot N \cdot ha^{-1}$ to only Canola	643	965	1287	583	905	1227
8. Canola-Pea in Same Row, $0 \ kg \cdot N \cdot ha^{-1}$	635	953	1270	635	953	1270
9. Canola-Pea in Same Row, $20 \ kg \cdot N \cdot ha^{-1}$	691	1036	1381	661	1006	1351
10. Canola-Pea in Same Row, $40 \ kg \cdot N \cdot ha^{-1}$	681	1022	1362	621	962	1302
$LSD_{0.05}$	88	132	176	88	132	176
SEM and Significance Level	30.3^{***y}	45.4^{***}	60.5^{***}	30.3^{***}	45.4^{***}	60.5^{***}

[z]The cost of N fertilizer was $1500 mg^{-1} of N; [y][***] refers to significant treatment effects in ANOVA at $P \leq 0.001$.

creased with application of $20 \ kg \cdot N \cdot ha^{-1}$ and then decreased beyond this N rate.

4. DISCUSSION

Earlier studies have explored the feasibility of intercropping grain legumes with various cereal [4,5,9] or oilseed [10] crops to improve land productivity, eco-

nomic returns and produce quality. The present study considered barley-pea or canola-pea intercrop mixtures as alternatives to barley, canola or pea sole crops in the Parkland region of western Canada under sub-humid climatic conditions. The following section discusses the effects of intercropping versus sole cropping in relation to crop production, land resource use efficiency, produce quality and N uptake, by providing possible explanations for any varying results from the published literature.

Previous research has established that pea-barley or pea-mustard intercrop produces greater seed yield than pure crops, in spite of the reduction in seed yield of the components [4,25]. This was attributed to a strong competition from cereals or oilseeds to pea in mixed intercropping [4,26]. Similarly, in our study barley or canola and pea intercrops produced greater seed yield than sole/pure crops, and seed yield of pea and barley or canola components decreased when grown in intercrop compared to grown as pure crops. Greater biomass production of intercrops, which in turn produced greater seed production, than sole crops could be possibly due to enhanced light interception because of greater canopy, especially for canola [27,28].

The LER values greater than 1 indicate higher crop yield and lower land requirement with intercropping compared to when barley and pea were grown as sole crops, especially when grown without applied N. The increase in seed yield of canola and decrease in seed yield of pea with application of N fertilizer compared to zero-N treatment could be most likely due to higher level of nitrate-N in soil which has been observed to favour competition of non-legume canola over legume pea [26, 29-31].

In our study, seed yields of the component crops in intercrop mixtures were much lower than their corresponding sole crop seed yields, but the total land productivity was usually higher in intercrop treatments as evidenced by higher LER values. For example, the LER values ranged from 1.15 to 1.50 for intercrops in alternate rows, and from 1.07 to 1.56 for intercrops in same row. In other words, sole cultivation of each crop would require more land than their cultivation in intercrop mixture, suggesting greater land-use efficiency of intercrops ((by 15 to 50% for alternate rows and by 7 to 56% for same row).) than sole crops. Because of different rooting pattern, intercropping legumes with oilseeds or cereals has the potential to improve the use of soil N resource compared to legumes grown as sole crops [2]. Therefore, it is also possible that higher land productivity under intercropping than sole cropping in our study was most likely due to more efficient use of available soil N with intercrops. Also, in our study there was more N uptake with intercrop combinations than sole crops. Similar results were reported previously in intercropping systems for mixed cultures of barley and pea [1,32] or oilseed rape/canola/mustard and pea [3,10]. The higher LER values of intercrops when grown in same row than alternate rows indicates that intercropping was superior when root systems of non-legume and legume crops are intermixed because of most likely the differences in the use of different sources of N and possibly other nutrients, as

also suggested by Martin and Snaydon [33].

The highest LER values when sole crop of canola was grown at zero-N compared with application of N fertilizer indicate greatest crop yield and lowest land requirement for canola-pea intercropping grown without any applied N fertilizer. This suggests increase in seed yield and decrease in land requirement for canola sole crop receiving increasing rate of applied N. This also suggests large advantage in favour of intercropping/mixed cropping, in spite of application of N fertilizer to oilseed or cereal crops. The increased in LER values with increasing N rate from 0 to 40 $kg \cdot N \cdot ha^{-1}$ in both canola-pea intercrop combinations suggests increase in seed yield and decrease in land requirement for canola-pea intercrop with increasing rate of applied N.

Knudson et al. [34] suggest that interspecies competetion can cause increase in the concentration total N in seed of barley when it is grown as a mixture with pea, due to lack of severe competition for N between the crops. In our study, the protein concentration (or total N) also increased in seed of barley-pea or canola-pea mixed intercrops, because more of soil N was available to barley or canola seeds, as pea fixes its own N requirement from the atmosphere.

Our findings suggest that if organic farming has limited possibilities for using manure, the inclusion of N-fixing legume grain crops can be introduced in the intercropping systems for preventing N deficiency in organic crops. Our results also show considerably more gain in protein yield with barley-pea or canola-pea intercrops than pure barley or canola crops, especially when no N fertilizer was applied. In summary, the barley-pea or canola-pea intercrops produced the highest seed yields, while also improving the sustainability of economic returns.

5. CONCLUSION

Compared to barley and pea as sole crops without applied N, crop yield, N uptake and LER improved with barley-pea intercrop. Seed yield, N uptake and LER also increased with canola-pea intercrop, particularly compared to canola as a sole crop. In barley-pea and canola-pea intercrop systems, seed yield of both crops together increased only slightly with application of N fertilizer. Net returns in both barley-pea and canola-pea intercrop systems improved greatly when no N fertilizer was applied, although the net returns were highest for pea grown as a sole crop without applied N in the barley-pea intercrop. In summary, intercrop of barley or canola with pea improved crop yield, N uptake and net returns, and reduced land requirements compared to barley, canola or pea as sole crops. The findings suggest the potential of barley-pea intercrop, canola-pea intercrop and pea for organic farming systems.

6. ACKNOWLEDGEMENTS

The author thanks Western Alfalfa for supplying alfalfa pellets amendment, International Compost Ltd., Calgary, Alberta, for supplying rock phosphate fertilizers for this study, and D. Leach, K. Strukoff and P. Boxall for technical help.

REFERENCES

[1] Jensen, E.S. (1996) Grain yield, symbiotic N_2 fixation and interspecific competition for inorganic N in pea-barley intercrops. *Plant and Soil*, **182**, 25-38.

[2] Hauggaard-Nielsen, H., Ambus, P. and Jensen, E.S. (2001) Temporal and spatial distribution of roots and competition for nitrogen in pea-barley intercrops—A field study employing 32P technique. *Plant and Soil*, **236**, 63-74.

[3] Aktar, M.S., Shamsuddin, A.M., Islam, N. and Rahman, A.R.M.S. (1993) Effects of mixed cropping lentil and mustard at various seeding ratios. *Lens*, **20**, 36-39.

[4] Hauggaard-Nielsen, H. and Jensen, E.S. (2001) Evaluating pea and barley cultivars for complementarity in intercropping at different levels of soil N availability. *Field Crops Research*, **72**, 185-196.

[5] Szczukowski, S. (1989) Yield and seed quality of field peas grown in mixtures with cereals and in pure stands. *Acta Academiae Agriculturae ac Technicae Olstenensis, Agricultura*, **47**, 40.

[6] Carr, P.M., Gardner, J.C., Schatz, B.G., Zwinger, S.W. and Guldan, S.J. (1995) Grain yield and weed biomass of a wheat-lentil intercrop. *Agronomy Journal*, **87**, 574-579.

[7] Paolini, R., Caporali, F. and Campiglia, E. (1993) Yield response, complementarity and competitive ability of bread wheat and pea in mixtures. *Agricoltura Mediterranea*, **123**, 114-121.

[8] Rauber, R., Schmidtke, K. and Kimpel-Freund, H. (2000) Competition and yield advantage in mixtures of pea (*Pisum sativum* L.) and oats (*Avena sativa* L.). *Konkurrenz und ertragsvorteile in gemengen aus erbsen (Pisum sativum L.) und hafer (Arena sativa L.)*, **185**, 33-47.

[9] Chalmers, S. and Day, S. (2009) Optimum seeding combinations for intercropping peas and canola: An investigation with surprising yields and other final stand effects. Manitoba Agronomists Conference, Winnipeg.

[10] Jetendra, K. and Mishra, J.P. (1999) Influence of planting pattern and fertilizers on yield and yield attributes and nutrient uptake in pea/mustard intercropping system. *Indian Journal of Pulses Research*, **12**, 38-43.

[11] Zielinska, A. and Rutkowski, M. (1988) Comparison of productivity of oats, barley and four cultivars of field peas in pure and mixed sowing. *Acta Academiae Agriculturae ac Technicae Olstenensis, Agricultura*, **46**, 113-124.

[12] Hiebesch, C.K. and McCollum, R.E. (1987) Area x time equivalency ratio: A method for evaluating the productivity of intercrops. *Agronomy Journal*, **79**, 15-22.

[13] de Wit, C.T. and van den Bergh, J.P. (1965) Competition between herbageplants. *Netherlands Journal of Agricultural Science*, **13**, 212-221.

[14] Szumigalski, A.R. and van Acker, R.C. (2008) Land equivalent ratios, light interception, and water use in intercrops in the presence or absence of in-crop herbicides. *Agronomy Journal*, **100**, 1145-1154.

[15] Poggio, S.L. (2005) Structure of weed communities occurring in monoculture and intercropping of field pea and barley. *Agriculture, Ecosystems and Environment*, **109**, 48-58.

[16] Helenius, J. (1991) Insect numbers and pest damage in intercrops vs. monocrops: Concepts and evidence from a system of faba bean, oats and *Rhopalosiphum padi* (Homoptera, Aphididae). *Journal of Sustainable Agriculture*, **1**, 57-80.

[17] Paras, N. and Chakravorty, S. (2005) Effect of intercropping on the infestation of chickpea pod borer *Helicoverpa armigera* (Hubner). *Journal of Plant Protection and Environment*, **2**, 86-91.

[18] Chen, Y., Zhang, F., Tang, L., Zheng, Y., Li, Y., Christie, P. and Li, L. (2007) Wheat powdery mildew and foliar N concentrations as influenced by N fertilization and belowground interactions with intercropped faba bean. *Plant and Soil*, **291**, 1-13.

[19] Danso, S.K.A. and Papastylianou, I. (1992) Evaluation of the nitrogen contribution of legumes to subsequent cereals. *Journal of Agricultural Science*, **119**, 13-18.

[20] Izaurralde, R.C., McGill, W.B. and Juma, N.J. (1992) Nitrogen fixation efficiency, interspecies N transfer, and root growth in barley-field pea intercrop on a Black Chernozemic soil. *Biology and Fertility of Soils*, **13**, 11-16.

[21] Hauggaard-Nielsen, H., Andersen, M.K., Jornsgaard, B. and Jensen, E.S. (2006) Density and relative frequency effects on competitive interactions and resource use in pea-barley intercrops. *Field Crops Research*, **95**, 256-267.

[22] Noel, R.J. and Hambleton, L.G. (1976) Collaborative study of a semi-automated method for the determination of crude protein in animal feeds. *Journal of Association of Official Analytical Chemists*, **59**, 134-140.

[23] Snedecor, G.W. and Cochran, W.G. (1967) Statistical methods. 6th Edition, The Iowa State University Press, Ames.

[24] SAS Institute Inc. (2004) Online documentation for SAS, version 8.

[25] Banik, P., Sasmal, T., Ghosal, P.K. and Bagchi, D.K. (2000) Evaluation of mustard (Brassica compestris var. Toria) and legume intercropping under 1:1 and 2:1 row-replacement series system. *Agronomy and Crop Science*, **185**, 9-14.

[26] Anderson, M.K., Hauggaard-Nielsen, H., Ambus, P. and Jensen, E.S. (2004) Biomass production, symbiotic nitrogen fixation and inorganic N use in dual and tri-component annual intercrops. *Plant and Soil*, **266**, 273-287.

[27] Kushwaha, B.L. and De, R. (1987) Studies of resource use and yield of mustard and chickpea grown in cropping systems. *Journal of Agriculture Science (Cambridge)*, **108**, 487-495.

[28] Morris, R.A. and Garrity, D.P. (1993) Resource capture and utilization in intercropping. *Water Field Crops Research*, **34**, 303-317.

[29] Francis, C.A. (1989) Biological efficiencies in multiple-cropping systems. *Advances in Agronomy*, **42**, 1-42.

[30] Cowell, L.E., Bremer, E. and van Kessel, C. (1989) Yield and N-2 fixation of pea and lentil as affected by inter-cropping and N application. *Canadian Journal of Soil Science*, **69**, 243-251.

[31] Waterer, J.G., Vessey, J.K., Stobbe, E.H. and Soper, R.J. (1994) Yield and symbiotic nitrogen-fixation in a pea mustard intercrop as influenced by N fertilizer addition. *Soil Biology and Biochemistry*, **26**, 447-453.

[32] Chen, C., Westcott, M., Neil, K., Wichmann, D. and Knox, M. (2004) Row configuration and nitrogen application for barley-pea intercropping in Montana. *Agronomy Journal*, **96**, 1730-1738.

[33] Martin, M.P.L.D. and Snaydon, R.W. (1982) Root and shoot interactions between barley and beans when intercropped. *Journal of Applied Ecology*, **19**, 263-272.

[34] Knudson, M.T., Hauggaard-Nielsen, H., Jørnsgaard, B. and Jensen, E.S. (2004) Comparison of interspecific competition and N use in pea-barley, faba bean-barley and lupin-barley intercrops grown at two temperate locations. *Journal of Agriculture Science*, **142**, 617-627.

6

Effects of broad-leaf crop frequency in various rotations on soil organic C and N, and inorganic N in a Dark Brown soil

Sukhdev S. Malhi[1*], R. L. Lemke[2], S. A. Brandt[3]

[1]Agriculture and Agri-Food Canada, Melfort, Canada; [*]Corresponding Author
[2]Agriculture and Agri-Food Canada, Saskatoon, Canada
[3]Agriculture and Agri-Food Canada, Scott, Canada

ABSTRACT

The objective of this study was to determine the impact of frequency of broad-leaf crops canola and pea in various crop rotations on pH, total organic C (TOC), total organic N (TON), light fraction organic C (LFOC) and light fraction organic N (LFON) in the 0 - 7.5 and 7.5 - 15 cm soil depths in autumn 2009 after 12 years (1998-2009) on a Dark Brown Chernozem (Typic Boroll) loam at Scott, Saskatchewan, Canada. The field experiment contained monoculture canola (herbicide tolerant and blackleg resistant hybrid) and monoculture pea compared with rotations that contained these crops every 2-, 3-, and 4-yr with wheat. There was no effect of crop rotation duration and crop phase on soil pH. Mass of TOC and TON in the 0 - 15 cm soil was greater in canola phase than pea phase in the 1-yr (monoculture) and 2-yr crop rotations, while the opposite was true in the 3-yr and 4-yr crop rotations. Mass of TOC and TON (averaged across crop phases) in soil generally increased with increasing crop rotation duration, with the maximum in the 4-yr rotation while no difference in the 1-yr and 2-yr rotations. Mass of LFOC and LFON in soil was greater in canola phase than pea phase in the 1-yr, 2-yr and 3-yr rotations, but the opposite was true in the 4-yr rotation. There was no consistent effect of crop rotation duration on mass of LFOC and LFON. The N balance sheet over the 1998 to 2009 period indicated large amounts of unaccounted N for monoculture pea, suggesting a great potential for N loss from the soil-plant system in this treatment through nitrate leaching and/or denitrification. In conclusion, the findings suggest that the quantity of organic C and N can be maximized by increasing duration of crop rotation and by including hybrid canola in the rotation.

Keywords: Broad-Leaf Crops; Canola; Frequency; Light Fraction Organic C; Light Fraction Organic N; Pea; Total Organic C; Total Organic N

1. INTRODUCTION

Crop rotation (the growing of different crops in the same field in a planned sequence) is often practiced to mitigate pests that often become unmanageable in monocultures (cultivation of the same crop year after year on the same field) [1,2]. In western Canada, research has indicated that rotations balanced between broad-leaf crops (canola and pea) with cereals (wheat and barley) tended to have less pest problems and lower production risk than rotations that were heavily cereal or broad-leaf based [3]. In many years, canola and/or pea provide the best economic return to producers compared to other field crops grown in western Canada. For this reason production of canola or pea is often intensive, meaning it is grown more than once every four years on the same field.

Research has shown that organic C in soil can be maintained or enhanced by combining reduced tillage or no-tillage with proper crop rotations that increase input of crop residues to soil and/or decrease their decomposition [4-8]. However, the magnitude of change in organic C in soil may vary with crop type/species/diversity/intensity, rooting characteristics (layer/volume/mass) of each crop, soil type and climatic conditions [9,10]. Our previous paper has discussed the impacts of frequency of broad-leaf crops canola and pea in various crop rotations on accumulation and distribution of nitrate-N and extractable P in the soil profile after 8 years [11]. The objective of this study was to determine the impact of frequency of broad-leaf crops canola and pea in various crop rotations (grown in monoculture or in rotation with wheat) on total organic C (TOC), total organic N (TON), light fraction organic C (LFOC), light fraction organic N (LFON) and pH in the 0 - 7.5 and 7.5 - 15 cm soil depths in autumn 2009 after 12 years (1998-2009) on a Dark Brown Chernozem (Typic Boroll) loam at Scott, Saskatchewan, Canada.

2. MATERIALS AND METHODS

A 12-yr field experiment was conducted from 1998 to 2009 on a Dark Brown Chernozem (Typic Boroll) loam at Scott, Saskatchewan. The field experiment was designed as split-plot of 12 crop rotations (main plots) and two fungicide (treated and untreated) treatments applied to sub-plots (**Table 1**), with all phases of each rotation present every year in four replications. For this area, the long-term average precipitation in the growing season (from May to August) is about 210 mm. Growing season precipitation was substantially below average in 1998, 2001 and 2008, fairly below average in 2002, 2003 and 2004, slightly below average in 2000, 2006 and 2009, above average in 1999 and 2007, and very wet in 2005. All crops were seeded with a Versatile Noble hoe drill set at 23-cm row spacing. Seeding rate was 245 kg·ha^{-1} for pea, 7 - 9 kg·ha^{-1} for canola, 100 kg·ha^{-1} for wheat and 50 - 62 kg·ha^{-1} for flax. All plots received monoammonium phosphate seed placed to supply adequate amounts of P (15 kg·P·ha^{-1} plus 7 kg·N·ha^{-1}) and nitrogen (N) as 46-0-0 midrow banded at the time of seeding at rates based on soil test recommendations (on average ranged from 11 to 60 kg·N·ha^{-1}, with no N fertilizer to pea) for optimum crop growth and yield. Appropriate herbicides were applied to control annual weeds. The crops were harvested for seed yield every year and straw was returned to each corresponding plot/treatment.

In the autumn of 2009, soil samples in selected treatments were obtained by taking 10 cores (about 2.4 cm diameter) from the 0 - 7.5, 7.5 - 15 and 15 - 20 cm layers. Bulk density of soil was determined by the core method

using soil weight and core volume [12]. The soil samples were air dried at room temperature after removing coarse roots and easily detectable crop residues, and ground to pass a 2-mm sieve. Sub-samples were pulverized in a vibrating-ball mill (Retsch, Type MM2, Brinkman Instruments Co., Toronto, Ontario) for determination of TOC, TON, LFOC and LFON in soil. Soil samples used for organic C and N analyses were tested for the presence of inorganic C (carbonates) using dilute HCl, and none was detected in any soil sample. Therefore, C in soil associated with each fraction was considered to be of organic origin. Total organic C in soil was measured by Dumas combustion using a Carlo Erba instrument (Model NA 1500, Carlo Erba Strumentazione, Italy), and Technicon Industrial Systems [13] method was used to determine TON in the soil. Light fraction organic matter (LFOM) was separated using a NaI solution of 1.7 Mg·m^{-3} specific gravity, as described by Janzen *et al.* [6] and modified by Izaurralde *et al.* [14]. The C and N in LFOM (LFOC, LFON) were measured by Dumas combustion. Soil samples (ground to pass a 2-mm sieve) from the 0 - 7.5 and 7.5 - 15 cm layers were also monitored for pH in 0.01 M CaCl$_2$ solution with a pH meter.

In autumn 2009, soil samples were also obtained by taking 4 cores (using 4-cm diameter coring tube) from the 0 - 15, 15 - 30 and 30 - 60 cm layers. The bulk density of soil was determined by the core method using soil weight and core volume [12]. The soil samples were air dried at room temperature, ground to pass a 2-mm sieve, and then analyzed for ammonium-N [15] and nitrate-N [16] by extracting soil in a 1:5 ratio of soil and 2 M KCl solution.

The cumulative amounts of crop residue (CR) input from 1998 to 2009 growing seasons were estimated as: above-ground residue (AGR) + belowground residue (BGR) returned to soil. The AGR was determined from the straw yield of each crop from 1998 to 2009 growing seasons. The BGR was estimated from grain dry weight

Table 1. Description of crop rotations in a field experiment from 1998 to 2009 at Scott Saskatchewan.

Crop rotation name (duration)	Crop rotation	Input of C from crop residue in 12 years (kg·C·ha^{-1})
Monoculture (1-yr)	Monoculture hybrid **canola**	20,847
Monoculture (1-yr)	Monoculture **pea**	9930
2-year rotation (2-yr)	Hybrid **canola**-wheat	26,697
2-year rotation (2-yr)	**Pea**-wheat	18,482
3-year rotation (3-yr)	Pea-hybrid **canola**-wheat	20,724
3-year rotation (3-yr)	**Pea**-hybrid canola-wheat	22,705
4-year rotation (4-yr)	Hybrid **canola**-wheat-pea-wheat	24,854
4-year rotation (4-yr)	Hybrid canola-wheat-**pea**-wheat	23,910

(GDW) and AGR, using the formula: BGR = a (GDW + AGR). The value of the constant "a" was 0.24 for wheat and 0.25 for canola [17], and 0.25 for pea. The amounts of crop residue C input were estimated by multiplying the concentration of total C by the amount of crop residue input in various rotation treatments. The estimated C concentration was 45% for wheat, 42% for canola [18], and 40% for pea [19]. The cumulative amounts of crop residue C were calculated as: CR-C = AGR-C + BGR-C; for 1998 to 2009 (12 years). The estimated amounts of N balance and unaccounted N over the 1998 to 2009 period for the 8 treatments [two crop phases (canola and pea) and four crop rotation durations (1-yr, 2-yr, 3-yr and 4-yr)]; **Table 7** was calculated as the difference between total of N applied as fertilizers in 12 years + N fixed by pea in rotations with pea + N added in seed at seeding in 12 years minus N removed in seed in 12 years (for N balance) + nitrate-N recovered in soil in autumn 2009 (for unaccounted N).

The data for each parameter were subjected to analysis of variance (ANOVA) using procedures as outlined in SAS [20]. Significant ($P \leq 0.05$) differences between treatments were determined using LSmeans (Proc GLM, SAS 6.1 for windows). The least significant difference ($LSD_{0.05}$) test was used to compare treatment means for various parameters. Correlations between mass of organic C or N in soil in autumn 2009 and the amount of crop residue C input from 1998 to 2009 growing seasons were calculated using the linear (REG) procedure.

3. RESULTS AND DISCUSSION

3.1. Soil pH

There was no significant effect of crop rotation duration and crop phase on pH in the 0 - 7.5 and 7.5 - 15 cm soil layers (data not shown). The soil pH ranged from 4.9 to 5.5 in the 0 - 7.5 cm layer and from 4.7 to 5.6 in the 7.5 to 15 cm layer among different treatments. Similar to a previous study at this site, there was no consistent effect of crop diversification/rotation on soil pH [21]. Because soil at this site is already fairly acid, acidification of soil from application of N fertilizer at moderate rates or from including pea legume in the rotation is not expected to be a serious problem in this soil site.

3.2. Organic C and N Fraction in Soil

The mass of TOC and TON in soil varied with the crop rotation duration and/or frequency of canola or pea in the rotation, but the crop rotation duration × crop phase interaction effect was significant only for TOC ($P \leq 0.10$) and TON ($P \leq 0.05$) in the 7.5 - 15 cm soil layer (**Table 2**). For example, mass of TOC and TON in the 0 - 15 cm soil tended to be greater in canola phase than pea

phase in the 1-yr (monoculture) and 2-yr crop rotations, while the opposite was true in the 3-yr and 4-yr crop rotations. Averaged across crop phases, mass of TOC and TON in soil generally increased with increasing crop rotation duration (but significant only for TOC ($P \leq 0.06$) and TON ($P \leq 0.05$) in the 7.5 - 15 cm soil layer, with the maximum in the 4-yr crop rotation while no difference in the 1-yr and 2-yr rotations. When ANOVA was conducted separately for canola and pea phases, the ANOVA for canola phase did not indicate any significant effect of crop rotation duration and/or frequency of canola in the rotation on TOC and TON. However, the ANOVA for pea phase showed significant effect of crop rotation duration and/or frequency of pea in the rotation on TOC and TON in the 7.5 - 15 ($P \leq 0.05$) soil layer and in the total 0 - 15 ($P \leq 0.14$) cm soil depth. Mass of TOC and TON (averaged across crop rotation duration) tended to be greater (but not significant) in canola phase than pea phase.

Of interest, despite markedly lower C inputs, TOC under the continuous pea treatment was not greatly dissimilar to other treatments. Similar results have been reported by other workers. In a long-term study on a Brown Chernozem, Campbell et al. [22] reported lower C inputs but similar SOC status for a wheat-lentil compared to a continuous wheat treatment. In an incubation study using ^{13}C-CO_2 to label C inputs from growing crops, Comeau [23] estimated significantly lower C inputs to soil under pea compared to canola, but similar amounts of ^{13}C "stabilized" in the soil at the end of two growing cycles. These results support the hypothesis that pea residues are more efficiently stabilized as SOC than wheat and particularly canola.

The mass of LFOC and LFON in soil varied greatly with the duration/frequency of canola/pea in the crop rotations, but the crop rotation duration × crop phase interaction effect was significant only for TOC ($P \leq 0.12$) in the 0 - 7.5 cm soil layer (**Table 3**). Like TOC and TON, mass of LFOC and LFON in soil tended to be greater in canola phase than pea phase in the 1-yr and 2-yr rotations, but the opposite was true in the 3-yr and 4-yr rotations. Compared to 1-yr rotation, mass of LFOC and LFON (averaged across crop phases) in soil decreased with 2-yr and 3-yr rotations, and then increased, with greater LFOC and LFON in soil for 4-yr than 1-yr rotations in the 0 - 7.5 cm layer, but again the effect was significant only for LFOC ($P \leq 0.12$) in the 0 - 7.5 cm soil layer. When ANOVA was conducted separately for canola and pea phases, there was no significant effect of crop rotation duration and/or frequency of canola in the rotation on LFOC and LFON for canola phase. However, for pea phase, the ANOVA showed significant effect of crop rotation duration and/or frequency of canola in the rotation on LFOC in the 0 - 7.5 cm soil layer ($P \leq 0.06$)

Table 2. Effect of broad-leaf crop phase and frequency over 12 years from 1998 to 2009 on mass of total organic C (TOC) and total organic N (TON) in soil in autumn 2009 at Scott, Saskatchewan, Canada.

Treatment		TOC mass ($Mg \cdot C \cdot ha^{-1}$) in soil layers (cm)			TON mass ($Mg \cdot N \cdot ha^{-1}$) in soil layers (cm)		
Crop rotation (duration)	Crop phase	0 - 7.5	7.5 - 15	0 - 15	0 - 7.5	7.5 - 15	0 - 15
Monoculture (1-year)	Canola	24.77	21.57	46.34	2.198	1.965	4.163
Monoculture (1-year)	Pea	22.33	21.95	44.28	2.007	1.960	3.967
Canola-wheat (2-year)	Canola	26.08	23.99	50.07	2.253	2.166	4.419
Pea-wheat (2-year)	Pea	21.19	20.00	41.19	1.858	1.787	3.645
Pea-canola-wheat (3-year)	Canola	23.55	24.03	47.58	2.045	2.103	4.148
Pea-canola-wheat (3-year)	Pea	23.14	26.81	49.95	2.086	2.388	4.474
Canola-wheat-pea-wheat (4-year)	Canola	24.75	24.95	49.70	2.152	2.208	4.360
Canola-wheat-pea-wheat (4-year)	Pea	26.00	25.56	51.56	2.222	2.256	4.478
	$LSD_{0.05}$	ns	4.74	ns	ns	0.387	ns
	SEM (significance)	1.589^{ns}	1.612^{*}	2.712^{ns}	0.1348^{ns}	0.1316^{*}	0.2330^{ns}
Crop rotation duration							
1-year		23.55	21.76	45.31	2.103	1.962	4.065
2-year		23.64	21.99	45.63	2.055	1.977	4.032
3-year		23.35	25.42	48.76	2.066	2.246	4.312
4-year		25.37	25.25	50.63	2.187	2.232	4.419
$LSD_{0.05}$		ns	3.44	ns	ns	0.291	ns
SEM (significance)		1.151^{ns}	1.178^{*}	2.015^{ns}	0.0965^{ns}	0.0997^{*}	0.1701^{ns}
Crop phase							
Canola		24.79	23.63	48.42	2.162	2.111	4.273
Pea		23.17	23.58	46.75	2.044	2.098	4.141
$LSD_{0.05}$		ns	ns	ns	ns	ns	ns
SEM (significance)		0.814^{ns}	0.833^{ns}	1.425^{ns}	0.0682^{ns}	0.0705^{ns}	0.1203^{ns}
Additional statistical comparisons							
Significance of four crop rotation durations for canola phase only							
$LSD_{0.05}$		ns	ns	ns	ns	ns	ns
SEM (significance)		1.109^{ns}	1.670^{ns}	2.515^{ns}	0.0829^{ns}	0.1458^{ns}	0.2108^{ns}
Significance of four crop rotation durations for pea phase only							
$LSD_{0.05}$		ns	5.00	10.18	ns	0.398	0.832
SEM (significance)		2.135^{ns}	1.564	3.183^{*}	0.1856^{ns}	0.1244^{*}	0.2599^{*}

*, * and ns refer to significant treatment effects in ANOVA at $P \leq 0.1$, $P \leq 0.05$ and not significant, respectively.

Table 3. Effect of broad-leaf crop phase and frequency over 12 years from 1998 to 2009 on mass of light fraction organic C (LFOC) and light fraction organic N (LFON) in soil in autumn 2009 at Scott, Saskatchewan, Canada.

Treatment		LFOC mass (kg·C·ha^{-1}) in soil layers (cm)			LFON mass (kg·N·ha^{-1}) in soil layers (cm)		
Crop rotation (duration)	Crop phase	0 - 7.5	7.5 - 15	0 - 15	0 - 7.5	7.5 - 15	0 - 15
Monoculture (1-year)	Canola	3283	889	4172	212	48	260
Monoculture (1-year)	Pea	2520	694	3214	168	42	210
Canola-wheat (2-year)	Canola	2733	831	3564	171	46	217
Pea-wheat (2-year)	Pea	2317	814	3131	154	49	203
Pea-canola-wheat (3-year)	Canola	2193	816	3009	150	47	197
Pea-canola-wheat (3-year)	Pea	2723	962	3685	176	54	230
Canola-wheat-pea-wheat (4-year)	Canola	3081	923	4004	186	51	237
Canola-wheat-pea-wheat (4-year)	Pea	3348	855	4203	201	48	249
	LSD$_{0.05}$	918	ns	ns	ns	ns	ns
	SEM (significance)	312.0$^{•}$	110.5^{ns}	365.9^{ns}	18.5^{ns}	6.3^{ns}	22.7^{ns}
Crop rotation duration							
1-year		2901	792	3693	190	45	235
2-year		2525	822	3347	162	47	209
3-year		2458	889	3347	163	50	213
4-year		3215	889	4014	193	49	242
LSD$_{0.05}$		677	ns	ns	ns	ns	ns
SEM (significance)		231.8$^{•}$	77.2^{ns}	273.1^{ns}	13.4^{ns}	4.3^{ns}	16.2^{ns}
	Crop phase						
	Canola	2823	865	3688	180	48	228
	Pea	2727	831	3558	175	48	223
	LSD$_{0.05}$	ns	ns	ns	ns	ns	ns
	SEM (significance)	163.9^{ns}	54.6^{ns}	193.1^{ns}	9.5^{ns}	3.0^{ns}	11.5^{ns}
Additional statistical comparisons							
Significance of four crop rotation durations for canola phase only							
	LSD$_{0.05}$	ns	ns	ns	ns	ns	ns
	SEM (significance)	334.5^{ns}	125.2^{ns}	430.8^{ns}	21.9^{ns}	7.2^{ns}	27.1^{ns}
Significance of four crop rotation durations for pea phase only							
	LSD$_{0.05}$	769	ns	818	ns	ns	ns
	SEM (significance)	240.3$^{•}$	112.4^{ns}	255.8*	13.2^{ns}	6.3^{ns}	17.6^{ns}

$^{•}$, * and ns refer to significant treatment effects in ANOVA at P ≤ 0.1, P ≤ 0.05 and not significant, respectively.

and in the total 0 - 15 cm soil depth (P ≤ 0.05), but for LFON it was significant at P ≤ 0.16 in the 0 - 7.5 soil layer. Mass of LFOC and LFON (averaged across crop rotation duration) in soil tended to be greater in canola phase than pea phase in most cases.

There were strong and highly significant positive correlation coefficients between TOC and TON, and between LFOC and LFON fractions in soil (**Table 4**). The correlations between TOC and LFOC (significant at P ≤ 0.12), and between TON and LFON (significant at P ≤ 0.10) were moderate. The crop residue C or N input over 12 growing seasons (**Table 1**) had significant correlation coefficients with TOC (significant at P ≤ 0.05) and TON (significant at P ≤ 0.06). For linear regressions between crop residue C input and TOC, TON, LFOC or LFON, the R^2 values were significant (significant at P ≤ 0.05) for TOC and TON (**Table 5**).

In an 8-yr study with 4-yr rotations on a Black Chernozem soil in northeastern Saskatchewan, previous research has shown no effect of broad-leaf frequency (1, 2 or 3 broad-leaf crops in 4-yr rotations) on organic C and N in soil [10]. In another study on a Brown Chernozem

soil in southern Saskatchewan, annually cropped rotations stored more organic C in soil than crop rotations with bare summer fallow, but there was no influence of crop phase on soil organic C under continuously cropped rotations [9]. However in our present 12-yr study, TOC, TON, LFOC and LFON all increased (although significant only in a few cases) with increasing duration of crop rotation, with the maximum soil organic C and N in the 4-yr rotations and also greater organic C and N in soil in canola phase than pea phase. It is possible that inclusion of other crops, such as wheat with high lignin content in straw which is relatively slow to decompose, may have resulted in this slow build-up of organic C and N in soil in the longer duration rotations, especially 4-year rotation with two wheat crops.

Because of relatively much smaller amounts of annual inputs of crop residue C or N compared to the amounts of organic C or N stocks in soil, there is generally a slow build-up of organic C or N in soil. This makes it very difficult to detect significant increase in storage of organic C or N in soil due to management practices, especially from crop rotations where crop residues (roots and

Table 4. Relationships among organic C or N fractions (TOC, TON, LFOC, LFON) in the 0 - 15 cm soil, or between crop residue input from 1998 to 2009 growing seasons and organic C or N stored in the 0 - 15 cm soil sampled in autumn 2009 at Scott, Saskatchewan, Canada.

Parameter	Correlation coefficients (r)			
	TOC	TON	LFOC	LFON
Relationships among soil organic C or N fractions				
TOC		0.986***	0.598*	0.463ns
TON			0.626*	0.516ns
LFOC				0.973***
LFON				
Relationships between crop residue C input and soil organic C or N fractions				
Crop residue C input	0.717*	0.678*	0.511ns	0.359ns

*, *, *** and ns refer to significant treatment effects in ANOVA at P ≤ 0.1, P ≤ 0.05, P ≤ 0.001 and not significant, respectively.

Table 5. Linear regressions for relationships between crop residue C input from 1998 to 2009 growing seasons and organic C or N (TOC, TON, LFOC, LFON) stored in the 0 - 15 cm soil sampled in autumn 2009 at Scott, Saskatchewan, Canada.

Crop parameter (X)	Soil C or N parameter (Y)	zLinear regression (Y = a + bX)	R^2
Crop residue C input	TOC	Y = 37.42 + 0.0005X	0.514*
	TON	Y = 3.408 + 0.00004X	0.458*
	LFOC	Y = 2640 + 0.047X	0.261ns
	LFON	Y = 192.4 + 0.002X	0.131ns

zY = Soil organic C or N fraction (TOC and TON as Mg C or N·ha^{-1}; and LFOC, LFON as kg C or N·ha^{-1}; a = Intercept on Y, origin of the line; b = Regression coefficient of Y on X, slope of line; X = Crop residue and/or swine manure C input (Mg·ha^{-1}); * and ns refer to significant treatment effects in ANOVA at P ≤ 0.05 and not significant, respectively.

straw) are returned to soil in all crop phases. Similarly, in our study there was an increase in storage of TOC or TON in soil with increasing length of crop rotation from monoculture pea (1-year) to canola-wheat-pea-wheat rotation (4-year) but the differences were significant only in a few cases even after 12 years.

Previous research has shown that dynamic organic C or N fractions are more responsive to management practices than total organic C or N under annual crops [24-27]. Similarly, in our study the changes in LFOC and LFON were more pronounced than TOC and TON, although the increases in LFOC and LFON in soil due to crop rotations were not significant.

3.3. Residual Nitrate-N and Ammonium-N in Soil

The crop rotation duration × crop phase interaction effect was significant in most soil layers (Table 6). This was due to usually greater amounts of nitrate-N in pea phase than canola phase, particularly in the 1-yr rotation. The mean effect of crop rotation length/duration on soil nitrate-N was significant (at $P \leq 0.1$) only in the total 0 - 60 cm depth, but the amounts of nitrate-N were usually highest in all soil layers for monocultures, with the lowest nitrate-N in the 4-yr rotation. When ANOVA was conducted separately for canola and pea phases, there was no significant effect of crop rotation duration and/or frequency of canola in the rotation on soil nitrate-N. However, for pea phase there was a significant (at $P \leq 0.1$) effect of crop rotation duration and/or frequency of pea in the rotation on nitrate-N in the 15 - 30 and 30 - 60 cm soil layers. The amount of nitrate-N (averaged across crop rotation duration) was significantly greater in pea phase than canola phase in all soil layers. The generally higher soil nitrate-N in most layers with monoculture was probably due to relatively lower cumulative crop yield and total N uptake in seed in 1-yr rotation compared to longer rotations [11]. In another study in Saskatchewan, nitrate-N in soil was also higher in 6-yr continuous rotations with low crop diversity compared to rotations with high diversity of annual grain crops [21]. Earlier research has also suggested that residual N in soil can be decreased with efficient cropping systems [28]. Similarly, in our study soil nitrate-N in most soil layers was usually lowest in the 3-yr and 4-yr rotations with pea and 4-yr rotation with canola, suggesting the importance of 3- or 4-year rotations in reducing residual nitrate-N in the soil profile.

The mean effect of crop phase on ammonium-N in soil was not significant in any soil layer, but the crop rotation duration × crop phase interaction effect was significant in the 0 - 60 cm depth most likely due to greater amounts of ammonium-N in soil in canola phase than pea phase

(Table 6). When ANOVA was conducted separately for canola and pea phases, there was no significant effect of crop rotation duration and/or frequency of pea in the rotation on soil ammonium-N. But, there was a significant effect of crop rotation duration and/or frequency of canola in the rotation on ammonium-N in the 15 - 30 and 30 - 60 cm soil layers. The tendency of greater amounts of ammonium-N (averaged across crop phases) in 1-yr or 2-yr rotation than 3-yr or 4-yr rotations in the 30 - 60 cm soil layer or 0 - 60 cm depth were most likely due to contribution through ammonification of N of recently added residue from crops with relatively longer taproots at deeper depth, particularly canola.

3.4. Amounts of N Uptake in Seed, Residual Nitrate-N in Soil and N Balance

The N balance over the 1998 to 2009 period for the 8 treatments (4 rotation durations × 2 crop phases) included amount of inorganic N applied as fertilizers, N added in seed at seeding time, estimated biologically fixed N (BFN) by pea when it was grown, amount of N recovered in seed over 12 years and mineral N (nitrate-N + ammonium-N) recovered in the 0 - 60 cm soil in autumn 2009 after 12 growing seasons, and the estimated amount of unaccounted N (Table 7). The estimated amount of N recovered in seed (which was removed from the land/field) plus the amount of mineral-N recovered in soil in various treatments ranged from 658 to 1103 kg·N·ha^{-1}. The corresponding values of N applied as inorganic fertilizers, plus N added in seed and BFN during the 12-year experimental period ranged between 556 and 1522 kg·N·ha^{-1}. The amounts of N that could not be accounted for ranged from 441 to −296 kg·N·ha^{-1}.

The amounts of unaccounted N were positive with monoculture pea, and the unaccounted N was much greater with pea than canola in the monoculture and 2-yr rotations, while the differences were small between the canola and pea phases in the 3-yr and 4-yr rotations. The positive unaccounted N reflects an excess of N that was applied to and/or fixed by pea compared to the N recovered in crop seed yield plus nitrate-N recovered in soil in the 1-yr rotation. The results of our study do not suggest any over-application of N, because little N was applied to pea and annual rates of applied N to other crops were moderate, and in canola phase the N balance was negative. It is possible that a portion of the fixed/applied N in 1-yr rotation in pea phase may have leached down below the 60 cm soil depth, as evidenced by greater amounts of nitrate-N recovered in the 30 - 60 cm layer in this study (Table 6), and also in the 60 - 90 cm layer in some treatments in the same experiment in autumn 2005 in our previous report [11]. Other researchers have reported an increase in the concentration of residual soil nitrate-N at

Table 6. Effect of broad-leaf crop phase and frequency over 12 years from 1998 to 2009 on nitrate-N and ammonium-N in soil in autumn 2009 at Scott, Saskatchewan, Canada.

Treatment		Nitrate-N (kg·N·ha⁻¹) in soil layers (cm)				Ammonium-N (kg·N·ha⁻¹) in soil layers (cm)			
Crop rotation (duration)	Crop phase	0 - 15	15 - 30	30 - 60	0 - 60	0 - 15	15 - 30	30 - 60	0 - 60
Monoculture (1-year)	Canola	12.7	0.0	0.0	12.7	6.9	6.8	14.0	27.7
Monoculture (1-year)	Pea	23.7	5.6	11.5	40.8	6.1	5.0	7.6	18.7
Canola-wheat (2-year)	Canola	12.7	1.0	0.5	14.2	4.8	4.2	9.1	18.1
Pea-wheat (2-year)	Pea	13.5	2.9	6.8	23.2	7.3	6.2	11.8	25.3
Pea-canola-wheat (3-year)	Canola	12.2	1.3	2.2	15.7	4.5	2.8	7.5	14.8
Pea-canola-wheat (3-year)	Pea	14.0	2.0	1.8	17.8	4.2	5.8	8.7	18.7
Canola-wheat-pea-wheat (4-year)	Canola	8.1	0.5	0.2	8.8	5.7	4.8	9.0	19.5
Canola-wheat-pea-wheat (4-year)	Pea	15.5	2.2	1.5	19.2	3.6	3.1	6.2	12.9
	LSD$_{0.05}$	ns	2.1	6.7	14.8	ns	ns	ns	10.4
	SEM (significance)	3.39^{ns}	0.72***	2.26*	5.04**	1.20^{ns}	1.13^{ns}	2.00^{ns}	3.54$^{•}$
Crop rotation duration									
1-year		18.2	2.8	5.6	26.6	6.5	5.9	10.8	23.2
2-year		13.1	2.0	3.7	18.8	6.1	5.2	10.5	21.8
3-year		13.1	1.7	2.0	16.8	4.4	4.3	8.1	16.8
4-year		11.8	1.4	0.9	14.1	4.7	4.0	7.6	16.3
LSD$_{0.05}$		ns	ns	ns	11.2	ns	ns	ns	ns
SEM (significance)		2.39^{ns}	0.61^{ns}	1.77^{ns}	3.86$^{•}$	0.86^{ns}	0.87^{ns}	1.51^{ns}	2.72^{ns}
	Crop phase								
	Canola	11.4	0.7	0.7	12.8	5.5	4.7	9.9	20.1
	Pea	16.7	3.2	5.4	25.3	5.3	5.0	8.6	18.9
	LSD$_{0.05}$	4.9	1.3	3.7	8.0	ns	ns	ns	ns
	SEM (significance)	1.69*	0.43***	1.25*	2.73**	0.61^{ns}	0.61^{ns}	1.06^{ns}	1.93^{ns}
Additional statistical comparisons									
Significance of four crop rotation durations for canola phase only									
	LSD$_{0.05}$	ns	ns	ns	ns	ns	3.5	4.6	9.6
	SEM (significance)	3.50^{ns}	0.47^{ns}	1.00^{ns}	3.89^{ns}	1.09^{ns}	1.09$^{•}$	1.45*	3.00$^{•}$
Significance of four crop rotation durations for pea phase only									
	LSD$_{0.05}$	ns	2.8	9.3	ns	ns	ns	ns	ns
	SEM (significance)	3.71^{ns}	0.88$^{•}$	2.92$^{•}$	6.33^{ns}	1.36^{ns}	1.14^{ns}	1.90^{ns}	3.39^{ns}

$^{•}$, * and ns refer to significant treatment effects in ANOVA at $P \leq 0.1$, $P \leq 0.05$ and not significant, respectively.

Table 7. Balance sheets of broad-leaf crop phase and frequency over 12 years from 1998 to 2009 at Scott, Saskatchewan, Canada.

Parameters	Crop rotation duration							
	1-yr (monoculture)		2-yr		3-yr		4-yr	
	Canola	Pea	Canola	Pea	Canola	Pea	Canola	Pea
Mineral-N recovered in soil (0 - 60 cm) after 12 years in autumn 2009 (kg·N·ha^{-1})	40	60	32	49	31	37	28	32
N recovered in seed in 12 years (kg·N·ha^{-1})	618	1021	820	1054	954	954	972	972
N recovered in soil after 12 years + N recovered in seed in 12 years (kg·N·ha^{-1})	658	1081	852	1103	985	991	1000	1004
Inorganic N applied in fertilizers to non-legume crops in 12 years (kg·N·ha^{-1})	552	84	534	300	384	384	477	477
Organic N added in seed in 12 years (kg·N·ha^{-1})	4	78	22	59	41	41	41	41
Organic N fixed by pea in 12 years (kg·N·ha^{-1})	0	1460	0	730	487	487	365	365
Total N added in fertilizers + in seed + fixed by pea in 12 years (kg·N·ha^{-1})	556	1522	556	1089	912	912	883	883
N balance (N applied in fertilizers/seed/fixed − N recovered in seed) (kg·N·ha^{-1})	−62	501	−264	35	−42	−42	−89	−89
Unaccounted N (N applied in fertilizers/seed/fixed − N recovered in soil + seed) (kg·N·ha^{-1})	−102	441	−296	−14	−73	−79	−117	−121

high N fertilizer rates [28-30] and nitrate leaching in the 90 - 240 cm soil profile [21], suggesting the need for deep soil sampling below the 60 or 90 cm depth in future in this long-term study. Soil nitrate-N below the effective root zone of crops is susceptible to leaching, and the loss of nitrate-N through leaching can result in N contamination of groundwater, and thus represents a potential risk to groundwater quality and soil health [31-32]. Furthermore, a portion of the applied N in these treatments may have been immobilized in soil organic N, as evidenced by large amounts of soil N in LFON in this study (**Tables 3** and **4**). In addition, a small portion of the applied N may have been lost from the soil-plant system through denitrification (e.g., nitrous oxide and other N gases) due to wet soil conditions which temporarily exist in the present study in some years after occasional heavy rainfall during summer and/or autumn [33,34]. The negative amounts of N balance and unaccounted N, especially in canola phase, suggest that N became available to the crop through mineralization of organic matter in the growing season, and possibly soil may be gaining N from wet deposition and/or non-symbiotic N fixation but this needs further research to verify any contribution of N from precipitation and non-symbiotic N fixation.

4. CONCLUSION

The findings suggest that the quantity of organic C and N can be maximized by increasing duration of crop rotation and by including hybrid canola in the rotation.

5. ACKNOWLEDGEMENTS

The authors are grateful to Brett Mollison, Colleen Kirkham, Kara Lengyel, Kimberly Martin, Don Gerein and Larry Sproule for technical support, Darwin Leach for preparing the poster, and Erin Cadieu for printing the poster.

REFERENCES

[1] Christen, O. and Sieling, K. (1995) Effect of different preceding crops and crop rotations on yield of winter oil-seed rape (*Brassica napus* L.) J. *Agronomy and Crop Science*, **174**, 265-271.

[2] Pearse, P.R., Morrall, A.A., Kutcher, H.R., Keri, M., Kaminski, D., Gugel, R., Anderson, K., Trail, C. and Greuel, W. (2001) Survey of canola diseases in Saskatchewan, 2000. *Canadian Plant Distribution Survey*, **81**, 105-107.

[3] Johnston, A.M., Kutcher, H.R. and Bailey, K.L. (2005) Impact of crop sequence decisions in the Saskatchewan Parkland. *Canadian Journal of Plant Science*, **85**, 95-102.

[4] Dalal, R.C. (1989) Long-term effects of no-tillage, crop residue, and nitrogen application on properties of a Vertsol. *Soil Science Society of America Journal*, **53**, 1511-1515.

[5] Havlin, J.L., Kissel, D.E., Maddux, L.D. and Long, J.H. (1990) Crop rotation and tillage effects on soil organic carbon and nitrogen. *Soil Science Society of America Journal*, **54**, 448-452.

[6]   Janzen, H.H., Campbell, C.A., Brandt, S.A., Lafond, G.P. and Townley-Smith, L. (1992) Light-fraction organic matter in soil from long term rotations. *Soil Science Society of America Journal*, **56**, 1799-1806.

[7]   Halvorson, A.D., Wienhold, B.J. and Black, A.L. (2002) Tillage, nitrogen and cropping system effects on soil carbon sequestration. *Soil Science Society of America Journal*, **66**, 906-912.

[8]   Liang, B.C., McConkey, B.G., Schoenau, J.J., Curtin, D., Campbell, C.A., Moulin, A., Lafond, G.P., Brandt, S.A. and Wang, H. (2003) Effect of tillage and crop rotations on the light fraction organic carbon and carbon mineralization in Chernozemic soils of Saskatchewan. *Canadian Journal of Soil Science*, **83**, 65-72.

[9]   McConkey, B.G., Liang, B.C., Campbell, C.A., Curtin, D., Moulin, A., Brandt, S.A. and Lafond, G.P. (2003) Crop rotation and tillage impact on carbon sequestration in canadian prairie soils. *Soil and Tillage Research*, **74**, 81-90.

[10] Malhi, S.S., Moulin, A.P., Johnston, A.M. and Kutcher, R.H. (2008) Short-term and long-term effects of tillage and crop rotation on some soil physical and biological properties in a Black Chernozem soil in northeastern Saskatchewan. *Canadian Journal of Soil Science*, **88**, 273-282.

[11] Malhi, S.S., Brandt, S.A. and Kutcher, H.R. (2011) Effects of broad-leaf crops frequency and fungicide application in various crop rotations on seed yield, and accumulation of nitrate-N and extractable P in soil after eight years in a Dark Brown Chernozem in Saskatchewan. *Communications in Soil Science and Plant Analysis*, **42**, 2795-2812.

[12] Culley, J.L.B. (1993) Density and compressibility. In: M.R. Carter, Ed., *Soil Sampling and Methods of Analysis*, Lewis Publishers, Boca Raton, 529-549.

[13] Technicon Industrial Systems (1977) Industrial/simultaneous determination of nitrogen and/or phosphorus in BD acid digests. *Industrial Method* 334-74W/Bt. Technicon Industrial Systems, Tarrytown.

[14] Izaurralde, R.C., Nyborg, M., Solberg, E.D., Janzen, H.H., Arshad, M.A., Malhi, S.S. and Molina-Ayala, M. (1997) Carbon storage in eroded soils after five years of reclamation techniques. In: Lal, R., Kimble, J.M., Follett, R.F. and Stewart, B.A., Eds., *Management of Carbon Sequestration in Soil, Advances in Soil Sciences*, CRC Press, Boca Raton, 369-385.

[15] Technicon Industrial Systems (1973) Ammonium in water and waste water. *Industrial Method* No. 90-70W-B. Technicon Industrial Systems, Tarrytown.

[16] Technicon Industrial Systems (1973) Nitrate in water and waste water. *Industrial Method* No. 100-70W-B. Technicon Industrial Systems, Tarrytown.

[17] IPCC (Integovernmental Panel on Climate Change) (2006) N_2O emissions from managed soils, and CO_2 emissions from lime and urea application. Chapter 11. In: *IPCC Guidelines for National Greenhouse Gas Inventories, Agriculture, Forestry and Other Land Use*, Vol. 4, Institute for Global Environment Strategies, Hayama.

[18] Lupwayi, N.Z., Clayton, G.W., O'Donovan, J.T., Harker, K.N., Turkington, T.K. and Soon, Y.K. (2007) Phosphorus release during decomposition of crop residues under conventional and zero tillage. *Soil and Tillage Research*, **95**, 231-239.

[19] Malhi, S.S. and Lemke, R. (2007) Tillage, crop residue and N fertilizer effects on crop yield, nutrient uptake, soil quality and greenhouse gas emissions in the second 4-yr rotation cycle. *Soil and Tillage Research*, **96**, 269-283.

[20] SAS Institute, Inc. (2004) SAS product documentation. Version 8, SAS Institute, Cary. http://support.sas.com/documentation/onlinedoc/index.html

[21] Malhi, S.S., Brandt, S.A., Lemke, R., Moulin, A.P. and Zentner, R.P. (2009) Effects of input level and crop diversity on soil nitrate-N, extractable P, aggregation, organic C and N, and N and P balance in the Canadian Prairie. *Nutrient Cycling in Agroecosystems*.

[22] Campbell, C.A., VandenBygaart, A.J., Grant, B., Zentner, R.P., McConkey, B.G., Lemke, R., Gregorich, E.G. and Fernandez,M. R. (2007) Quantifying carbon sequestration in a conventionally tilled crop rotation study in southwestern Saskatchewan. *Canadian Journal of Soil Science*, **87**, 23-38.

[23] Comeau, L. (2012) The influence of lentil, canola, pea and wheat on carbon and nitrogen dynamics in two Chernozemic soils. M.Sc. thesis, University of Saskatchewan, Saskatoon.

[24] Gregorich, E.G. and Janzen, H.H. (1995) Storage of soil carbon in the light fraction and macroorganic. In: Carter, M.R. and Stewart, B.A., Eds., *Structure and Organic Matter Storage in Agricultural Soils. Advances in Soil Science*, Lewis Publishers, CRC Press, Boca Raton, 167-190

[25] Malhi, S.S., Nyborg, M., Goddard, T. and Puurveen, D. (2010) Long-term tillage, straw and N rate effects on quantity and quality of organic C and N in a Gray Luvisol soil. *Nutrient Cycling in Agroecosystems*.

[26] Malhi, S.S., Nyborg, M., Goddard, T. and Puurveen, D. (2011b). Long-term tillage, straw management and N fertilization effects on quantity and quality of organic C and N in a Black Chernozem soil. *Nutrient Cycling in Agroecosystems*, **90**, 227-241.

[27] Malhi, S.S., Nyborg, M., Solberg, E.D., McConkey, B., Dyck, M. and Puurveen, D. (2011) Long-term straw management and N fertilizer rate effects on quantity and quality of organic C and N, and some chemical properties in two contrasting soils in western Canada. *Biology and Fertility of Soils*, **47**, 785-800.

[28] Guillard, K., Griffin, G.F., Allinson, D.W., Yamartino, W.R., Rafey, M.M. and Pietryzk, S.W. (1995) Nitrogen utilization of selected cropping systems in the US northeast. II. Soil profile nitrate distribution and accumulation. *Agronomy Journal*, **87**, 199-207.

[29] Malhi, S.S., Harapiak, J.T., Nyborg, M. and Flore, N.A.

(1991) Soil chemical properties after long-term N fertilization of bromegrass: Nitrogen rate. *Communications in Soil Science and Plant Analysis*, **22**, 1447-1458.

[30] Malhi, S.S., Harapiak, J.T., Gill, K.S. and Flore, N. (2002) Long-term N rates and subsequent lime application effects on macroelements concentration in soil and in bromegrass hay. *Journal of Sustainable Agriculture*, **21**, 79-97.

[31] Zhang, W.L., Tian, Z.X., Zhang, N. and Li, X.O. (1996) Nitrate pollution of groundwater in northern China. *Agriculture, Ecosystem and Environment*, **59**, 223-231.

[32] Yuan, X., Tong, Y., Yang, X., Li, X. and Zhang, F. (2000) Effect of organic manure on soil nitrate accumulation. *Soil Environmental Science*, **9**, 197-200.

[33] Heaney, D.J., Nyborg, M., Solberg, E.D., Malhi, S.S. and Ashworth, J. (1992) Overwinter nitrate loss and denitrification potential of cultivated soils in Alberta. *Soil Biology and Biochemistry*, **24**, 877-884.

[34] Nyborg, M., Laidlaw, J.W., Solberg, E.D. and Malhi, S.S. (1997) Denitrification and nitrous oxide emissions from soil during spring thaw in a Malmo loam, Alberta. *Canadian Journal of Soil Science*, **77**, 53-160.

7

Isolation and pathogenicity of fungi associated to ambrosia borer (*Euplatypus segnis*) found injuring pecan (*Carya illinoensis*) wood

Ramón Alvidrez-Villarreal[1], Francisco Daniel Hernández-Castillo[1*], Oswaldo Garcia-Martínez[1], Rosalinda Mendoza-Villarreal[2], Raúl Rodríguez-Herrera[3], Cristóbal N. Aguilar[3]

[1]Department of Agricultural Parasitology, Universidad Autónoma Agraria Antonio Narro, Saltillo, México;
*Corresponding Author
[2]Department of Basic Sciences, Universidad Autónoma Agraria Antonio Narro, Saltillo, México
[3]Department of Food Research, School of Chemistry, Universidad Autónoma de Coahuila, Saltillo, México

ABSTRACT

***Euplatypus segnis* is an insect pest of economic importance in pecan (*Carya illinoensis*) trees grown at Parras, General Cepeda and Torreón Coahuila, Mexico. The objectives in this study were to identify the fungal strains associated to ambrosia borer body and diseased pecan wood and determine their pathogenicity. The results showed that the associated fungi to *Euplatypus segnis* and damaging the pecan wood were identified as: *Helminthosporium sp.*, *Aspergillus sp.*, *Penicillium sp.*, *Phoma sp.*, *Ascochyta sp.*, *Phaecylomices sp.*, *Umbeliopsis sp.*, *Torula sp.*, *Fusarium solani*, *Alternaria alternata*, *Fusarum oxysporum*, and *Lasiodiplodia theobromae*. The pathogenicity tests on healthy 3 year old pecan trees cv. western using *Fusarium oxysporum*, *Fusarium solani*, *Alternaria alternata* and *Lasiodiplodia theobromae* suspension conidia shown die back tree branches after 84 days inoculation. The insect in combination with the fungal invasion eventually cause the death of trees. Additionally, the insect contributes to the spread of fungi in pecan nut orchards.**

Keywords: Pathogenicity; Pecan Nut; *Euplatypus segnis*; Ambrosia Borer; *Carya illinoensis*

1. INTRODUCTION

In Mexico, *Euplatypus segnis* also known as trunk and branch borer is distributed in the states of Coahuila, Durango, San Luis Potosi, Jalisco and Chihuahua [1-3]. In the Coahuila Southwest region there are approximately 525 thousand pecan (*Carya illinoensis*) trees on produc-

tion, and annually are harvested about 19,345 tons of nuts. However, in more that 20 percent of the tree plantations the presence of ambrosia borer *Euplatypus segnis* has been reported causing yield losses of up to 4% (773 tons of nuts) [4]. In addition, *E. segnis* (Chapuis) (Coleoptera: Platypodinae), attacks wild trees such as: *Bursera copallifera*, *Delonix regia*, *Ficus sp.* *Ficus lyrata*, *Ficus microcarpa*, *Ficus elastica* and *Ficus cotinifolia* [5], fruit trees such as avocado, apple, peach, pomegranate, apricot, and mango, and some forestry trees like silver poplar, blackberry and tropical ash [6]. There are reports on symbiotic associations between ambrosia insects and fungus, this relation help both of them to survive, per example *Xyleborus ferrugineus* and *Lasiodiplodia theobromae* in avocado (*Persea americana* Mill) [7], *Xyleborus ferrugineus* and *Fusarium solani*, *Cephalosporium sp.* and *Graphium sp.* [8], *Ips typographus* and *Ophiostoma polonicum* on *Picea abies* [9], *Hypotenemus hampei* and *Fusarium solani* on Coffea genus trees [10]. *Platypus quercivorus* and *Raffaelea quercivora* on *Quercus crispula* and *Q. serrata* [11], *Hypothenemus hampei* and *Penicillium brocae* on *Coffea arabiga* [12], *Platypus mutatus* associated with a symbiotic fungus in commercial plantations of poplar [13]. In this study, fungi associated to ambrosia borer isolated from diseased pecan wood were morphologically and molecularly identified and incidence of diseases induced by these fungal species was determined using pathogenicity tests.

2. MATERIALS AND METHODS

2.1. Study Area

The sample collections of *E. segnis* specimens and pecan damaged tissues were obtained from Parras de la Fuente (25°22'N 102°11'W 1520 m), General Cepeda

(25°22'N 101°28'W 1470 m) and Torreon (25°42'N 103°27'W 1120 m) all of these counties located in Coahuila, Mexico. Three orchards were sampled by locality and at each orchard were selected three 20-years old damaged trees cv. Western, the damaged trees were in the phase 3 (50% damaged leaf area, 25 to 50 holes entries and presence of sawdust at the stem base) [14]. Wood and insects samples were placed in plastic bags and were properly coded and carried to the entomology and phytopathology laboratories for examination.

2.2. Fungal Isolates

The fungi were obtained from xylem, phloem and wood contained entry points and borer insects. The borer insects were sectioning into head, thorax and abdomen. Each section was disinfected used sodium hypochlorite (1.5%) by immersion for 3 min, after those insect sections were washed with sterile distilled water for 1 min and dried on sterile paper towels. Wood and insect sections were placed on Petri dishes containing potato dextrose agar (PDA: 20 g potato, 20 g of dextrose, 18 g agar and 1000 mL distilled water) as a culture medium, placing 4 pieces per plate. The Petri dishes were incubated and a continuous black light lamp 40 W to 25°C ± 1°C was using during the day. The isolates were purified by monospore cultures on water-agar (AA: 18 g agar in 1000 mL distilled water), and increased on PDA. The strains were spread on liquid medium (Pontecorbo) in order to obtain spores and mycelium production for pathogenicity test, and freeze drying and preservation.

2.3. Identification of Pathogenic Fungi

2.3.1. Fungal Morphological Characterization

After 7 - 10 days, the isolated and purified fungi from wood and insects were observed under a composed microscope and identified by morphology using taxonomic keys for ascomycetes and imperfect fungi, in this case characterization was based on characteristics of isolate reproductive structures using Barnnet and Hunter [15], Hanlin [16], Boot [17], Rotem [18], Neergaard [19], Sutton [20] keys for genus and Wei *et al.* [21] key for species. To identify *Lasiodiplodia theobromae* were used Punithalingam [22], Sutton [20] and Burgess *et al.* [23] keys.

2.3.2. Molecular Identification of Isolated Fungi

To confirm the morphological identification of those fungi colonies that showed the highest pathogenicity on pecan trees under field conditions. DNA was isolated from *Fusarium oxysporum*, *Fusarium solani*, *Alternaria alternata* and *Lasiodiplodia theobromae* fungi at the Molecular Biology laboratory, CIIDIR-IPN Oaxaca. The DNA was isolated using the methodology proposed by Ahrens and Seemüller [24]. Amplification of internal transcribed spaces of ribosomal genes (rRNA) was performed by PCR using the primers ITS4 (TCC TCC GCT TAT TGA TAT GC) and ITS5 (GGA AGT AAA AGT CGT AAC AAG G) according to the methodology proposed by Ahrens and Seemüller [24] with minor modifications, PCR reaction was composed of sterile ultrapure water (13.22 µL), 10X TBE buffer (2.5 µL), $MgCl_2$ at 2.5 mM (2.08 µL), dNTPs at 0.2 mM (2 µL), primers ITS4 and ITS5 to 20 pmol (2 µL of each), DNA polymerase (biogenic®) 1 U (0.2 µL) and 80 ng of DNA sample (1 uL). The amplified product was purified with Wizard kit (Promega®) and sequenced with the Genetic Analyzer® Model 3100 (Applied Biosystems). The data base sequences with the highest value of similarity in were considered for comparison with the sequences obtained in this study. The ITS sequences obtained were analyzed using the software Lasergene® 2001, V5 (DNASTAR®, Inc.) and were aligned in the database of the National Gene Bank Center for Biotechnology Information (NCBI), by the BLAST program (Basic Local Alignment Search Tool)

2.4. Pathogenicity Tests

Inoculum was incremented in PDA culture medium which was autoclaved three times continues with an interval of 24 hours among sterilizations. Then were inoculated each of the twelve fungi isolated from pecan damaged wood and borer insect body: *Phoma*, *Fusarium oxysporum*, *Ascochyta*, *Phaecylomices sp Lasiodiplodia theobromae*, *Alternaria alternata*, *Umbeliopsis sp*, *Torula sp*, *Fusarium solani*, *Helminthosporium sp*, *Aspergillus sp* and *Penicillium sp*. The isolates were multiplied in the culture media described above, then were kept in an incubator at 27°C with a photoperiod of 12:12 and a relative humidity of 50% ± 10%. Once sporulation was observed, the conidia from each isolate were harvested adding to each Petri dish 50 mL of a solution of Tween 80 (0.05%) and sterile distilled water, and removing the mold with a spatula. Subsequently, the mold solution was liquefied for 60 sec and placed in a beaker of 250 mL (stock suspension). Then, were made three successive dilutions of 100 uL of the stock suspension in 900 uL of sterile distilled water and were labeled 10^{-1}, 10^{-2} and 10^{-3}. Finally, spore concentration was determined using a Neubauer's camera and microscope, the dilution samples were observed in 40× to calculate spore concentration of the stock solution which was adjusted to a final concentration of 10^9 conidia/mL.

Pathogenicity tests were performed *in vivo* in twenty eight three years-old healthy pecan trees cv western (height (150 cm) and stem diameter (4 - 5 cm)) established in a high technology greenhouse at a temperature of 24°C ± 2°C, at the Universidad Autonoma Agraria

Antonio Narro located in Saltillo, Coahuila. The inoculation technique used in this step was that proposed by Kuroda [11], briefly; trees bark at a height of 100 cm was sterilize with sodium hypochlorite (1.5%), then washed with sterile distilled water and dried on using sterile absorbent paper. To each tree, two holes were made on the stem until xylem with a sterile 3-mm drill diameter. Twenty four trees were inoculated with 1 mL of the conidial suspension at 1×10^9 conidia·ml^{-1} of each one of the twelve identified fungi and two trees were inoculated with a mixed suspension of all fungi spores and two trees with sterile distilled water (control). The inoculations were added using a manual sterile pipette, each wound was covered with parafilm to prevent rapid drying and possible contamination by other microorganisms. According to this technique was favored the low density against mass inoculations in the inoculated tree to prevent wilting and rapid reactions to characterize tissues in a cytological way in response to the inoculated fungi activities. The evaluations were done every 7 days, watching symptoms and disease progression for 84 days after inoculation. Later, from the infected tissues of inoculated stems was isolated the fungus in pure culture and compared morphologically with the fungus inoculated to confirm Koch's postulates.

2.5. Statistical Analysis

The response variable was the presence of the fungal specie each case, which was evaluated in a quality way (presence and absence). For results analysis a categorical data analysis was performed. In this case Sx2 tables were used, where row and columns were nominally classified. Statistical analysis was performed using SAS (8.1 versions).

3. RESULTS

3.1. Disease Symptoms in Pecan Plantations under Field Conditions

In 20 years-old pecan trees were observed the characteristic symptoms of regressive dead and insect presence mainly in weakened and/or stressed (because age, water deficiency, injuries, fires, intense cold, snow or prolonged drought) trees. In trunk and branches were detected 2 mm-diameter entry holes and sawdust at the trees base, crystal reddish exudates that eventually became dark brown and pungent, after, taking off the bark, necrotic spots were observed with the typical form of diamond. On the other hand one cans heard a sound in trunks and branches; in addition one can see the sap out of the branches when phloem and xylem are piercing by the borer. Also the presence of adults, especially during morning and evening is detected, chlorosis is observed in

the leaf area, wilting, premature abscission of stems and leaves, partial death of branches and downward total death of trees.

3.2. Morphological and Molecular Identification of Plant Pathogenic Fungi Associated with Plant Damage

The morphological characterization based on reproductive structures of the fungal isolates was performed using the taxonomic keys proposed by Barnnet and Hunter [15], Hanlin [16], Boot [17], Rotem [18], Neergaard [19], and Sutton [20] for genus and Wei *et al.* [21] keys were used for the specie identification. *Lasiodiplodia theobromae* was identified using the Punithalingam [22], Sutton [20] and Burgess *et al.* [23] keys. In **Table 1** are shown the fungal species associated to die-back in pecan damaged wood.

The identification to genera and specie level by sequencing of the regions ITS1 and ITS2 of ribosomal genes (rRNA) was performed only to confirm identity of those fungal more common as pathogens. *L. theobromae*

Table 1. Genus and species identified in samples of pecan with dieback symptoms and presence of ambrosia insects.

Name	Morphology
Lasiodiplodia theobromae	Hyphae septate, dark 1.5 to 2.8 microns in diameter. Dark-colored pycnidia 493.04 × 622.82 microns, with ostiole at the apex, formed both in artificial medium and in plant tissue. Hyaline septate conidia in immature, dark and septate with longitudinal striations in mature state 11.35 26.54 × microns.
Fusarium oxysporum	Microscopically *Fusarium oxysporum* isolates showed abundant macroconidios, hyaline, oval, elliptical and occasionally kidney-shaped, with one or two septa, which measured 4 - 5 × 2 - 4 um, formed in false heads. The thin-walled hyaline macroconidia are just curved (canoe) or nearly straight, pointed at both ends, in the form of basal cell foot. Presented 3 - 5 septa and measure 20 - 59.5 × 2 - 6 microns. It has also marked by abundant spherical chlamydospores, large, hyaline, with smooth or rough walls. Can be seen individually or in pairs at intervals along the hyphae or branched short lateral sections.
Fusarium solani (Mart.) Sacc.	*Fusarium solani* (Mart.) Sacc. Abundant microconidia hyaline, ovoid, with 0 - 1 septum and occasional 2. Born in monophialides long with collar, a false head, measuring 8 - 16 × 3 - 5 microns. Macroconidia slightly curved, with ventral and dorsal plane parallel in most of its length, with 5 - 6 septa, 23 - 67 × 4 - 6 um in, slightly curved apical cell and basal cell pedicellate, cream-colored sporodochia at the beginning, later turning blue or blue.
Alternaria alternata (fr,) Keissler	*Alternaria alternata* (fr) Keissler presents colony dark green to olive, conidia (dictyosporae) in chains on short conidiophores and branching. The strings appear in dense cultures and isolated as bundles of 1 to 9 transverse septa and several longitudinal or oblique.

Isolation and pathogenicity of fungi associated to ambrosia borer (Euplatypus segnis) found injuring pecan (Carya illinoensis) wood

79

genome sequence was compared with the genetic sequences of *L. theobromae* deposited in the gene bank (NCBI) with accession numbers EF622073 (542 bp) and EU564805 (540 bp) in 98% and AY662402.1 in 99%; the *Fusarium solani* sequences was lineated with *Fusarium solani* sequences deposited in gene bank (NCBI) with accession numbers AF440567.1 (480 bp) in 94%, and AY310442.1 (554 bp) in 100%. The *Fusarium oxysporum* was aligned with *F. oxysporum* accession number EF495235.1 with a similarity index of 99%, *Alternaria alternata* gene sequence was compared and showed to be homologous to *A. alternata* with access numbers AF397236.1 with a similarity index of 100%, all these fungal species were confirmed as causal agents of dieback in pecan grown in Coahuila Southeast Region.

3.3. Fungi Associated to Insects and Damaged Pecan Trees in the Coahuila Southwest Region

There were isolated 459 fungal strains from insects and diseased pecan wood on artificial culture medium (PDA), distributed as follows: In Parras county were isolated 165 strains, in General Cepeda county 161 strains, and in Torreón county 131 strains. The genera isolated were described according to their morphological structures based on microscopic observations and literature guides.

In the **Table 2** are displayed the statistics for fungi by incidence. The Mantel-Haenszel Chi-Square (Qcs), is valid for these data since fungi is ordinally scaled, and the response is dichotomous; it indicates that there is a strong association between fungi and incidence (Qcs =

73.53) which is highly significant. By looking at the percentages in the **Figure 1**, we can see that fungi *F. solani* and *F. oxysporum* had the highest frequency of incidence.

In the **Table 3** are displays the Mantel-Haenszel results for the stratified analysis where the strata are all combinations of county and tissue used (insect o wood) is Qcs_{MH} = 73.4498, which is strongly significant, in this case some fungal species where observed only in some counties and a specific tissue (wood or insect) but no in others (**Figures 2** and **3**). The association of fungi and incidence controlling for counties is Qcs_{MH} = 73.4878, which is highly significant. In this case some fungal species appeared only in samples taken in specific counties (**Figures 3** and **4**). On the other hand, the Mantel-Haenszel results for the stratified analysis where the strata are all combinations of tissue used (insect o wood) is Qcs_{MH} = 73.5103, respectively, which is highly significant. In

Table 2. Statistics for fungi by incidence.

Statistic	DF	Value	Prob
Chi-Square	11	820.44	<0.0001
Likelihood Ratio Chi-Square	11	770.72	<0.0001
Mantel-Haenzel Chi-Square	1	73.53	<0.0001
Phi Coefficient		0.33	
Contigency Coefficient		0.31	
Cramer's V		0.33	

Sample size = 7202.

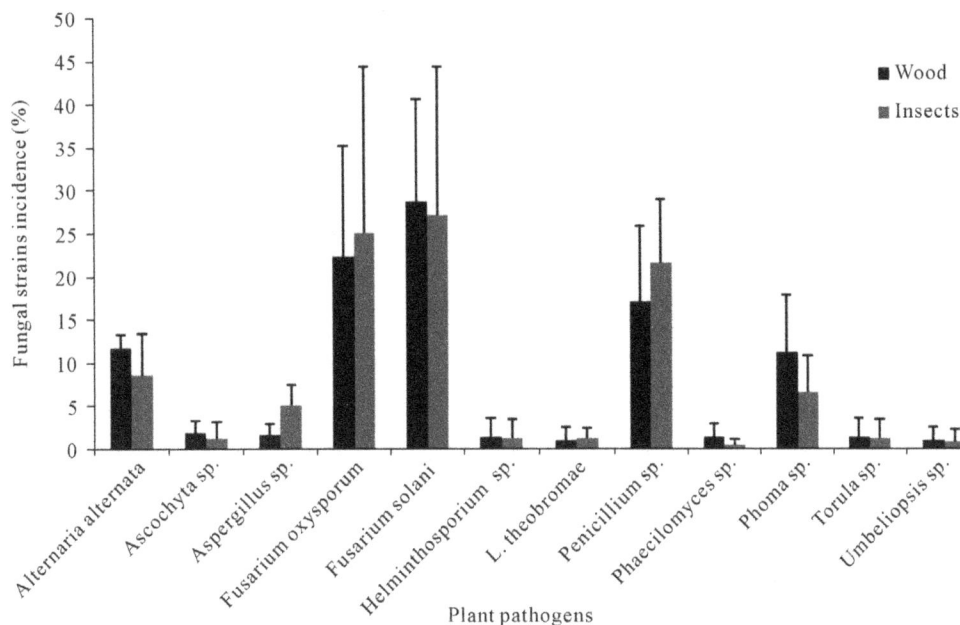

Figure 1. Fungi species isolated from the *Euplatypus segnis* collected attacking pecan branches and trunks at three Coahuila counties.

Table 3. Statistics for stratified by county and tissue used, stratified by county and stratified by tissue used.

Alternative Hypothesis	DF	Value			Probability
		County and tissue	County	Tissue	
Nonzero Correlation	1	73.4498	73.4878	73.5103	<0.0001
Row mean Scores Differ	11	819.7609	820.1022	820.216	<0.0001
General Associations	11	819.7609	820.1022	820.216	<0.0001

Cochran—Mantel-Haenszel Statistics (Modified Ridit Scores).

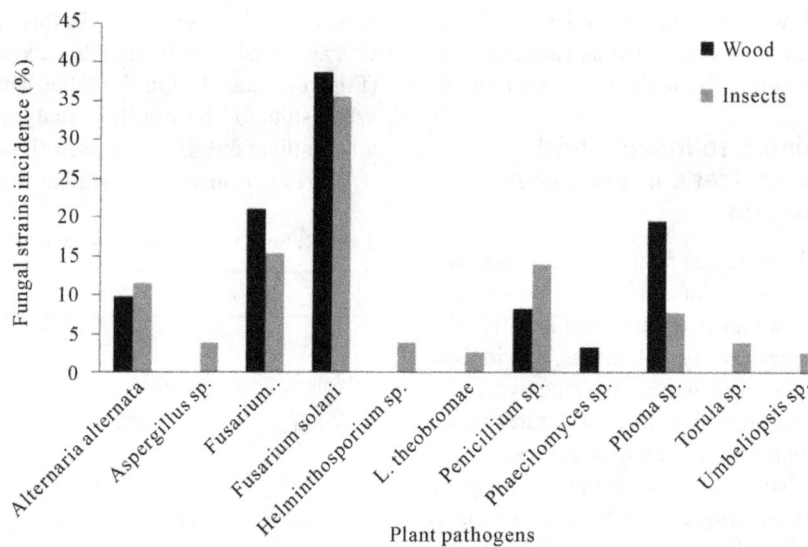

Figure 2. Fungal species found associated with *Euplatypus segnis* and diseased pecan (*Carya illinoensis*) wood collected at Parras Coahuila, Mexico.

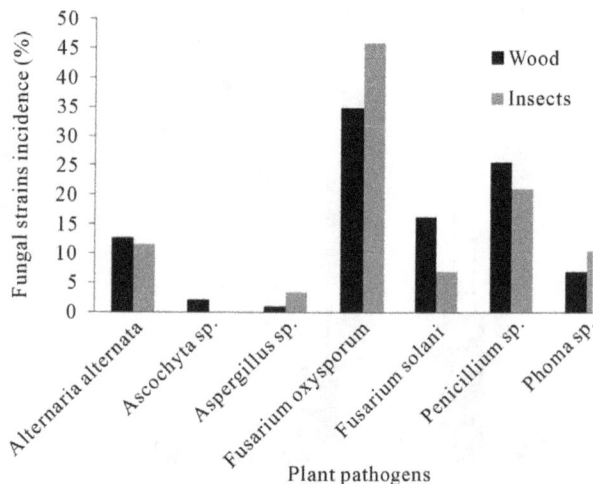

Figure 3. Fungal species found associated with *Euplatypus segnis* and diseased pecan (*Carya illinoensis*) wood collected at General Cepeda, Coahuila, Mexico.

this case some fungal species appeared only in samples taken in a specific tissue but not in the others (**Figure 2**). The significant association between incidence and fungal specie remains evident. Q_{GMH} and Q_{SH} are significant too in all the analyses.

Eighty nine fungal strains were isolated from borers insect body (*Euplatypus segnis*) collected in pecan branches and trunks at Parras county, 35% of the isolates belongs to *Fusarium solani*, 15.2% to *Fusarium oxysporum*, 13.9% to *Penicillium*, 11.4% to *Alternaria alternata*, 7.6% to Phoma *sp.*, 3.8% to *Helminthosporium sp.*, *Torula sp.*, *Aspergillus sp.*, and 2.5% to *Lasiodiplodia theobromae* and *Umbeliopsis sp.*, (**Figure 2**), also 76 fungal strains were isolated from diseased pecan wood, 38.7% belongs to *Fusarium solani*, 21% to *Fusarium oxysporum*, 19.4% to *Phoma sp.*, 9.7% to *Alternaria alternata*, 8.1% to *Penicillium sp.*, and 3.2% to *Phaecylomices sp.* (**Figure 2**).

Eighty five fungal strains were isolated from borer insect (*E. segnis*) body collected in pecan branches and trunks at General Cepeda county, these strains were distributed as follow: 45.9% belonging to *Fusarium oxysporum*, 21.2% to *Penicillium*, 11.4% to *Alternaria alternata*, 10.6% to *Phoma sp.*, 7.1% to *Fusarium solani* and 3.5% to *Aspergillus sp.* (**Figure 3**). On the other

Isolation and pathogenicity of fungi associated to ambrosia borer (Euplatypus segnis) found injuring pecan (Carya illinoensis) wood

81

hand 76 fungal strains were isolated from pecan (*C. illinoensis*) diseased wood, distributed as follow 38.7% belongs to *Fusarium oxysporum*, 21.2% to *Penicillium sp.*, 16.3% to *Fusarium solani*, 11.8% to *Alternaria alternata*, 7% to *Phoma sp.*, 2.3% to *Ascochyta sp.* and 1.2% to *Aspergillus sp.* (**Figure 3**).

From the samples collected in the Torreon county, sixty nine fungal stains were isolated from borer insect (*E. segnis*) body collected attacking pecan branches and trunks. These strain were distributed as follow: 38.5% belongs to *Fusarium solani*, 28.6% to *Penicillium sp.*, 10.3% to*Fusarium oxysporum*, 7.7% to *Aspergillus sp.*, 4.0% to *Helminthosporium sp.* And *Alternaria alternata*, 3.3% to *Ascochyta sp.*, 2.2% to *Phoma sp.* and 1.1% to *Phaecylomices sp.* and *Lasiodiplodia theobromae* (**Figure 4**), also were isolated 62 fungal strains from diseased pecan (*C. illinoensis*) wood distributed as follow 34.7% belongs to *Fusarium solani*, 14.7% to *Penicillium sp.*, 9.3% to *Fusarium oxysporum*, 7.7% to *Aspergillus sp.*, 12.0% to *Alternaria*, 9.3% to *Phoma sp.*, 4.0% to *Torula* and *Helminthosporium sp.*, and 2.7% to *Lasiodiplodia theobromae*, *Umbelliopsis sp.*, *Ascochyta sp.* and 1.3% to *Aspergillus sp.* and *Phaecylomices sp.* (**Figure 4**).

3.4. Pathogenicity of Fungi Strains Isolated from Pecan Trees with Dieback

At 14 days after inoculation were observed the first signs of disease in some trees (**Table 4**). In trees inoculated with *Fusarium solani* had discharge in the drilling at 21 and 28 days, chlorosis at 42 and 49 days, wilting at 63 and 70 days and dieback after 77 days after inoculation, the inoculated trees with *Fusarium oxysporum* shown exudates at the inoculation point in the two treatments at 14 and 21 days, and brown leaves at 35 and 42 days, chlorosis at 45 and 56 days, at 63 and 70 days caused wilting and dieback at 77 and 84 days after inoculation. The inoculated trees by *Alternaria alternata* showed chlorosis at 63 days, wilting to 70 days and dieback to 84 days after inoculation. The trees infected with *Lasiodiplodia theobromae* presented leaves chlorosis at 56 days, premature abscission at 70 days, defoliation at 77 days and dieback at 84 days, the other tree shoots inoculated only presented reddish tip at 77 and chlorosis at 84 days; *Torula sp.* Showed brown leaves at 77 days and defoliation at 84 days, the second treatment presented only brown leaves at 84 days, the two treatments behaved similarly. *Phoma sp.*, caused wilting only 70 days to complete the evaluation at 84 days, the trees inoculated with *Phaecylomices sp.* presented chlorotic symptoms at the end of the evaluation, *Ascochyta sp.*, caused brown leaves at 63 days, wilting from 77 to 84 days, in the second treatment was observed brown leaves at 70 days and defoliation at 84 days; *Helminthosporium sp.*, caused chlorosis at 77 days and wilting at 84 days, the other treatment only showed a slight chlorosis; The treatments that received a mixture of spores showed the most damaged where trees were killed at 77 and 84 days in both treatments, *Penicillium sp.* and *Aspergillus sp.* *Umbeliopsis sp.* and the control treatment did not cause any outward symptoms.

4. DISCUSSION

4.1. Insect-Fungi Association

Euplatypus segnis is is an ambrosia specie with monogenic reproductive behavior [25], polyphagous in its degree of specificity with respect to the host and xylomicetophagus by feeding habit, consuming the conidia producing by

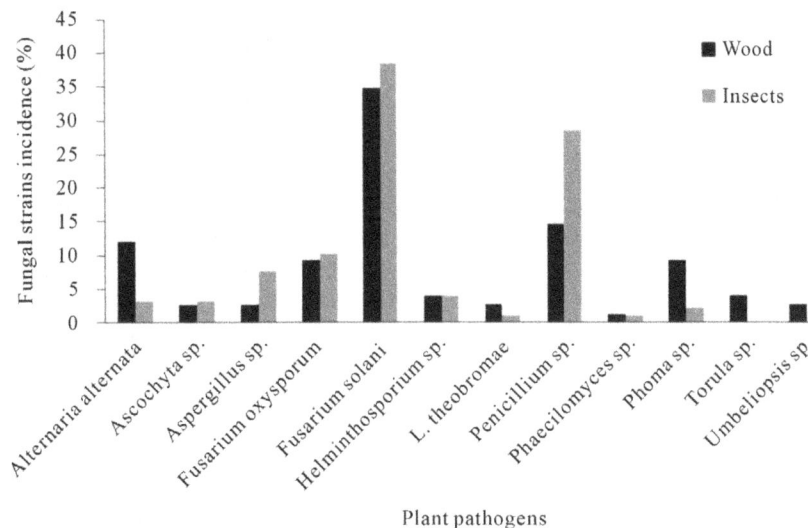

Figure 4. Fungi species found associated with *Euplatypus segnis* and diseased pecan (*Carya illinoensis*) wood collected at Torreon Coahuila, Mexico.

Table 4. Development of symptoms and signs in pecan trees inoculated with fungal strains isolated from *E. segnis* and damaged pecan wood.

Inoculated fungi	Symptoms after different days of inoculation												
	0	7	14	21	28	35	42	49	56	63	70	77	84
Helmintosporium sp.	S											C	C, M
Botryodiplodia sp.	S						C			AP		M, Ar	MR
Torula sp.	S											Hc	D
Umbeliopsis sp.	S												NS
Ascochyta sp.	S									Hc	Hc	M	D, M
Phaecylomices sp.	S												C
Fusarium oxysporum	S		E	E	E	Ar	Ar	C	C	M	M	MR	MR
Fusarium solani	S		E	E				C	C	M	M		MR
Phoma sp.	S										M		M
Alternaria alternata	S									C	M		MR
Penicillium sp.	S												NS
Aspergillus sp.	S												NS
Spore mixture	S							C	C	M	M	MR	MR
Control (Sterile water)	S												NS

S: Health, NS: No symptoms, E: trunk exudates, Ar: reddish tip, Hc: Brown Leaves, C: Chlorosis, M: wilt, AP: Premature Abscission, D: Defoliation. MR: dieback.

the fungi inoculated to the trees [26]. Ulloa [27] mentions that borers have the habit of feeding on fungi introduced by them in the gallery system built into the wood. These fungi appear to be relatively non-specific and need basically moisted wood for their development. Insect species that grow fungus in galleries are called beetles or ambrosia borers and they are dominant in tropical areas [28]. Most of the buildings in the tunnel systems of ragweed borer (galleries) are in the vigorous host tissues, weakened or recently dead, though some species are specialize in colonize the bone marrow, large seeds, fruits and petiole leaves [26,29]. The term ambrosia refers to the fungi that are cultivated by beetles in the gallery walls, which are feed exclusively for insects. Beetles are necessarily dependent on fungi, from which acquired vitamins, amino acids, and essential sterols [30, 31]. The fungi are the main source of food for larvae and adults and are essential to complete insect life cycle [10]. In some insect tribes, only females perform tasks inside the culture, while males are short lived and little flight [32]. After mating, females disperse to new host substrate, carrying fungi in specialized pockets called mycangia. Once inside, the founder females planted the fungi on the walls of tunnels, lay eggs, and serve the growing brood [32]. In a way not understood, the female beetles can control fungal growth and degree of multi-species composition [30,33-35]. If the female dies, the culture is quickly contaminated by other fungi and bacte-

ria, which ultimately leads to the brood death [32,36]. The ambrosia fungi cultured by borers are not pure monocultures, also which consist of a mixture of fungi mycelium, yeasts, and bacteria [37,38]. Norris [39] calls these complex multi-species mixtures. However, recent work has shown that a fungus always dominates in primary cultures of the insect [37,40-42]. In addition, beetles carry only the primary fungus in the mycangia (although sometimes secondary fungi are isolated from the mycangia), females tend to favor the primary fungal culture, which gives them a nutritional benefit [32,41,43, 44], some auxiliary fungi also support beetle development [32]. These observations imply the primary fungus as the primary crop, while fungi, yeasts and bacteria may be secondary "weeds" or with additional auxiliary roles to play in the cultures. The fungus growing transmission among generations of farmer's beetles is transmitted vertically from parent to offspring generations [38,43,45]. Reproductive females acquire original inoculum from their crops, they take with them in specialized pockets during dispersal flights and use this as seed inoculum for their new crops. Ambrosia borers are only associated with a particular kind of primary cultures within a specific geographic region [37,41]. However, although most of the beetles are associated with a specific species of fungus in their primary area, ambrosia beetle species distantly related are sometimes associated with cultivating the same fungus, involving the exchange, direct or indirectly. The fungal exchange between and within species of beetles can occur when several females colonize the same tree and fungal associations contaminate adjacent galleries [41]). The primary fungi are primarily asexual [46,47], while less specific auxiliary fungi are often sexual [43].

4.2. Fungi Associated with Pecan in the Coahuila Southwestern Region

The identification of different fungal isolates from both wood and insect body, one may infer that diebackmay be associated with *Fusarium* since it is the fungal specie most frequently isolated from pecan tissue and insect body (**Figure 1**).

This association of *Fusarium solani* and *Fusarium oxysporum* as a primary agent on pecan tree is due to exudates or gummosis presented at 14 days, later inoculated trees showed marked chlorosis and red apices, finally trees declined to dead. Inside of stalks were initially observed brown areas and when these areas were cut transversely, staining was observed in the vascular system in ring forms, this situation was observed in leaf petiole. When the fungal pathogen completely invaded the conducting vessels caused widespread of wilt and tree death. From all the isolated fungal species, *Fusarium solani* and *Fusarium oxysporum* are highlighted in this study because they have a recognized importance as ambrosia fungus [30]. *Fusarium solani* had the highest percentage of occurrence at Parras county, 35% in insects and 38.7% in diseased wood and in Torreon county, 38.5% in insects and 34.9% in diseased wood, by this reason it is considered as the primary ambrosia fungus, in contrast *Fusarium oxysporum* was the most frequent fungal specie at General Cepeda county, 45.9% in insects and 38.7% in diseased wood. As genera *Fusarium* are a ubiquitous necrotic parasite [48], and may tolerates pH changes [49], and host susceptibility to this fungus is directly influenced by temperature and osmotic pressure [50], Fusarium is widely distributed in soils and organic substrates, and it has been isolated from permafrost in the Arctic to Sahara sands, it presence is abundant in cultivated soils from temperate and tropical areas and it is the most frequently isolated fungi from plants by pathologists [17]. *Fusarium* species are commonly associated with some group of insects like Scolytidae and Platypodidae families habit in wood [51]. Regarding the association of *Fusarium* species and order Coleoptera insects in woods, Norris [32] mention *Fusarium solani* as a primary ambrosia fungus associated with species from *Xyleborus* genus, this report is consistent with the results obtained in this study, since 27.1% of our samples from the three different locations showed damage associated with insects of the Coleoptera order. Gil *et al.*, [52] identified

Corthylus sp. (Coleoptera: Scolytidae), causing damage to alder (*Alnus acuminata*) trees, which showed wood rot and death, in this case the inset was associated to some ambrosia fungi like *Fusarium solani*, *Fusarium sp. Ceratocystis sp.*, *Verticillium sp.* and a yeast identified as *Pichia sp.* Bonello *et al.* [53] found that the ambrosia beetle *Xylosandrus germanus* associated with *Fusarium lateritium* and *F. oxysporum*, is related with vascular stained, cancers, wilting and dieback in black walnut (*Juglans nigra*). Tisserat *et al.* [54] found that the mortality of black walnut (*Juglans nigra*) in Colorado, USA, is the result of association of *Pityophthorus juglandis* (Coleoptera: Scolytidae) with *Fusarium solani* and *Geosmithia sp.* this fungus was latter isolated from trunk marginal cancer in the final stages of decline. *F. solani* has been mentioned as a primary symbiotic with the coffee berry borer (*Hypothenemus hampei*) [55,56]. The immature stages of *X. ferrugineus* are unable to complete the development in absence of *F. solani* [34,57-59]. Ergosterol is essential for development and reproduction of *X. ferrugineus* [60,61]. This sterol is absent in wood, but *F. solani* offer it in sufficient quantities [31]. This fungus also provides to *X. ferrugineus* essential fatty acids and phospholipids [62]. Norris [32] found that *X. ferrugineus* offspring was highly associated to the primary ambrosia fungus *Fusarium solani*, but with *Cephalosporium sp.* association was reduced by 50% and with *Graphium sp.* up to 70%. *Pityophthorus juglandis* is associated with *F. solani* in black walnut (*Juglans nigra*) trees in North America and South Africa; it is believed that trees are predisposed to cancer formation by stress factors such as optimum site conditions, improper pruning and adverse weather conditions [54], such as abandoned orchards and poor management. In our study, *F. solani* and *F. oxysporum* were consistently isolated from bone, xylem, phloem, cambium and galleries, as well as all *Euplatypus segnis* body parts. This observation and pathogenicity studies suggest an important role of *F. oxysporum* and *F. solani* in the pecan tree mortality of all ages in our region.

In the pathogenicity tests, *Lasiodiplodia theobromae* induced necrotic lesions approximately 10 cm long and 1.0 cm wide, that caused necrosis in the cortical tissue and deepened to the bone in the site of inoculation, at 56 days showed a slight chlorosis followed by early abscission, leaves with red tips, wilting and dieback tree at 84 days, the slow progression of the disease coincides with Milholl and [62] who mentioned that this type of fungus grow slowly and progress in living tissue host due to its saprophytic condition. Optionally, *L. theobromae* is a parasite that usually infects its host plants by penetrating through wounds and decaying tissue. Occurrence of this fungus is common in tropical and subtropical regions and in different ecological areas, where it has been identified as the cause of disease in approximately 280 species of vascular plants, including avocado, apple, mango, grapes, pine, rose, rubber, cotton, cocoa, coffee, sugar cane, peanuts, tobacco, etc. [63]. Umezurike [64] mentions that *Botryodiplodia theobromae* has cellulolytic activity in *Bombax buonopozense* wood. The fungus attacks plant cells in a similar manner to soft rot fungi, use starch and other saccharides present in the initial wood substrates before degradation of cellulose and hemicellulose, but not degrade lignin. Rajput *et al.* [65] mention that decline and death of mango (*Mangifera indica* L.) trees in India, are related to the presence of *Botryodiplodia theobromae*, *Alternaria alternata*, *Fusarium moniliforme*, *Cephalosporium sp.*, *Chaetomium sp. Aspregillus ellipticus*, *A. niger*, *Penicillium sp.*, *Curvularia lunata*, *Gloeosporium mangiferae*, *C. globosum*, *Daldinia sp* associated to the attack of ambrosia beetles, termites, mechanical damage to trunk, branches and pruning. Rondon [7] established the relationship between the wood borer *Xyleborus ferrugineus* F. (Coleoptera: Scolytidae) and *Botryodiplodia theobromae* Pat attacking avocado trees in Venezuela, this author found that *B. theobromae* is the most frequently isolated fungi from the body of this wood borer insect. Flores *et al.* [66] detected the association of *Scolytus sp.* and *Botryodiplodia theobromae* fungus, which causes circular lesions to elongated oval up to 15 cm long, dark color that can be covered by large masses of spores in dry periods, causing the death of tips, and branches of *Tectona grandis* trees in Ecuador. Masood *et al.* [67] identified the association of *Hypocryphalus mangiferae* (Coleoptera: Scolytidae) with *Lasiodiplodia theobromae*, *Phomopsis fimbriata* and *Ceratocystis sp.* in the sudden death syndrome in Pakistan associated with symptoms of gummosis, dry rot and vascular discoloration. Tress inoculated with *Alternaria alternata* showed at 63 days marked chlorosis, wilting leaf area at 70 days and dieback at 84 days as external symptoms and necrotic areas, browning of tissues and a deep dehydration of inoculated area. The genus *Alternaria* contains cosmopolitan species found in a wide range of materials and products, can damage food and feed, producing biologically active compounds such as mycotoxins. As pathogens reduce crop yields or affect stored plants therefore requires a precise identification of the species because each has specific characteristics (preferences for growth, pathogenicity, and production of secondary metabolites) that predict the behavior of this fungus [68]. Armengol *et al.* [69] mentioned that *A. alternata* is normally a saprophytic fungus, but in Spain some strains have been described as pathogenic on different citrus cultivars like Fortune, Nova, and Mineola. Negron *et al.* [70] described *Alternaria* as parasites of annual and bi-annual plants or saprophytes on organic substrates, however found that *Alternaria alternata* associated with *Scolytus*

schevyrewi (Coleoptera: Scolytidae) and other fungi affect some species of elms (*Ulmus* spp) in Colorado and Utah, which showed necrotic areas, browning of xylem tissue, wilting of foliage and general death of elms, symptoms consistent with the disease progression in the pecan trees of this study inoculated with *A. alternata* in our pathogenicity tests. Reyes [71] found that 8% of fungal strains isolated from wood samples and insect remains of adult weevils (Coleoptera: Platypodidae) belong to *Alternaria alternata* and 7% to *Phoma sp.*

The fungus *Paecilomyces sp.* is a pathogen of wide host range and wide geographical distribution, which has been isolated from soil and insects of various orders such as Coleoptera, Homoptera and Collembola, Lepidoptera, Diptera, Homoptera, Hymenoptera and spiders [72]. *Paecilomyces farinosus* may inhibit some strains of pathogenic fungi [73]. Finally, *Aspergillus*, *Paecilomyces*, and *Penicillium*, have been isolated from insect gut *Triatoma sp.* [74], which has been found sporadically in pecan orchards at the Coahuila Southeast region, Mexico.

5. CONCLUSIONS

Twelve genera of fungi were isolated from the body of *E. segnis* and diseased pecan wood. These fungi were identified as *Phoma sp.*, *Fusarium sp.*, *Ascochyta sp.*, *Phaecylomices sp.*, *Lasiodiplodia sp.*, *Alternaria sp.*, *Umbeliopsis sp.*, *Torula sp.*, *Helminthosporium sp.*, *Aspergillus sp.*, *and Penicillium sp.*, and inoculated into healthy pecan trees, only *Fusarium oxysporum*, *Fusarium solani*, *Alternaria alternata* and *Lasiodiplodia theobromae* were highly pathogenic causing dieback at 87 days of inoculation, they were characterized morphologically and molecularly to the specie level.

In Mexico, this is the first study that determines to *Fusarium solani*, *Fusarium oxysporum*, *Alternaria alternata* and *Lasiodiplodia theobromae*, as the causal agent of pecan (*Carya illinoensis*) dieback in association with *Euplatypus segnis* ambrosia borer.

Fusarium solani and *F. oxysporum* were the most prevalent fungi and isolated from insects and diseased wood at all studied locations. It has been suggested that some species of fungi associated with ambrosia borer are in a symbiotic relationship with the insect. Although, there may be a symbiotic relationship between fungi and insects, there is no conclusive evidence to show that this relationship is sufficient, due to the great diversity of fungi found in association with *E. segnis*.

REFERENCES

[1]   Equihua-Martínez, A. and Burgos-Solorio, A. (2007) Platypodidae y Scolytidae (Coleoptera) de Jalisco, México. *Dugesiana*, **14**, 59-82.

Isolation and pathogenicity of fungi associated to ambrosia borer (Euplatypus segnis)
found injuring pecan (Carya illinoensis) wood

85

[2] Galván, L.O.A. (2000) *Euplatypus segnis* (Chapuis): Fluctuación poblacional y magnitud de daño a nogales en Parras, Coahuila. In: Vázquez, N.J.M., Ed., *Memoria del II Curso de Actualización Fitosanitaria en Nogal. 10 y 11 de Marzo.* Instituto Tecnológico de Estudios Superiores de Monterrey (ITESM), Torreón, 45-47.

[3] García, M.O. (1999) El barrenador ambrosia *Euplatypus segnis* (Chapuis) del tronco y ramas del nogal (*Carya illinoensis*). Memorias del Séptimo Simposium Internacional Nogalero NOGATEC. 23, 24 y 25 de septiembre. Instituto Tecnológico de Estudios Superiores de Monterrey, Torreón, 39-42.

[4] Cesaveco (2010) Comité estatal de sanidad vegetal del estado de Coahuila. *Boletin*, **86**, 20.

[5] Atkinson, T.H., Fernández, E.M., Céspedes E.S. and Burgos, A.S. (1986) Scolytidae y platypodidae (coleoptera) asociados a selva baja y comunidades derivadas en el estado de Morelos, México. *Folia Entomologica Mexicana*, **69**, 41-82.

[6] Samaniego-Gaxiola, J.A., Ramírez-Delgado, M., Pedroza-Sandoval, A. and Nava-Camberos, U. (2008) Association between cotton root rot (Phymatotrichopsis omnivore) and borer insects of pecan tree (*Carya illinoensis*). *Agricultura Técnica en México*, **34**, 21-32 (in Spanish).

[7] Rondon, A. and Guevara, Y. (1984) Algunos aspectos relacionados con la muerte regresiva del aguacate (*Persea Americana* Mill). *Agronomía Tropical*, **34**, 119-129.

[8] Kok, L.T. and Norris, D.M.J. (1972) Symbiotic interrelationships between microbes and ambrosia beetles. VI Amino-acid composition of ectosymbiotic fungi of *Xyleborus ferrugineus*. *Annals of the Entomological Society of America*, **65**, 598-602.

[9] Brignola, C., Lacroix B., Lieutier F., Sauvard D., Drouet A., Claudot, C., Yart, A., Berryman, A. and Christiansen, E. (1995) Induced Responses in phenolic metabolism in two norway spruce Inoculations with *Ophiostoma polonicum* and bark beetle-associated fungus. *Plant Physiology*, **109**, 821-827.

[10] Morales, R.J.A., Rojas, M.G., Bhatkar, H.S. and Saldaña, G. (2000) Symbiotic relationships between *Hypothenemus hampei* (coleoptera: scolytidae) and *Fusarium solani* (moniliales: tuberculariaseae). *Annals of the Entomological Society of America*, **93**, 541-547.

[11] Kuroda, K. (2001) Responses of *Quercus* sapwood to infection with the pathogenic fungus of a new wilt disease vectored by the ambrosia beetle *Platypus quercivorus*. *Japan Wood Society*, **47**, 425-429.

[12] Peterson, W.S., Pérez, E.J., Vega, F. and Infante, F. (2003) *Brocae Penicillium*, a new species associated coffee berry borer with the in Chiapas, Mexico. *Mycologia*, **95**, 141-147.

[13] Alfaro, R. (2003) El taladrillo grande de los forestales, *Platypus mutatus* (=*sulcatus*): Importante plaga de la populicultura Argentina. *Forestal*, **28**, 17.

[14] Rosas, R.E., Avila, G.M.R. and Cano, R.P. (2004) El metodo Laguna tecnica para conbatir el barrenador del tronco del nogal (Euplatypus segnis Chapuis). Memorias de la XVI Semana Internacional de Agronomia FAZ-UJED. Gomez Palacios, del 6-10 de Septiembre 2004.

[15] Barnett, H.L. and Hunter, B.B. (2006) Illustrated genera of imperfect fungi. 4th Edition, The American Phytopatological Society, St. Paul Minnesota.

[16] Hanlin, R.T. (1990) Illustrated genera of ascomycetes I. APS Press, Saint Paul.

[17] Booth, C. (1971) The genus *Fusarium*. Commonwealth Mycological, Kew.

[18] Rotem, J. (1988) The genus *Alternaria*, biology, epidemiology and pathogenicity. APS Press, St. Paul.

[19] Neergaard, P. (1977) Seed pathology. The McMillan Press Ltd., Surrey.

[20] Sutton, B.C. (1980) The fungi imperfecti with pycnidia coellomycetes, acervuli and stromata. Commonwealth Mycological Institute, Surrey.

[21] Wei, R.R., Sorger, P.K. and Harrison, S.C. (2005) Molecular organization of the Ndc80 complex, an essential kinetochore component. *Proceedings of the National Academy of Sciences*, **102**, 5363-5367.

[22] Punithalingam, E. (1976) *Botryodiplodia theobromae*. CMI description of pathogenic fungi and bacteria No. 519. Commonwealth Mycological Institute, Surrey.

[23] Burgess, T.I., Barber, P.A., Mohali, S., Pegg, G., De Beer W. and Wingfield, M.J. (2006) Three new *Lasiodiplodia* spp. From the tropics, based on DNA sequence recognized comparisons and morphology. *Mycologia*, **98**, 423-435.

[24] Ahrens, U. and Seemüller, E. (1992) Detection of DNA of plant pathogenic mycoplasmalike organisms by polymerase chain reaction amplifying a sequence of 16S rRNA gene. *Phytopathology*, **82**, 828-832.

[25] Kirkendall, L.R. (1983) The evolution of mating system in bark and ambrosia beetles (Coleoptera: Scolytidae and Platypodidae). *Zoological Journal of the Linnean Society*, **77**, 293-352.

[26] Wood, S.L. (1982) The bark and ambrosia beetles of north and central America (Coleoptera: Scolytidae), a taxonomic monograph. *Great Basin Naturalist Memoirs*, **6**, 1359.

[27] Ulloa, M. (1991) Illustrated dictionary of mycology. UNAM, Mexico.

[28] Flechtmann, C.A.H. (1995) Scolytidae em reflorestamentos com pinheiros tropicais. IPEF, Piracicaba.

[29] Harrington, C.T. (2005) Ecology and evolution of mycophagous bark beetles and their fungal partners. In: Vega, F.E. and Blackwell, M., Eds., *Ecological and Evolutionary Advances in Insect-Fungal Associations*, Oxford University Press, Oxford, 257-291.

[30] Beaver, R.A. (1989) Insect-fungus in the bark relationships and ambrosia beetles. Academic Blackwell, London.

[31] Kok, L.T, Norris, D.M and Chu, H.M. (1970) Sterol metabolism as a basis for a mutualistic symbiosis. *Nature*,

225, 661-662.

[32] Norris, D.M. (1979) The mutualistic fungi of xyleborine beetles. Halsted Press, Chichester.

[33] French, J.R.J. and Roeper, R.A. (1972) Interactions of the ambrosia beetle *Xyleborus dispar* (Coleoptera: Scolytidae) with STI symbiotic fungus, *Ambrosiella hartigii* (Fungi Imperfecti). *Canadian Entomologist*, **104**, 1635-1641.

[34] Kingsolver, J.G. and Norris, D.M. (1977) External morphology of *Xyleborus ferrugineus* (Fabr.) (Coleoptera: Scolytidae) I. Head and prothorax of adult male and female. *Journal of Morphology*, **154**, 147-156.

[35] Roeper, R.A, Treeful, L.M., O'Brien, K.M., Foote, R.A. and Bunce, M.A. (1980) Life history of the ambrosia beetle *Xyleborus affinis* (Coleoptera: Scolytidae) from *in vitro* culture. *Great Lakes Entomologist*, **13**, 141-144.

[36] Borden, J.H. (1988) The striped ambrosia beetles. In: Berryman, A.A., Ed., Dynamics of Forest Insect Populations, Plenum, New York.

[37] Batra, L.R. (1966) Ambrosia fungi: Extent of specificity to ambrosia beetles. *Science*, **153**, 193-195.

[38] Haanstad, J.O. and Norris, D.M. (1985) Microbial symbionts of the ambrosia beetle *Xyloterinus politus*. *Microbial Ecology*, **11**, 267-276.

[39] Norris, D.M. (1965) The complex of fungi essential to growth and development of *Xyleborus sharpie* in wood. *Material und Organismen Beiheft*, **1**, 523-529.

[40] Baker, J.M. (1963) Ambrosia beetles and their fungi, with particular reference to *Platypus cylindricus* Fab. *Symposium of the Society for General Microbiology*, **13**, 232-265.

[41] Gebhardt, M., Bergerow, D. and Oberwinkler, F. (2004) Identification of the ambrosia fungus of *Xyleborus monographus* and *X. dryographus* (Curculionidae, Scolytinae). *Mycologia*, **3**, 95-102.

[42] Kinuura H. (1995) Symbiotic fungi associated with ambrosia beetles. *Japan Agricultural Research Quarterly*, **29**, 57-63.

[43] Francke-Grosmann, H. (1967) Wood-inhabiting ectosymbiosis in insects. Academic, New York.

[44] Morelet, M. (1998) Une nouvelle raffaelea spec. isolee cylindrus platypus, coleoptera des chenes xylomycetophage. Extr. *Annales de la Societe des Sciences Naturelles*, **50**, 185-193.

[45] Fernandez-Marin, H., Zimmerman, J.K. and Wcislo, W.T. (2004) Ecological traits and evolutionary sequences of nest establishment in fungus growing ants (Hymenoptera, Formicidae, Attini). *Biological Journal of the Linnean Society*, **81**, 39-48.

[46] Jones, K.G. and Blackwell, M. (1998) Phylogenetic analysis of ambrosia species in the genus *Raffaelea* based on 18S rDNA sequences. *Mycological Research*, **102**, 661-665.

[47] Rollins, F., Jones, K.G., Krokene, P., Solheim, H. and

[48] Blackwell, M. (2001) Phylogeny of asexual fungi associated with bark beetles and ambrosia. *Mycologia*, **93**, 991-996.

[48] Nicholson, P. (2001) Molecular Assays as aids in the detection, diagnosis and Quantification of *Fusarium* species in plants. APS Press, St. Paul, 176-192.

[49] Carrillo, L. (1990) Micotoxinas de *Fusarium spp* en frutos deteriorados de *Cucurbita ficifolia*. *Revista Argentina de Microbiología*, **22**, 212-215.

[50] Doohan, F.M., Brennan, J. and Cooke, B.M. (2003) Influence of climatic factors on *Fusarium* species pathogenic to cereals. *European Journal of Plant Pathology*, **109**, 755-768.

[51] Cooke, R. (1977) Mutualistic symbioses with insects. John Wiley & Sons, London.

[52] Gil, P.Z.N., Bustillo, P.A.E., Gómez, D.D.S. and Marín, M.P. (2004) *Corthylus novo sp.* (Coleoptera: Scolytidae), plaga del aliso en la cuenca del rio Blanco en Colombia. *Revista Colombiana de Entomología*, **30**, 171-178.

[53] Bonello, R., McNee, W.R., Storer, A.J., Wood, D.L. and Gordon, T.L. (2001) The role of olfactory stimuli in the location of weakened hosts by twig-infesting *Pityophthorus* sp. *Ecological Entomology*, **26**, 8-15.

[54] Tisserat, N., Cranshaw, W., Leatherman, D., Utley, C. and Alexander, K. (2009) Mortality in colorado black walnut caused by the walnut twig beetle and thousand cankers disease. *Plant Health Progress.*

[55] Morales-Ramos, J.A., Rojas, M.G. and Harrington, T. (1999) Association between the coffee berry borer, *Hypothenemus hampei* (Coleoptera: Scolytidae) and *Fusarium solani* (Moniliales: Tuberculariaceae). *Annals of the Entomological Society of America*, **92**, 98-100.

[56] Morales-Ramos, J.A., Rojas, M.G., Sittertz-Bhatkar, H. and Saldaña, G., (2000) Symbiotic relationship between *Hypothenemus hampei* (Coleoptera: Scolytidae) and *Fusarium solani* (Moniliales: Tuberculariaceae). *Annals of the Entomological Society of America*, **93**, 541-547.

[57] Norris, D.M. and Baker, J.K. (1967) Symbiosis effects of a mutualistic fungus upon the growth and reproduction of *Xyleborus ferrugineus*. *Science*, **156**, 1120-1122.

[58] Norris, D.M. and Chu, H.M. (1971) Maternal *Xyleborus ferrugineus* transmission of sterol or sterol-dependent metabolites necessary for progeny pupation. *Journal of Insect Physiology*, **17**, 1741-1745.

[59] Norris, D.M., Baker, J.K. and Chu, H.M. (1969) Symbiotic interrelationships between microbes and ambrosia beetles. III. Ergosterol as the source of sterol to the insect. *Annals of the Entomological Society of America*, **62**, 413-414.

[60] Chu, H.M., Norris, D.M. and Kok, L.T. (1970) Pupation requirement of the beetle, *Xyleborus ferrugineus:* Sterols other than cholesterol. *Journal of Insect Physiology*, **16**,

Isolation and pathogenicity of fungi associated to ambrosia borer (Euplatypus segnis) found injuring pecan (Carya illinoensis) wood

87

1379-1387.

[61] Kok, L.T. (1979) Lipids of ambrosia fungi and the life of mutualistic beetles. In: Batra, L.R., Ed., *Insect-Fungus Symbiosis*, Halsted Press, Sussex.

[62] Milholland, R.D. (1970) Histology of *Botryosphaeria* canker of highbush blueberries susceptible and resistant. *Phytopathology*, **60**, 70-74.

[63] Riva, R. (1996) Tecnología del cultivo de camu camu en la Amazonía Peruana. INIA, Pucallpa-Perú.

[64] Umezurike, G.M. (1969) Cellulolytic activities of *Botryodiplodia theobromae* pat. *Annals of Botany*, **33**, 451-462.

[65] Rajput, K.S. and Rao, K.S. (2007) Death and decay in the trees of Mango (*Mangifera indica* L.). *Microbiological Research*, **162**, 229-237.

[66] Flores, T.V., Crespo, R.G. and Cabezas, G.F. (2010) Plagas y enfermedades en plantaciones de Teca (*Tectona grandis* L.F) en la zona de Balzar, Provincia del Guayas. *Ciencia y Tecnología*, **3**, 15-22.

[67] Masood, A., Saeed, S., Silveira, S.F., Akem, C.N., Hussain, N. and Farooq, A.M. (2011) Of mango quick decline in Pakistan: Survey and pathogenicity of fungi isolated from bark beetle and mango tree. *Pakistan Journal of Botany*, **43**, 1793-1798.

[68] Andersen, B., Krøger, E. and Rodney G.R. (2001) Chemical and morphological segregation of *Alternaria alternata*, A. *gaisen* and A. *longipes*. *Mycological Research*, **105**, 291-299.

[69] Armengol, J., Sales, R., Garcia-Jimenez, J. and Alfaro-Lassala, F. (2000) First report of *Alternaria* brown spot of citrus in Spain. *Plant Disease*, **84**, 1044.

[70] Negron, J.F., Witcosky, J.J., Cain, R.J., LaBonte, J.R., Duerr, D.A., McElwey, S.J., Lee, J.C. and Seybold, S.J. (2005) The banded elm bark beetle: A new threat to elms in North America. *American Entomologist*, **51**, 84-94.

[71] Capurro, M. and Reyes, S. (2007) Fungi, insect borers Association present in samples of wood entered into the regional laboratory of the agricultural and livestock service. Bachelor Thesis, Universidad Austral de Chile.

[72] Chan-Cupul, W., Ruiz-Sánchez, E., Cristóbal-Alejo, J., Pérez-Gutiérrez, A., Munguía-Rosales, R. and Lara-Reyna, J. (2010) *In vitro* development of four *Paecilomyces Fumosoroseus* isolates and their pathogenicity on immature whitefly. *Agrociencia*, **44**, 587-597.

[73] Gemma, J.N., Hartmann, G.C. and Wasti, S.S. (1984) Interaction between inhibitory *Ceratocystis ulmi* and several species of entomogenous fungi. *Mycologia*, **76**, 256-260.

[74] Moraes, A.M.L. Junqueira, A.C.V., Costa, G.L., Celano, V., Oliveira, P.C. and Coura, J.R. (2000) Fungal flora of the digestive tract of triatomines of 5 species vectors of *Trypanosoma cruzi*, Chagas 1909. *Mycopathologia*, **151**, 41-48.

8

Effects of environmental factors on *Sparganium emersum* and *Sparganium erectum* colonization in two drainage ditches with different maintenance

Korehisa Kaneko[1], Hiroshi Jinguji[2]

[1]Hokuso Creature Association, Tokyo, Japan
[2]School of Food, Agricultural and Environmental Sciences, Miyagi University, Sendai, Japan

ABSTRACT

In the Niheishimizu and Ooshimizu sections of the town of Misato in the Akita Prefecture, Northern Japan, there are many abundant spring water areas. *Sparganium* (*Sparganium emersum* and *Sparganium erectum*) species are widely distributed in the irrigation water that fed by spring water. The irrigation waters were divided the natural type ditch and the maintained ditch that connect with nearby natural ditch to promote environmentally friendly agriculture. This study was conducted in both sections to support the maintenance of the irrigation water fed by the abundant spring water. A vegetation survey was conducted in September of 2005. The survey collected data on the amount of vegetation cover and the stem lengths of the plant species found in selected locations of the study area. The water depths and the flow velocities were also measured in these locations. As for the growth situation of *S. emersum* and *S. erectum*, the submerged form of *S. emersum* was found in water approximately 15 cm deep with a surface flow velocity of approximately 7 cm/s. This species was characterised by a relatively fast flow and relatively shallow water. The emergent and submerged growth forms of *S. emersum* were found in waters having flow velocities faster than those associated with *S. erectum*. The emergent form of *S. emersum* grew in relatively deep water. *S. emersum* is more capable of adjusting to the conditions of stream habitats than *S. erectum*.

Keywords: *Sparganium*; Flow Velocity; Water Depth; Emergent Growth Form; Submerged Growth Form; Natural Type Ditch; Maintained Ditch

1. INTRODUCTION

[1] has established the "Ministry of Agriculture, Forestry and Fisheries (MAFF) biodiversity strategy" to promote approaches to agriculture that recognise the value of biodiversity, reduce damage to the populations of local birds and other animals and protect agricultural resources. The MAFF is promoting types of environmentally friendly agriculture that can coexist with many living organisms. It is necessary to maintain and restore rice fields, ditch and the habitats of wild flora and fauna.

Spring water areas occur in some parts of the alluvial fans located on the plains in the northern area of the Senboku District in the Yokote basin, Akita Prefecture, Northern Japan. The spring water is primarily used to supply an irrigation water. Species included on the Red List of the [2] and [3] are found in the area, including *Hippuris vulgaris*, *Sparganium erectum*, *Sparganium japonicum*, *Sparganium simplex*, and *Pungitius pungitius* [4]. In particular, *Sparganium* (*S. emersum* and *S. erectum*) has a very extensive distribution in the ditch of Niheishimizu and Ohshimizu sections of the town of Misato. *Sparganium* grows to water area where spring water is abundant, and these breeding seasons are August and September [5]. We consider that *Sparganium* is suitable for growth with the above-mentioned ditch. However, recent years, the habitat of *Sparganium* (*i.e.*, *S. erectum*, *S. emersum* and *S. emetsum*) shows the tendency to decrease according to the influence of the development of the river maintenance [6,7]. In addition, basic information on the life history and habitat characteristics of *Sparganium* is lacking. Moreover, few studies have investigated the differences in distribution between *Sparganium* species (*i.e.*, *S. emersum* and *S. erectum*) found in the same aquatic areas or in neighbouring areas.

In this study, we examined the differences in growth situation between different species *Sparganium* (*S. emersum* and *S. erectum*) by comparing the state of the vegetation and the growth environment (water depth, flow

Effects of environmental factors on Sparganium emersum and Sparganium erectum colonization
in two drainage ditches with different maintenance

89

velocity and t he maintenance) of these s pecies in the abundant ditch of spring water.

2. MATERIALS AND METHODS

2.1. Study Site

The study sites were located in a spring water area in the Niheishimizu and Ooshimizu sections of the town of Misato, Akita prefecture, Northern Japan (**Figure 1**).

We investigated areas located upstream (Site 1) a nd downstream (Site 2) of th e weir in the Ooshimizu, upstream (Site 3) and downstream (Site 4) of the weir in the Niheishimizu.

The ditch in the Niheishimizu section is maintained by protecting the banks with ston es placed in the channel and was originally created and maintained by a farmland consolidation project in 2003. The bottom of the ditch is lined with sling stone, and *Sparganium* plants which were grown in the neighbourhood have been transplanted to the bottom of the ditch. The ditch in the Ooshimizu section has been left in nearby natural form. For t he purposes of this study, Niheishimizu was referred as to be a maintained ditch, and Ooshimizu was referred as to be a natural type ditch.

2.2. Sampling and Identification Methods

The investigation was conducted during September of 2005. The vegetation was sur veyed using a quadrant frame (50 × 5 0 cm^2) in eac h section. The numbers of quadrant frames were 15, 23 in upstream (Site 1) and downstream (Site 2) of the weir in the Ooshimizu, and were 18, 26 in upstream (Site 3) and downstream (Site 4) of the weir in the Niheishimizu. The amount of vegetation cover for each plant species summed the c over of a quadrant for these. T he stem length was measured in same quadrant in which vegetation cover was recorded. The stem length included the above-ground portion of the plant in the central of quadrant frame and the sub-merged portion of th e plants in the stream area was included in th is measurement. The flow velocity (surface and bottom), the water depth and the vegetation cover were measured at the sam e time. The fl ow velocity was measured with a KE NEK VP-3000 three-dimensional electromagnetic current meter. The surface measured approximately 3cm under water surface and the bottom measured approximately 1 cm above the stream bed. The following diagnostic characteristics were used to identify *Sparganium* species: *S. erectum* has a divided scape and 3 or more branches, and th e upper part of th e seed extends above the dome; *S. emersum* has an undivided scape and 4 or m ore staminate heads, and th e pistillate heads above the second are ax illary. Moreover, as for *S.erectum*, the dorsal crest of the leaf develops, and the section is triangular. The leaf is 7 - 20 mm in width. As

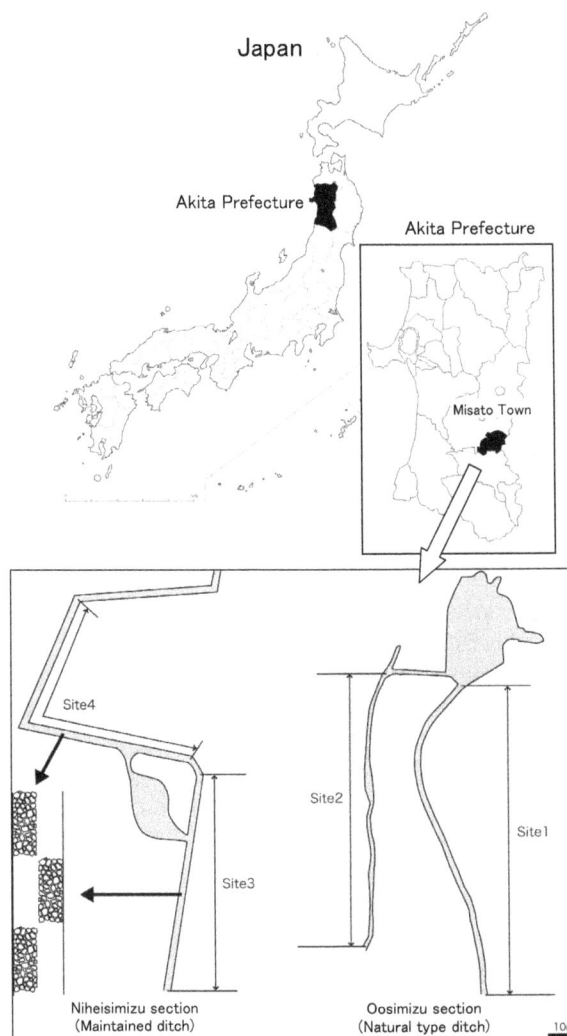

Figure 1. Study site.

for *S. emersum*, the emerged leaf is 5 - 16 mm in width, dorsal crest develops a little. The submerged leaf is 6 - 9 mm in width, the dorsal crest is not remarkable [5].

3. RESULTS

3.1. Vegetation Coverage

Upstream of the weir in the maintained ditch, *S. emersum* and *S. erectum* covered areas of 2.58 m^2 and 0.94 m^2, respectively. Their coverage downstream of the weir were 1.83 m^2 and 3.7 m^2, respectively, and *Typha latifolia* covered 0.35 m^2 downstream of the wei r. In the natural type ditch, *S. emersum* covered 2.64 m^2 upstream of the weir, and *S. emersum* and *S. erectum* covered 1.46 m^2 and 1. 12 m^2, respectively, downstream of the weir (**Table 1**).

3.2. Flow Velocity and Water Depth

The flow velocity was the hig hest upstream of the weir

Table 1. The vegetation cover areas (m^2) of aquatic botany founded in the investigation ground.

Species	Maintained ditch Niheishimizu section		Natural type ditch Ooshimizu section	
	Upstream of the weir	Downstream of the weir	Upstream of the weir	Downstream of the weir
Sparganium emersum	2.58	1.83	2.64	1.46
Sparganium erectum	0.94	3.70	-	1.12
Typha latifolia	-	0.35	-	-
Total	3.52	5.88	2.64	2.58

in the natural type ditch. T he flow velocity was successively lower downstream of the weir in the natural type ditch, downstream of the weir in th e maintained ditch and upstream of the weir in th e maintained ditch. The water was the deepest upstream of the weir in the maintained ditch. The water depth was found to be successively reduced downstream of the weir in the maintained ditch, downstream of the weir in the natural type ditch and upstream of the weir in the natural type ditch.

The relative coverage of the emergent and submerged growth forms of the plants differed among the sampling sites. The emergent form showed a coverage of 100% upstream of the weir in th e maintained ditch, and the submerged form showed a c overage of 100% upstream of the weir in the natural type ditch. Overall, the relative abundance of the s ubmerged form increased with increasing flow velocity and decreasing water depth (**Figure 2**).

3.3. The Relative amounts of Vegetation Cover according to Species, Stem-Length Class and Growth Form

The emergent form of *S. emersum* was found both upstream and downstream of the weir in th e maintained ditch. The relative amount of vegetation cover represented by the short stem lengths was higher in these locations than in the location downstream of the weir in the natural type ditch. The submerged growth form was found upstream of the weir in the natural type ditch, and both growth forms were found downstream of the weir in the natural type ditch. In the natural type ditch, the relative coverage of t he shortest stem-length class was higher upstream of the weir than downstream of the weir.

The emergent form of *S. erectum* upstream of the weir in the maintained ditch showed the highest relative amount of vegetation cover in the stem-length class of 1.5 - 2.0 m. Both growth forms of this species were found downstream of the weir in the maintained ditch and downstream of the weir in the natural type ditch. *S. erectum* was not found upstream of the weir in the natural type ditch (**Figure 3**).

3.4. The Growth Environment (Flow Velocity and Water Depth) of the Species and Growth Forms of *Sparganium*

The surface and bottom flow velocities were nearly equal in each case studied. The submerged form of *S. emersum* was found in th e fastest-flowing, shallowest water. The emer gent form of *S. erectum* was found in even slower and even deeper water. The emergent form of *S. emersum* was found in water having flow velocities similar to those associated with the submerged form of *S. erectum* and in water deeper than that associated with the other growth forms of both species (**Figure 4**).

4. DISCUSSION

The Growth Situation of *Sparganium* in Each Sampling Location

The emergent form of *S. erectum* was only found upstream of the weir in the maintained ditch. The relative vegetation cover was the highest for the stem-length class of 1.5 - 2.0 m. Both growth forms of *S. erectum* were found downstream of the weir i n the maintained ditch and downstream of the weir in   the natural type ditch. *S. erectum* was n ot found upstream of the weir in the natural type ditch (**Figure 3**).

S. erectum grows in lotic and l entic environments. Many species of this genus grow as a submerged form in running water [7]. However, the emergent of *S. erectum* occur commonly at the margins of low- and medium-energy river systems across the northern temperature zone [8]. *S. erectum* grows densely in gradually flowing waters that have a flow velocity of less than 5 cm/s, it is rarely found in rapidly flowing waters that have a velocity of 5 cm/s or m ore. Indeed, it h as been reported to grow tall in a deep, gently flowing stream [9].

Upstream and downstream of the weir in the maintained ditch, an extremely gradual flow of less t han 5 cm/s was rec orded, and the water depth was 30 cm or greater. Downstream of the weir in the natural type ditch, the flow velocity exceeded 5 cm/s, and the water depth was (16.73 ± 0.34) cm, a relatively low value.

Effects of environmental factors on Sparganium emersum and Sparganium erectum colonization
in two drainage ditches with different maintenance

91

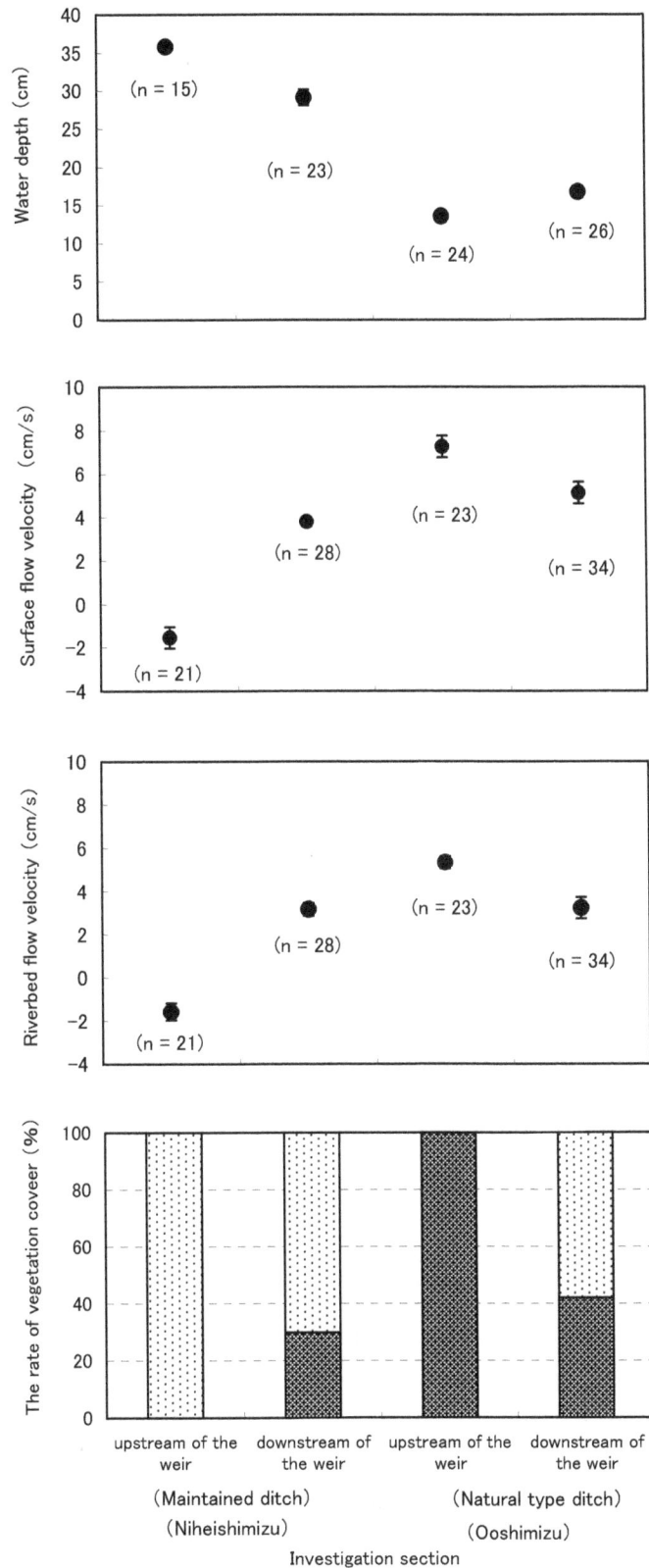

Figure 2. The rate of vegetation cover on each life form of aquatic macropyhtes (Total values), flow velocity and water depth in each investigation section. ※The vertical bars indicate a standard error. N shows the number of measurement.

【 *Sparganium emersum* 】   【 *Sparganium erectum* 】

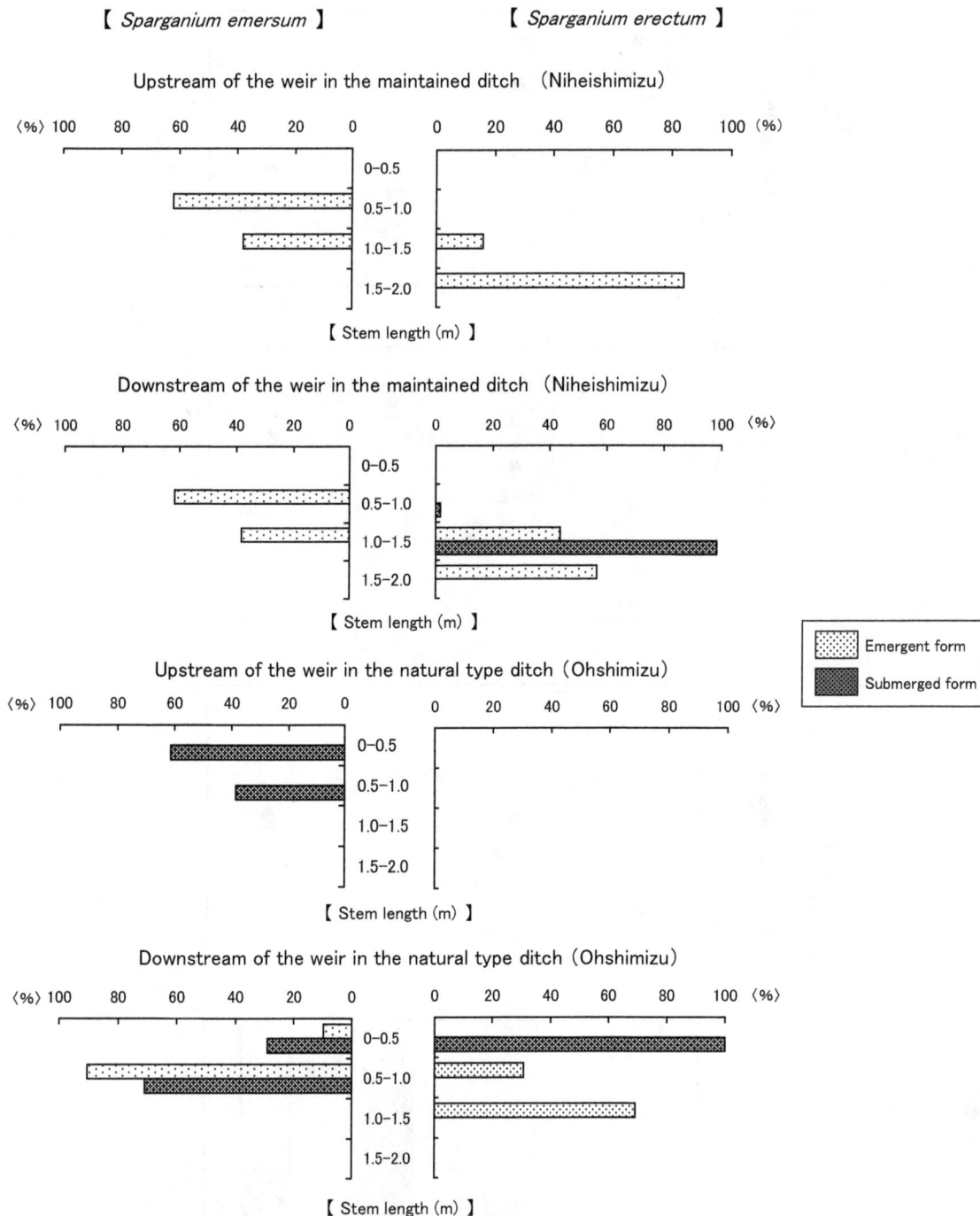

Figure 3. The relative amounts of vegetation cover of stem length class of life form on *Sparganium* (*S. emersum*, *S. erectum*) in investigation section (%).

We hypothesised that the emergent form of *S. erectum* grows by extending its stems and expanding its distribution in waters whose flow velocity is less than 5 cm/s and whose depth exceeds 30 cm. If the flow velocity exceeds 5 cm/s and the water depth is approximately 15 cm or greater, the plant must develop a submerged growth form. Upstream of the weir in the natural type ditch, only the submerged form of *S. emersum* was found. The relative

vegetation cover was highest for the shortest stem-length class of 0 - 0.5 m.

S. emersum is dominated in waters where a fast flow and disturbance are strong, and adapted by the submerged type as these influences strengthens in comparison to *S. erectum* [10]. The submerged form of *S. emersum* grows in running waters. It grows densely at bottom flow velocities (measured approximately 1 cm above the

Effects of environmental factors on Sparganium emersum and Sparganium erectum colonization
in two drainage ditches with different maintenance

93

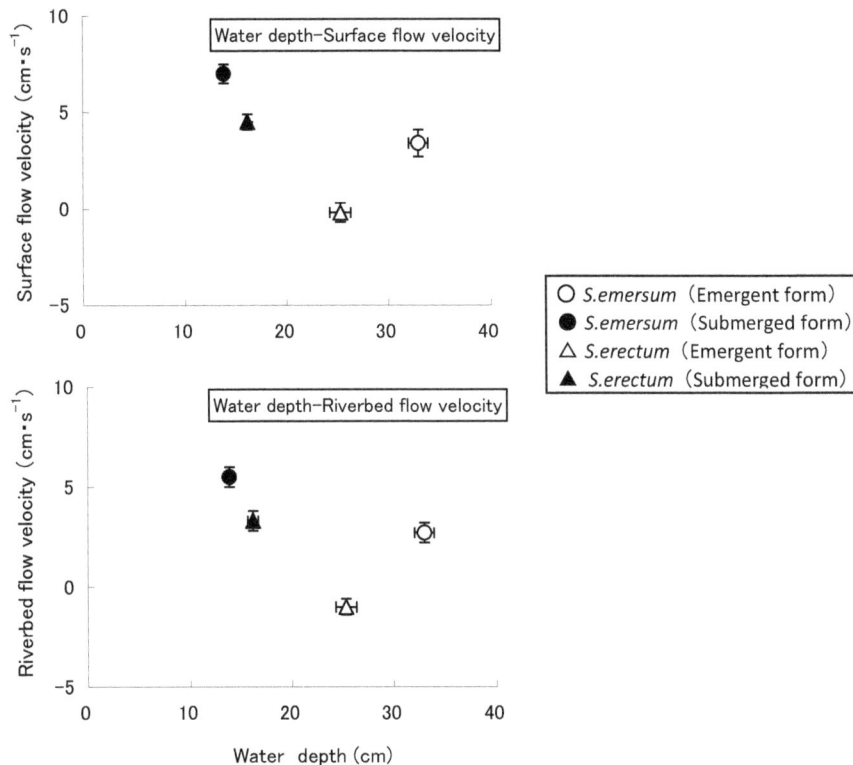

Figure 4. The relationship flow velocity, water depth and stem length in h abitat of *Sparga-nium* (*S. emersum, S. erectum*). ※The vertical and horizontal bars indicate a standard error.

stream bed) of (3.9 ± 0.4) cm/s to (5.9 ± 2.4) cm/s [11]. In this study, the stream bed flow velocity upstream of the weir in th e natural type ditch was (5 .3 ± 0.3) cm/s, and the water depth was (13.6 ± 0.15) cm (**Figure 2**).

We hypothesised that the submerged form of *S. emer-sum* is su itable for these environmental conditions be-cause the plant controls its growth as t he flow velocity increases and the depth decreases. Although the stems of *S. emersum* tend to be s horter than those of *S. erectum*, the short stem might enhance the capability to resist strong drag in fast flow.

As other study cases, *S. erectum* grows in waters where the flow is comparatively gradual and sand and silt accumulate [12]. *S. emersum* grows well in a gradual flow, mud, and argillaceous soil [13], and it also grows in sandy, muddy sediment with constant water flow veloci-ties and is found in eutrophic conditions [14]. To clarify the factors affecting the growth of the two species, future research on *Sparganium* habitats should include investi-gations of the water quality and the soil substrate.

REFERENCES

[1] Ministry of Agriculture, Forestry and Fisheries (2007) MAFF's biodiversity strategy. Ministry of Agric ulture, Forestry and Fisheries, Tokyo, Japan.

[2] Ministry of the Environment of Japan (2007) Threatened wildlife of Japan, red list plant I (Vascular plant). Minis-try of the Environment of Japan, Tokyo, Japan.

[3] Akita Prefecture (2002) Threatened wildlife of Akita Prefecture 2002—Red data bo ok of Akita Pr efecture-Plants. Akita Prefecture, Akita, Japan.

[4] Kaneko, K. and Jinguji, H. (2 011) Biota (h ydrophyte, ichthyofauna) of the main spring water ground in Sen-boku-gun and Minamiakita-gun, Akita Prefecture, Japan. *Japanese Journal of Landscape Ecology and Manage-ment*, **15**, 63-70.

[5] Kadono, Y. (1994) Acuatic plants of Japan. Bun-ichi Sogo Shuppan, Tokyo, Japan.

[6] Ichikawa, K., Nishigami, D., Sato H. and Morimoto, Y. (2002) Fundamental stud y on restoration of *Sparganium fallax Gradebn* community. *Journal of the Japanese So-ciety of Revegetation Technology*, **27**, 574-581.

[7] Ishii, T., Nakayama, Y. and Yamaguchi, H. (2005) A note on phenology and seed-germination behavior in two natu-ral populations of the endangered aquatic macrophytes, *Sparganium erectum var. erecutum* and *S. erectum var. macrocarpum* (*Sparganiaceae*). *The Weed Science Society of Japan*, **50**, 82-90.

[8] Pollen-Bankhead, N., Thomas, R. E., Gurnell, A. M., Liffen, T., Simon, A. and O'Hare, M.T. (2011) Quantify-ing the potential for flow to remove the emergent aquatic macrophyte *Sparganium erectum* from t he margins of low-energy rivers. *Ecological Engineering*, **37**, 1779-1788.

[9] Cook, C.D.K. (1962) *Sparganium erectum L. Journal of Ecology*, **50**, 47-55.

[10] Naden, P., Rameshwaran, P., Mountford, O. and Robertson, C. (2006) The inf luence of macroph yte growth, typical of eutrophic condition, on river f low velocities and turbulence production. *Hydrological Processes*, **20**, 3915-3938.

[11] Jensen, K. (199 8) Influence of submerged macrophytes on sediment co mposition and near-bed flow in lowland streams. *Freshwater Biology*, **39**, 663-679.

[12] Kaneko, K., Fukawa, H. and Fujisaku, M. (20 10) Influence of d ifferences in maintained and management of river on the growth of h ydrophytes such as *Sparganium L.*—Case study of Uzuma River (Tochigi City, Tochigi, Japan). *Japanese Journal of Landscape Ecology and Management*, **15**, 11-17.

[13] Riis, T., Sand-Jensen, K. and Vestergaad, O. (2000) Plant communities in lowland Danish streams: Species composition and environmental factors. *Aquatic Botany*, **66**, 255-272.

[14] Takahashi, H., Volotovskyi, K, A. and Toshiyuki, S. (2001) A quantitative comparison of distribution patterns in four common *Sparganium* species in Yakutia, Eastern Siberia. *Acta Phytotax Geobot*, **51**, 155-167.

9

A case research on economic spatial distribution and differential of agriculture in China

—An application to Hunan province based on the data of 1999, 2006 and 2010

Jian Wang[1], Zhenghe Zhang[2], Baozhong Su[2], Liyang Zhang[3]

[1]College of Economics and Management, China Agricultural University, Beijing, China
[*]Corresponding Author
[2]College of Economic and Management, China Agricultural University, Beijing, China
[3]China Finance and Economy News, Beijing, China

ABSTRACT

This paper is to provide an empirical work for agricultural spatial distribution of agriculture. We consider the spatial location pattern in order to offer spatial views on the agricultural economic research and how Chinese agricultural economic spatial location pattern is forming, we also tested the agglomeration situation of agriculture and the process is going on in the future. The results indicate that the periphery areas exist significant differential among regions in Hunan province, China. It really presents some kinds of agglomeration pattern of agriculture and characteristic spatial autocorrelation; the biggest rate of contribution to the region agriculture economic gap is productivity per agriculture worker.

Keywords: Geography Economic; Agriculture location; Economic differential; Moran'l; Core-Periphery

1. INTRODUCTION

A new trend in regional economic analysis is introduces the spatial views to establish the new economic methodology. As we can see, the spatial pattern for the whole economy is frequently found in many researches on industrial cluster and agglomeration [1-4]. As in Mori's research, most studies of cluster focus on the overall degree of agglomeration in industry. If we turn our mind back and take focus on the developing country, just as China, we will find that the industrial agglomeration really did a great contribution for economy development. But for the whole country, there is more than 80% of total population in China lives and works in rural areas,

which we usually call countryside society. But the role of agriculture in the spatial analysis is ignored in many researches, as key factor in spatial analysis, the agriculture had been took as "periphery" [5-8] and serves for the "core" in the economy system. What we must accept is those most regional economists' interests are not in the agriculture, and most of recent literatures focus on the manufacturing and industrial sectors rather than the rural and agriculture sectors. In the common, they have no hesitation to place agriculture on the outside of their core research and take it as "remote" and "periphery" factor, just like the Krugman's New Geography Economic (NEG) method [7]. "Very frequently, a peripheral region is also a rural region, *i.e.* having a greater-than-average share of agricultural employment and a lower-than-average contribution to GDP [9]." We usually acknowledge that the persons in the rural areas mainly work in the agriculture, so there exits a balance rate between the total population and the population of agriculture workers in rural region, but the fact is that the proportion of agriculture workers is rapidly declining in many developed countries and the population in rural areas are constant, and this trend is continuing in developing countries right now. Many sufficient labors are moving to the "Core" city from agriculture "Peripheral". This movement breaks the hypothesis "The agriculture labor is immobile" down in the NGW basic model "Core-Peripheral Model". It seems like that this movement of agriculture labor is due to exogenous reasons, in fact, the reason is endogenous. The excessive economic cluster in "City Core" will plunder the resources of "agriculture periphery" and make it weaker for developing. On the other side, the exogenous impact, such as new technology and innovation in agriculture production will improve the productivity and relax many fix agriculture workers and turn them to mobility workers for manufactory. The economy system will effect by the combination force from exogenous and endogenous (Mobility agriculture worker and new technology, etc.). For instance, in China, the urbanization processes have established a city net cross the whole country, and there are three main city clusters in the east coastline, Zhujiang Delta, Yangzi river Delta and Bohai Economic Rim. Many immobile agriculture workers turn to mobile during the process of urbanization and Industrialization. That is wh y we can see th e huge workers flow from the rural to cities in China during the past two decades. The new work force really offers big cities continuingly growing and the "Core-Periphery" structure inner cities have been emerged, such as the CBD in Beijing, Shanghai and Guangzhou, etc. If we turn our attention to agricultural areas, we can notice that the big city and city net just like a Black Hole, conce ntrates all resources, worker, funding, knowledge, etc., to the "core" for city. It makes the agriculture (we can call it Agriculture Periphery) development lags far away fro m cities clusters. Pierre and Zeng's research shows that agricultural productivity improvement is asso ciated to so me re-dispersion of economic activity. Price subsidies strengthen dispersion forces and export subsidies weaken them. The regional differential and economic gap between urban and rural will be a key barrier for Chinese economic development process in the coming future. "The spatial evolution of rural-economic activities when the markets for agricultural and m anufactured goods become more accessible has become key issue for Chinese planners [10]". However, there has not been much empirical work for agriculture spatial distribution pattern and rarely working on introducing spatial views to agriculture economic research. The purpose of this paper is to provide an empirical work for agri culture spatial distribution. Based on th e county level data, we consider the spatial location pattern in order to offer spatial views on the agriculture economic research.

2. PERIPHERIES AND LOCATION FOR AGRICULTURE SECTOR

There are rarely researches in the agriculture spatial analysis, but still have few researchers give some discuss on the roles of agriculture in their work on spatial economic research [5,8], the roles of agriculture in their research is the p roblem of transport cost and agricultural market in core-periphery models. Fujita showed that agricultural transport costs act as a brake on urban development. A rise in agricultural transport costs fosters dispersion as strongly as a r ise in manufacturing transport costs. The basic model of New Geography Economic place the hypothesis as "Both goods are tradable across regions, where trade in agricultural goods is costless, and trade in manufactured goods is subject to some iceberg transport costs [7]", so "This result is the outcome of a

process of cumulative causation, where additional firms in the (prospective) core attract ad ditional workers from the (prospective) periphery as a result of higher wages, which in turn attracts more firms as a result of increasing demand [9]". In the coreperiphery model, agricultural products are abundant and agricultural sector can satisfy all the core city need. It's not a real world. This hypothesis makes many economists pay rarely attentions to the agriculture cluster and agglomeration. In Krugman's NGE model, the immobility of agricultural workers is a centrifugal force for the process of agglomeration, "because they consume both types of goods". In contrary, "the centripetal force is more complex, involving a circular causation [6]". We can image that if many firms locate in one area, there will be a greater number of varieties are produced, then, "workers (who are consumers) in that region have a better access to a greater number of varieties in comparison with workers in the other region [6]". So, workers in those areas will get a higher income level and it will induce more workers to come. On the other hand, many workers getting together will increase the demand for goods and lead to higher production, which Krugman call it "Home Market Effect (HME)". The cluster and agglomeration is the conclusion of games between centripetal force and centrifugal force, "the centripetal force is generated through a circular causation of forward linkages (the incentive of workers to be close to the producers of consumer goods) and backward linkages (the incentive for producers to co ncentrate where the market is larger) [6]". "If forward and backward linkages are strong enough to overcome the centrifugal force generated by immobile farmers, the economy will end up with a core-periphery pattern in which all manufacturing is concentrated in one region [8]". In most of the basic NEG models, the primary role of the agriculture sector is to serve as a "numéraire sector", "producing under constant returns to scale and perfect competition [11]".

At the beginning, before the industrial revolution, the natural economy sh ows a disperse location in the whole region (See **Figure 1(a)**) due to the immobile inputs in agriculture produce, like land, worker, water resources, etc.). The human social action concentrated small group and this small group disperse in the whole region, each small group has their central, the economic actives surrounding in the small central (We can image this central as a village in d eveloping country). In this period, the transportation cost is too high and there are rarely communications among those small groups, the economy was thus diverse at household level. T he few economic exchange activities just existing between two regions which adjacent. In the **Figure 1**, we use dotted line in "A" to present the two regions economic linkage. In this situation, the centrifugal force is stronger than the centripetal force, the economic activates can't be cluster to

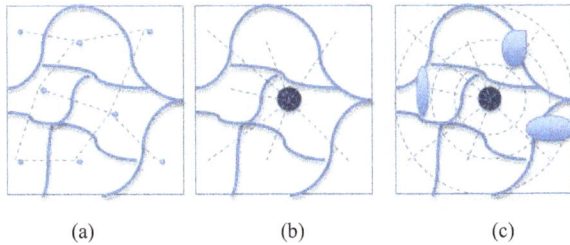

Figure 1. The process of city current and agriculture location form.

one place. As the transportation cost decline (May be due to new road, may be due to new vehicle), the economic location pattern happened to mutation, low cost m ake worker easy commute to the high income region and the home market effect will make the economic end up to a cluster point (See **Figure 1(b)**). In this process, the centripetal force is stron ger than the ce ntrifugal force, "a larger number of firms locate in a re gion, a greater number of varieties are produced there", "workers in that region has a better access to a greater number of varieties in comparison with workers in the other region [5]". "The resulting increase in the number of workers creates a larger market than the other region, which in turn yields the home market effect familiar in international trade [7]". This agglomeration process is a circular causation.

The new geo graphy economics or spat ial analysis stopped on this step and defines the outside city areas as agriculture periphery. We need to know however, how the agricultural spatial location pattern evolves for next steps. Thünen gives us a view on his monocentric economic theory. Thünen showed that "competition among the farmers will lead to the gradient of lan d rents that declines from a maximum at the town to zero at the outermost limit of cultivation [5,12]". In Thünen's theory, each farmer is faced with a trade-off between land re nt and transport cost. "Since transport costs and yield differ among crops, a pattern of concentric rings of production will result. In equilibrium, the land-rent gradient must be such as to induce farmers to grow just enough of each crop to meet the demand, and this condition together with the condition that rents be zero for the outermost farmer fully determines the outcome [6,10,13]". However, what we miss is agriculture in Thünen ring that will show some kinds of cluster and agglomeration (See C in **Figure 1**). The new economic geography leads researchers to focus on the sp atial impact of con tinuing areas, and hence spatial location patterns more generally for Industry. But this combination of po litical impetus and theoretical development is not sufficient to solve the problem for most regions in developing countries due to that there are great lands which belong to the agriculture in these countries. Our work in this paper is to try to find out the agriculture (may be in the Thünen ring) spatial lo cation

pattern and the process of this pattern evolution through the empirical work.

Before introducing the method we use, we need to do a definition for the agricultural areas or periphery. In general, the area that ends up as the so-called "agricultural periphery" is a region serving the core, which is due to the fact that the peripheral region acts as the supplier of agricultural products for all the workers in the "Core" and as the importer of manufactured goods from the there [14]. Inevitable, the differentiation inner the agriculture exists due to agricultural resources, society system, etc. But we can p ick out factors that will characterize agricultural areas: Transport cost and Resources capacity. Just like the definition by Steve Wiggins and Sharon Proctor [9,15]. Cost of movement and the relative abundance of land and other natural resources, the structure of classification of agriculture location will be change, "patterns of activ ity are af fected by c hanging technologies, market liberalization, improved communications, and rising population [15]".

3. STANDARD METHODOLOGY

3.1. The Exploratory Spatial Data Analysis (ESDA)

3.1.1. Globe Moran' I Index

Before our analysis, we usually assume that the agriculture economic activities have relationship in a certain region. This relationship distributes to goods price, information gap and so on. Also, the economic correlation shall have some kinds of geographical effect. How to measure this effect is reduced by location, and many researchers offer kinds of methods. But the popular and most acceptable method is Globe Moran' I Index, which defined by:

$$Moran'I = \frac{n}{S_0} \frac{\sum_i \sum_j w_{ij}(X_i - \bar{X})(X_j - \bar{X})}{\sum_i (X_i - \bar{X})^2}$$

where X_i and X_j denote observed value from region i and j, then $\sum_i (X_i - \bar{X})$ and $\sum_j (X_j - \bar{X})$ are deviation of the variable of X in different regions with respect to the mean. $\bar{X}$ is the average value of X; w_{ij} denotes a n space weight matrix, and be used to show the relationship between regions. The matrix w_{ij} is requ ired because in order to address spatial autoc orrelation and also model spatial interaction, we need to impose a structure to constrain the number of neighbors to be considered. There are two rules to establish the w_{ij} matrix: neighboring rule and distance rule, in th is paper, we use the neighboring rule due to t hat each county has a clea r boundary. The factor of diagonal in w_{ij} matrix is 0, and

$w_{ij} = 0$ if region i neighboring to region j, otherwise, $w_{ij} = 0$.

We can give the Z and P value to inspect the significance of Globe Moran' I Index. Of course, it will automatically show in the result if calculating the Moran index using statistical software. In general, the significance level we selected is 0.05, on this level. If P value is less than this level, we can accept that the Moran' I index shows a significant effect in the economic activities, otherwise, if P value is greater than the significance level, the effect showed by Moran' I index is weakened.

3.1.2. Local Moran' I Index and Local Indicator of Spatial Association

The Globe Moran' I index can tell us whether there is economic spatial autocorrelation in a certain region and the significance level through the Z value and P value. The analysis of globe Moran' I index yields only one statistic to summarize the whole study area. In other words, global analysis assumes homogeneity. If that assumption does not hold, then having only one statistic does not make sense as the statistics should differ over spaces. Besides that, what we have more interest is the specific place if there is high correlation. The Local Moran' I index can help us. The fact that Local Moran's I is a summation of individual is exploited by the "Local Indicators of Spatial Association (LISA)" to evaluate the clustering in those individual units by calculating Local Moran's I for each spatial unit and evaluating the statistical significance for each region. From the previous equation we then set the Local Moran' I as following:

$$\text{L'Moran'I} = \left(X_i - \overline{X}\right)\frac{\sum_j\left(X_j - \overline{X}\right)}{\sum_i\left(X_i - \overline{X}\right)^2 \big/ n}$$

where n is the number of regions (observations). The other variables are the some with the globe Moran' I index.

3.2. Theil Index

The Theil index is a statistic used to measure economic inequality. So, to see how the deferential in spatial regions, most traditionally, we can use Theil Index. The use of the dissimilarity Theil index to assess relative concentration is subject to a straightforward economic interpretation. In this paper, we will do some changes for the basic Theil index: separate the effect; we will discuss this in the next section. The basic Theil index formula is:

$$T = \sum_{i=1}^{n} r_i \log\left(\frac{g_i}{g}\right)$$

r_i denotes the proportion of population in region i, g_i denotes the average output per person in region i.

4. DATA AND EMPIRICAL WORK

4.1. Data

In this section, we consider the data that are available for studying the agriculture spatial distribution of economic activities. There are 31 provinces, 600+ cities and more than 2000 counties in China. As the basic unit, we choose county as the research unit due to their independent policy-making system. You can find a full economy structure in a single county: agriculture sector, industry sector, center city, etc. The county is a small integration economic system. The relevant notion of a basic region for this paper is the 88 counties in Hunan province area (See **Figure 2**), which locates on the south middle in China. There are 88 counties in Hunan province and these 88 counties are our data unit. These 88 unit counties cover over more than 95% of total land areas of Hunan province. The data we use is source from the public data from statistical book, it include: "Chinese county social and economic statistical year book" 2011, 2007 and 2000; "China Statistical Yearbook" 2011, 2007 and 2000; "China agricultural yearbook" 2011, 2007 and 2000. The data set used in this paper contains the main economic statistical data, such as population, total agriculture output, main agricultural products (Food, Meat, Cotton and Oilseed), of 88 counties in Hunan Province.

4.2. Agriculture Economic Distribution Pattern of Hunan Province

Figure 3 shows us a glance of distribution of total economic output of agriculture in Hunan province in 1999, 2006 and 2010. The top pic in the **Figure 3** shows years' total economic output bar chart (From left to right is 1999, 2006 and 2010). The agriculture output in 2010 doubles that in 1999, and also has a high increasing from 2006. The bottom pictures shows total agricultural economic output spatial location spread in Hunan on the corresponding periods (1999, 2006 and 2010). The distribution pattern is preformed as little cluster in west, northeast and mid-southwest. The increasing of produc-

China   Hunan Province

Figure 2. Hunan location and the county unit.

tivity in the high output areas is continuing to highly grow, and we can explain this phenomenon as the inner scale effect. High output areas trends to have higher output in next period due to the potential input of new resource and "Home Market Effect". In high output areas, capital and technology skills have some kinds of advantage than other places, and we can call this advantage as "Advantage of Producer (AOP)". AOP effect can be active only when there are abundant input factors, such as worker, land and nature resource.

The cluster pic at the bottom of **Figure 3** shows us a change trend of agriculture economic output location that the form of agricultural spatial is from dispersing in northeast areas to clustering surrounding the triangle periphery circle gradually, which set by "Zhangjiajie-Changsha-Shaoyang" as a city triangle shape (See the third pic at the bottom of **Figure 3**). It really presents some kind of agglomeration for agriculture outside the areas of cities in Hunan province. But the cluster level is slight and the agglomeration process is still going on. We can see little change of agriculture spatial location pattern from 2006 to 2010, but we also should notice that the cluster steps from 1999 to 2010 is obviously happening. The maps in

Figure 4 highlight the key stylized facts concerning the spatial distribution of four main agricultural products output across regions in Hunan province between 2006 and 2010, where from up to down are Food, Meat, Cotton and Oilseed.

The specific classification of agricultural products shows more significance in cluster and agglomeration in Hunan province. For Food, the output of Food products has a similar spatial location pattern with the total agriculture economic output location (See the first row pics in **Figures 3** and **5**). The similar spatial location pattern between Food and total output of agriculture contributed to high proportion of food supply of the whole agricultural products in Hunan. The Food supply presents a ring form spatial location pattern in the north of Hunan (See **Figure 5**). For Meat, the location pattern is more remote from the center and similar with Food. There also are some kinds of cluster for meat product in the southeast areas (See **Figure 5**). High productive area for meat clusters in the north and east of Hunan, but compared to the 2006, what differentiates with the food's pattern is that most of counties with high productivity show some decline in output. The level of agglomeration for meat is

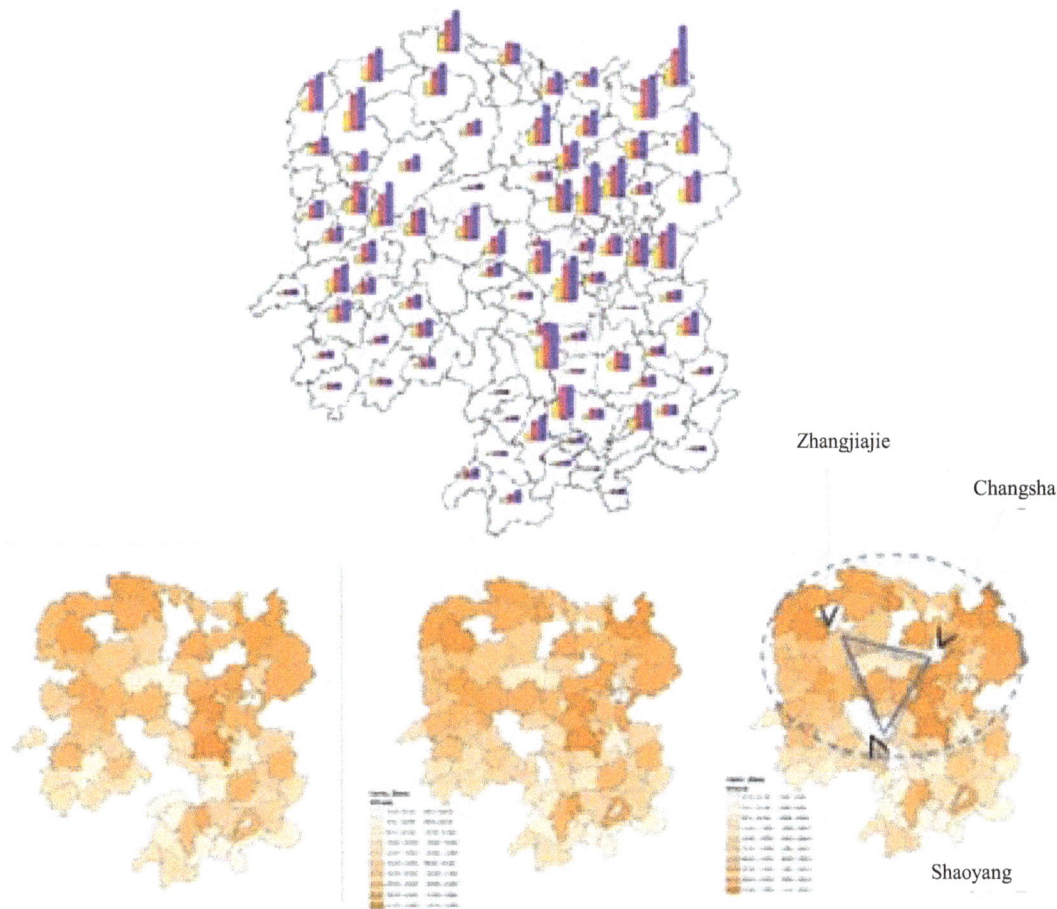

Figure 3. Hunan agricultural economic spatial location agglomeration and trends.

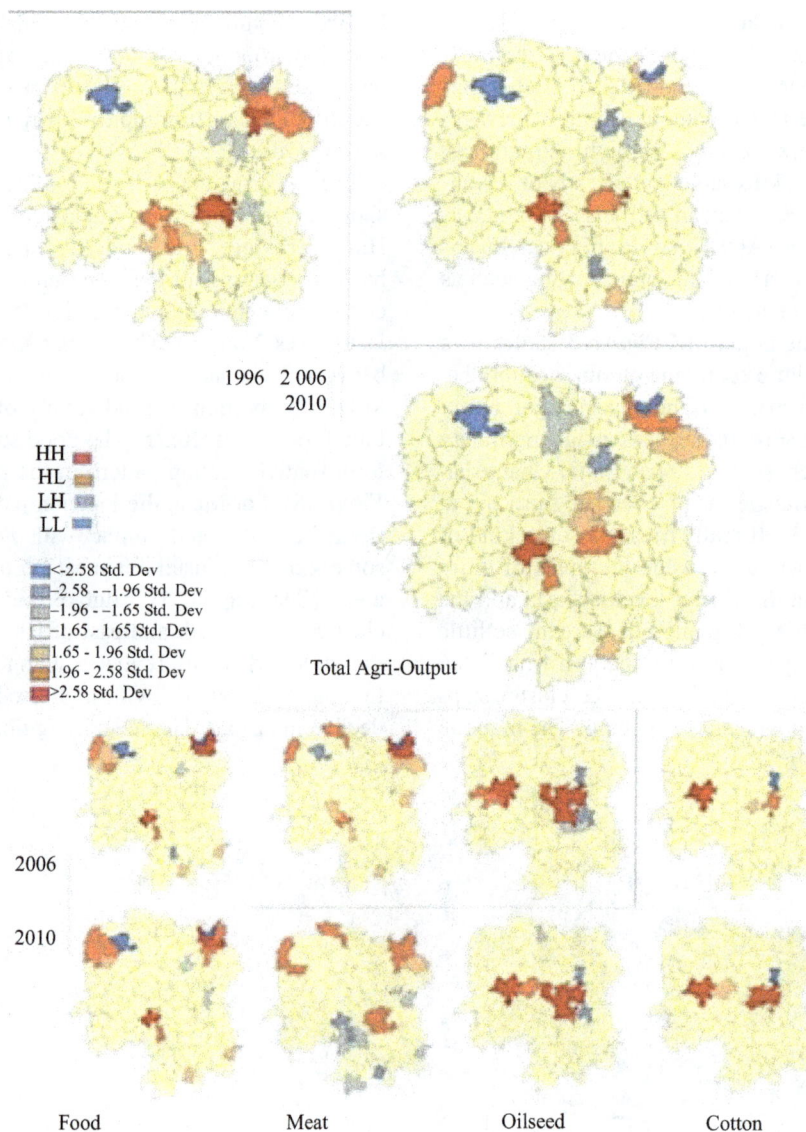

Figure 4. Local moran' I index with LISA for agricultural output in Hunan province.

more obvious in 2006 than 2010.

The spatial location patterns of Cotton and Oilseed are more obvious (See **Figure 5**). There are high cluster patterns in the center of Hunan, which present a Line shape. Compared with that in 2006, the agglomeration level is more significant in 2010 for both Cotton and Oilseed.

The pattern of agriculture in Hunan, which shows that the spatial location cluster of agricultural products, characters with describing as exhibiting pattern of "Ring pattern in the north" for Food and Meat, "Line Pattern in the center" for Cotton and Oilseed (See **Figure 5**). The form of total agricultural output spatial is from dispersing in northeast areas to clustering surrounding the triangle periphery circle gradually, which is set by "Zhangjiajie-Changsha-Shaoyang" as a city triangle shape (See **Figure 3**). There is some evidence that this core-periphery

pattern in agriculture ec onomic output may be weakening among counties, while stable within counties.

In this case, Food and Meat follow a slight "Core-Periphery" pattern in Hunan, and we can figure it out on visual map. For Food, the wide core periphery pattern has more significance since 2006 as high output areas are more cluster. In contrast, the Meat core periphery pattern has some kind of declining slightly since 2006 as t he high output areas are more dispersed. Cotton and Oilseed in Hunan province follow a clear pattern. We can identify a high output area for them and the line in center pattern in Hunan remained stable from 2006 to 2010.

4.3. The Relation Analysis of Agriculture Economic Spatial Location

In the previous section, we described the spatial loca-

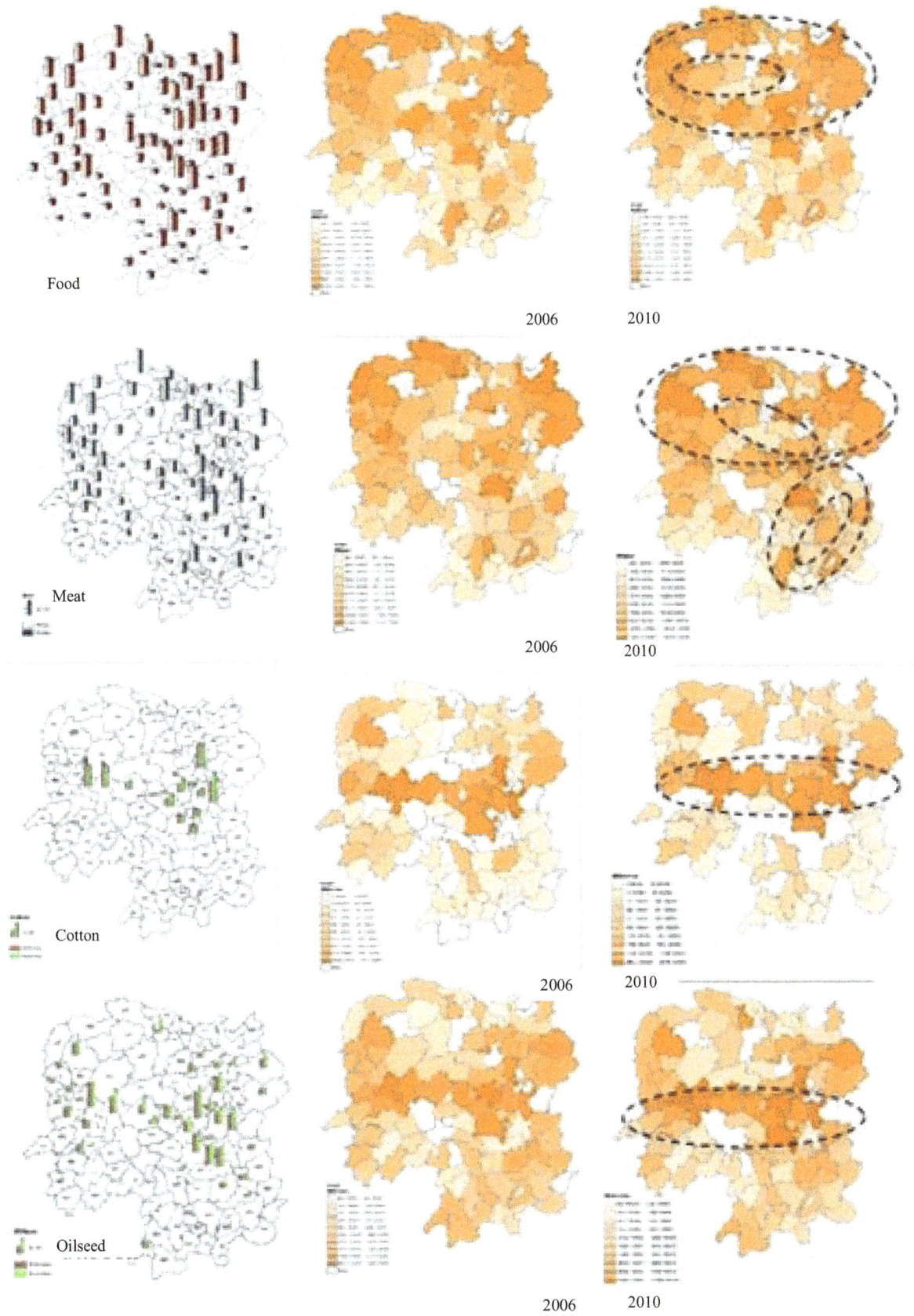

Figure 5. The spatial location for food, meat, cotton and oilseed in Hunan (up to low: food, meat, cotton, oilseed).

tion pattern for agriculture economic output and specify products (Food, Meat, Cotton and Oilseed) output by county level in Hunan. The problem emerges: are agriculture production areas having spatial autocorrelation in county level? Whether there are economic spatial autocorrelation in a certain region and how is their significance level? In this section, we will apply the spatial autocorrelation to identify tool to answer it. For the first question, whether the spatial autocorrelation exists, the Globe Moran' I Index will be applied and for the following questions, whether there is a certain region with high cluster and how the significance level is, we will use the Local Moran' I index. The fact that Local Moran's I is a summation of individual is exploited by the "Local Indicators of Spatial Association (LISA)" to evaluate the clustering in t hose individual units by calculating L ocal Moran's I for each spatial unit and evaluating the statistical significance for each region.

The Globe Moran' I Index showed in the **Table 1** presents the spatial autocorrelation level in 1999, 2006 and 2010 for total agriculture output, 2006 and 2010 for Food, Cotton, Meat and Oilseed by county level in Hunan province.

The conclusion of **Table 1** summarizing the total agriculture output of county in Hunan shows some kind of spatial autocorrelation, even though the spatial autocorrelation is low, but stable from 1999 to 201 0. Globe Moran' I Ind ex for total agriculture output are listed as 0.14, 0.11 and 0.13 in 1999, 2006 and 2010. The corresponding Z score and P value is Z(1.92), P(0.05); Z(1.72), P(0.08); Z(1.95), P(0.05). For total agricultural economic output, the significance set for Globe Moran' I i ndex presents more stable. The process of clustering total agriculture economic output is standing at slight agglomeration level (See **Figure 6(b)**).

The spatial autocorrelation of Co tton and Oilseed is higher than Food a nd Meat, and we can read the estimated result from **Table 2**. For the Cotton and Oilseed, the Globe Moran' I index are 0.27 and 0.30 in 2006; 0.31 and 0.31 in 2010. The level of spatial autocorrelation is similar with ot her between Cotton and Oilseed. We need to no tice that the cotton's spatial autocorrelation

level shows some kind of increasing in 2010 compared to that in 2006. The Z score and P value to Globe Moran' I Index for Cotton and Oilseed listed are: Z-co tton (3.83), P-cotton (0.00); Z-oilseed (4.47), P-oilseed (0.00) in 2006 and Z-cotton (3.90), P-cotton (0.00); Z-oilseed (4.71), P-oilseed (0.00) in 2010. The significance set for Globe Moran' I ind ex to Cotton and Oilseed presents higher in 2010 than that in 2006. The pr ocess of clustering for Cotton and Oilseed is getting to stronger agglomeration (**Figure 4**).

The spatial autocorrelation level of outputs for Food and Meat is sli ghter than Cotton and Oilseed, but it still presents positive effect in spatial autocorrelation. For the Food and Meat, th e Globe Moran' I index are 0.12 and 0.13 in 2006; 0.13 and 0.16 in 2010, which shows some kind of increasing in the digital level from 2006 to 2010. The level of spatial autocorrelation is similar with each other between Food and Meat. The Z score and P value to Globe Moran' I Inde x for Food and M eat listed are: Z-food (1.90), P-food (0.06); Z-meat (2.02), P-meat (0.04) in 2006 and Z-food (1.93), P-food (0.05); Z-meat

Table 1. The globe Moran' I index for agriculture output of county in Hunan Province.

	Time	Moran I	Z-Score	P-value
	2010	0.13	1.95	0.05
Agri-GDP	2006	0.11	1.72	0.08
	1999	0.14	1.92	0.05
FOOD	2010	0.13	1.93	0.05
	2006	0.12	1.90	0.06
COTTON	2010	0.31	3.90	0.00
	2006	0.27	3.85	0.00
MEAT	2010	0.16	2.48	0.01
	2006	0.13	2.02	0.04
OILSEED	2010	0.31	4.71	0.00
	2006	0.30	4.47	0.00

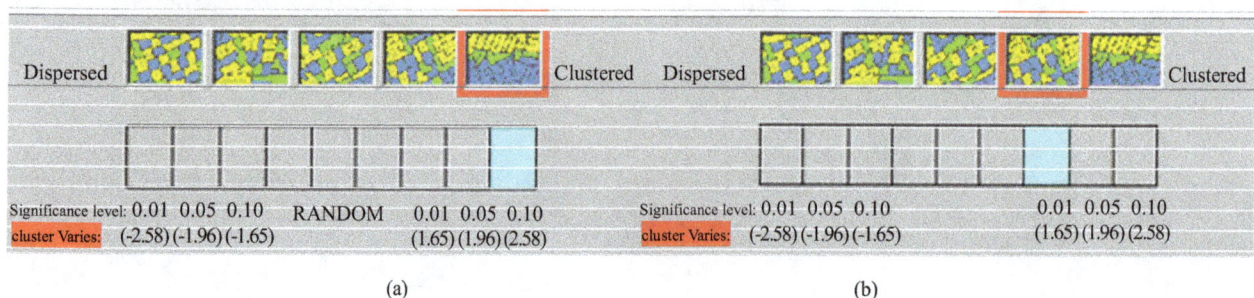

Figure 6. Spatial autocorrelation and cluster level.

Table 2. Theil index: Decomposition ef fect of population structure.

	T	T1	T2	T3	T1CR	T2CR	T3CR
2010	0.0274	0.0199	0.0059	0.0016	72.69%	21.59%	5.72%
2006	0.0263	0.0197	0.0052	0.0014	74.84%	19.65%	5.51%
1999	0.0231	0.0179	0.0037	0.0015	77.42%	15.98%	6.60%

CR: Contribution Rate.

(2.48), P-meat (0.01) in 2010. Similar to Cotton and Oil-seed, the significance set for Globe M oran' I index to Food and Meat presents higher in 2010 than 2006. The process of cluster for Food and Meat is getting to slight agglomeration and getting to higher in future.

The analysis of globe Moran' I index yields on ly the statistics to summarize the whole area sp atial autocorre-lation level in Hunan. But what we are m ore interest is the specific place if there is high correlation for agricul-ture output. In th is section, we will apply the Local Mo-ran' I index tool exploited by the "Local indicators of spatial association (LISA)" t o evaluate the clustering in those individual county uni ts by calculating for each county unit and evaluating the statistical significance. **Fig-ure 6** is mapping the result of estimate of Local Moran' I Index with LISA method.

The three t op maps in th e **Figure 6** show the local cluster effect for total agricultural economic output in Hunan province by county unit. It present the high output of agriculture locates in cen ter and northeast. The dark area in the map m eans there are hi gh output and also high output area surrou nding it. Th ere is po sitive scale effect for high output of total agriculture. Also, these high product positive scale effect areas are slig ht decli-ning from 1999 to 2010 (See **Figure 6**). The high output areas with high spatial autocorrelation for agriculture in Hunan are locating surrounding the Changsha city (In the northeast of Hunan) and Shaoyang city (In the center near to southwest of Hunan), this phenomenon presents the significant effect from city d evelopment for ag ricul-ture development. We can no tice that the High output hotpot areas are relative to "Peri-urban zones". The bot-tom maps summarize the four main products (Food, Meat, Cotton and Oilseed) output local cluster effect and the comparable between 2006 and 2010. Food and Meat have similar distribution of local cluster effect, and it show this effect is weak en in the south more than in the north. Especially for meat output, the high output areas spatial correlation in the south is weaken in 20 10 than it in 2006. In contract, cotton and oilseed show a different side. The high output areas effect of cotton and oilseed are higher in 2010 than 2006 and we can image this process still be continue, especially for oilseed (See **Fig-ure 6**). We can set this phenomenon for cotton and oil-seed as high output spillover effect.

For the low spatial output effect areas, the low output can't be change even thought there are high output areas surrounding it. This may be cause by lack of agriculture resource, such as water, land or insufficient on wo rker and so on. These ar eas can be set as "poor natural re-sources" in **Table 1**. On the other hand, the low product of agriculture may be cau se by the land be occupy for other using, especially for tourism, these areas can be set as "the m iddle countryside" or "remote rural areas" in **Table 1**. Theoretically, for the explaining of phenomenon of agriculture distribution pattern can be focused mainly on the role of demand externalities in determining agri-cultural locations. In particular, do not like the industrial agglomeration, the types of demand externalities that induce agriculture cluster are often set just only for one certain agriculture products, but for so me product, there still exist their spatial market overlap between neighbor regions. In such case, it' s natural for thos e agriculture products show some kind of "ring pattern" surrounding the marketplace. More over, in terms of nature environ-ment and transportation situation, it is also natural for agglomerations in certain pl ace as more concentrated to coincide with those of less co ncentrated, and leading to the type of synchronization predicated by the hierarchy principle.

4.4. Spatial Inequality Measured by Theil Index

Obviously, the agriculture output show sig nificant gap between counties in Hunan province, the analysis o f lo-cation pattern in previous can prove this. In this settion, we will apply the Theil Index to an alysis the inequality of agriculture product in Hunan province. What we focus on is worker efficient and the structure of population. After this work, we need to decomposition the Theil In-dex with population style.

The Theil index is a statistic used to measure econo-mic inequality. It can help us to figure out the factor and their contribution for inequality (See th e original Theil Index formula in section 3). We split the Theil Index ac-cording the deviation of human factor: R. We divided population R to three overlap part: agricultural employed population: l, rural population: k, and total population: n. The relationship of them is: $l \subset k \subset n$ (See **Figure 7**). Then for th e average agriculture output per per-son g can be di vided as $g = \frac{G}{l} \cdot \frac{l}{k} \cdot \frac{k}{n}$, where $\frac{G}{l}$ denotes labor productivity for agriculture output, $\frac{l}{k}$ denotes the agriculture employed rate and $\frac{k}{n}$ denotes the agriculture population proportion. So we can reset the

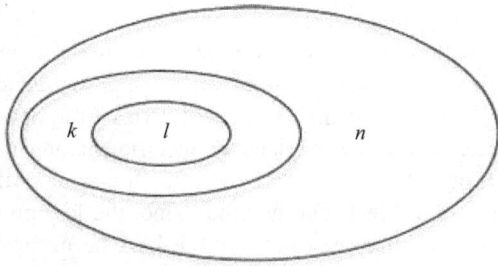

Figure 7. The relationship of population structure.

Theil Index as:

$$T = \sum_{i=1}^{n} r_i \log\left[\left(\frac{G_i}{l_i}\cdot\frac{l_i}{k_i}\cdot\frac{k_i}{n_i}\right)\Big/\left(\frac{G}{l}\cdot\frac{l}{k}\cdot\frac{k}{n}\right)\right]$$

Decomposed to:

$$T = \sum_{i=1}^{n} r_i \log\left(\frac{G_i l}{l_i G}\right) + \sum_{i=1}^{n} r_i \log\left(\frac{l_i k}{k_i l}\right) + \sum_{i=1}^{n} r_i \log\left(\frac{k_i n}{n_i k}\right)$$

And we can set:

$$T_1 = \sum_{i=1}^{n} r_i \log\left(\frac{G_i l}{l_i G}\right)$$

$$T_2 = \sum_{i=1}^{n} r_i \log\left(\frac{l_i k}{k_i l}\right)$$

$$T_1 = \sum_{i=1}^{n} r_i \log\left(\frac{G_i l}{l_i G}\right)$$

So:

$$T = T_1 + T_2 + T_3$$

The previous section considers the a griculture economic spatial location pattern by using county data in Hunan province, the pattern map show us di rectly that there exist some kind of differentials among different counties. The Theil Index for counties agricultural output show that the inequality between 88 counties in Hunan is increasing slightly from 1999 to 2010. The to tal Theil Index list as: T $_{1999}$(0.0231), T$_{2006}$(0.0263) and T$_{2010}$(0.0274). The reason for in equality are v ery complex, in this paper, we take our focuses on the cause from person. From the **Table 2**, it seems that the labor productivity is th e biggest cause the agricu ltural economic inequality between counties in Hu nan province, the contribution rate of labo r productivity to those inequality is more than 70%. Bu t the influence of labor productivity to agricultural shows a d eclining during 1999 to 2010. The agriculture employed rate show a in creasing influence to the agricultural inequality between regions. From 1999 to 2010, the agriculture employed rate contribution rates to ag ricultural inequality between 88 counties in Hunan are list as: 15.98%, 19.65% and 21.59%. The agriculture population proportion do some kind of contri-

bution to the regional agricultural inequality, the influence to agricultural output inequality in Hunan is stab le (See **Table 2**).

5. CONCLUSIONS

This paper provides an empirical work for a griculture spatial distribution and considers what we know about the spatial location pattern in order to offer spatial views on the agriculture economic research. Also, we know that the periphery areas exist significant differential among regions. Set county as basic unit to analysis shows us that: It really presents some kind of agglomeration for agriculture outside the areas of city in Hunan province; The pattern of ag riculture in Hunan shows that th e spatial location cluster of agriculture products c haracters with describing as exhibiting pattern of "Ring pattern in the north" for Food and meat, "Line Pattern in the center" for cotton and oilseed. In additional, for total agricultural economic output, the significance set for Globe Moran' I index are presents more stable. T he process of cluster agricultural economic total output is stand ing at sli ght agglomeration level. The process of cluster for cotton and oilseed is getting to more strong agglomeration and food and meat is getting to slight agglomeration and getting a process to higher in future. Theil Index result shows us th e labor productivity is th e biggest cause the agricultural economic inequality between counties in Hunan province.

We have to accept that m any economists take their emphasis on industrial sector, especially for the economic geography, many researches neglected to study the role of agricultural sector in the process of cluster and how the agriculture sector agglomeration by itself. We can jump out agriculture to analysis the whole econom ic or even only a city, still, agriculture take an important part f the economy in most developing country, like China, and it seem s that the cluster of agricu lture has give some significant phenomenon in those areas. In additional, developing countries agriculture is often subj ect to important policy issues.

6. ACKNOWLEDGEMENTS

The research works for this paper was supported by the National Philosophy and Social Science Foundation of China (No. 11 & ZD009).

REFERENCES

[1] Brülhart, M. (1998) Economic g eography, industry location, and trade: The evidence. *The World Economy*, **21**, 775-801.

[2] De Lucio, J., Herce, J. and Goicolea, A. (2002) The effects of externalities on productivity growth in Spanish industry. *Regional Science and Urban Economics*, **32**, 241-

258.

[3] Venables, A.J. (1996) Equilibrium locations of vertically linked industries. *International Economic Review*, **37**, 341-359.

[4] Fujita, M., Krugman, P. and Venables, A. (1999) The spatial economy. MIT Press, Cambridge.

[5] Fujita, M. and Thisse (2002) Econom ics of agglomeration: Cities, industrial location, and regional growth. Cambridge University Press, Cambridge.

[6] Fujita, M. and Thisse, J.F. (2002) Economics of agglomeration: Cities, industial location, and regional growth. Cambridge University Press, Cambridge.

[7] Gruber, S. and Soci, A. (2010), Agglomeration, agriculture, and th e perspective of th e periphery. *Spatial Economic Analysis*, **5**.

[8] Ottaviano, G.I.P. and Puga, D. (1998) Agglomeration in the global economy: A survey of the new economic geography. *World Economy*, **21**, 707-731.

[9] Ottaviano, G.I.P. and Rob ert-Nicoud, F. (2006) The "genome" of NEG m odels with vertical linkages: A positive and normative synthesis. *Journal of Economic Geography*,

6, 113-139.

[10] Hanson, G.H. (2001) Scale economies and the geographic concentration of industry. *Journal of Economic Geography*, **1**, 255-276.

[11] Wiggins, S. and Proctor, S. (200 1) How special are rural areas? The economic implications of lo cation for rural development. *Development Policy Review*, **19**, 427-436.

[12] Baldwin, R.E. (2001) Core-periphery model with forward-looking expectations. *Regional Science and Urban Economics*, **31**, 21-49.

[13] Bosker, M., Brakman, S., Garretsen, H. and Schramm, M. (2007) Looking for m ultiple equilibria when geography matters: German city growth and the W WII shock. *Journal of Urban Economics*, **61**, 152-169.

[14] Chan, K.W. (1994) Urbanization and rur al-urban migration in China since 1982: A new baseline. *Modern China*, **20**, 243-328.

[15] Duranton, G. and Overman, H. (2005) Testing for localisation using micro-geographic d ata. *Review of Economic Studies*, **72**, 1077-1106.

10

Long term effects of treated wastewater irrigation on calcisol fertility: A case study of Sfax-Tunisia

Nebil Belaid[1,2*], Catherine Neel[2,3], Monem Kallel[4], Tarek Ayoub[5], Abdelmoneim Ayadi[1], Michel Baudu[2]

[1]National School of Engineers of Sfax, Radio Analyzes and Environment laboratory (LRAE), Sfax, Tunisia;
[*]Corresponding Author
[2]University of Limoges, Research Group, Water, Soil and Environment (GRESE), Limoges, France
[3]CETE Lyon-DLCF, Clermont-Ferrand, France
[4]National School of Engineers of Sfax, Water, Energy and Environment Laboratory (L.3E), Sfax, Tunisia
[5]CRDA-Sfax, Sfax, Tunisia

ABSTRACT

The use of treated wastewater (TW) for irrigation is increasingly being considered as a technical solution to minimize soil degradation and to restore nutrient content of soils. Indeed, TW usually contain large amounts of nutrient elements. The objective of this study is to evaluate the impact of long-term irrigation by TW on soil fertility under real field conditions. In the vicinity of the city of Sfax, a semi-arid region, a calcisol field has been irrigated for more 15 years with organic sodic TW; soil was modeled at three different depths (0 - 30, 30 - 60 and 60 - 90 cm) and along soil pits in the TW irrigated zone and in a nearby non-irrigated zone (control). Several parameters have been measured: Soils pH, CEC, exchangeable cations, nitrate and ammonia, total contents of nitrogen, phosphorus and other essential macro and micro nutrients, electrical conductivity, soil organic carbon and dissolved organic carbon. C/N ratio and SUVA were calculated for each soil layer. The calculation of the isovolumic mass balance on soil profile scale was used to measure macro and micro nutrients supply. The TW irrigation has led to important supply in organic carbon (+100%), phosphorus (+80%) and in most essential nutrients (N, Mn, Zn). Due to the high rate of irrigation and low CEC of the studied soil, the added nutrient cations and nitrate are removed with leaching waters compared to the non-irrigated control soil. Moreover, Sfax's TW bring about important amounts of salts and Na. Therefore the beneficial addition of nutrients could quickly be inhibited by the excessive supply of salts and available nitrogen. Apart from future crops produc-

tion risk, groundwater degradation quality and soil fertility will be endangered over the long term.

Keywords: Arid Region; Wastewater; Irrigation; Fertility; El Hajeb-Sfax

1. INTRODUCTION

In arid and semi-arid regions of countries such as Tunisia which are facing rising serious water shortage problems, the reuse of urban wastewater for non potable purposes, such as agriculture [1-3] has became an usual practice. Indeed, wastewater reuse for irrigation has been the largest field of application because it usually offers some attractive environmental and socio-economic benefits, mainly due to the reduction of effluent disposal in receiving water bodies, to the supply of nutrients as fertilizers, and to the improvements in crops production during the dry season [4,5]. Benefits apart, planners are aware of the potential disadvantages of wastewater reuse for irrigation which are, aside from pathogenic contamination of irrigated crops, mainly related to the specific chemical composition of wastewater being somewhat different from most natural waters used in irrigation [6]. Over time, wastewater irrigation affects some soil parameters. Increase in soil pH is observed in acidic soils [7,8], whereas slight decrease of pH is mentioned for alkaline soils [5,8-11]. In alkaline calcarous soils, the sligh acidification is due to the leaching of limestone by the leaching water [10]. Wastewater can also supply ammonia anion to soil which is another source of soil acidification [11]. In general, the decrease of pH is also explained by the low C/N ratio of effluents and the subsequent enhancement of the organic mineralization substances [7,10]. Wastewater irrigation thus usually leads to macro and micro nutrients supply [5,11] which stimulates the microbial activity [12,13] and promotes the mineralization of the soil organic matter. This can lead to the decrease of the soil cation exchange capacity [7,10] and mitigate the soil buffer capacity. By the same way, wastewater usually contains elevated concentrations in metal elements such as Mn, Cu and Zn which constitute essential micro nutrients for plants. Over time, the same elements can accumulate in the organic topsoil layer in such manner that they reach their critical level for plant growth.

Hence, the use of wastewater for irrigation of crops requires assessment of the balance in supply of macro and micro nutrient over the long-term. Most of previous mentioned studies recorded impacts on contents in macro and micro nutrients without any regard for changes in the total store of these elements in soil. Moreover, specific studies dedicated to highlight long-term impacts (along several decades) have involved soils that have been irrigated by untreated domestic effluents or municipal wastewaters (Rattan *et al.* [8]: 10 and 20 yr; Yadav *et al.* [5]: 30 yr; Solis *et al.* [10]: 50 and 100 yr). Thus, the treatment of wastewater has been generalized after the heighties on the spur of the F.A.O. guidelines for application in agriculture [4]. The treatment of urban effluents modifies their nutritional value, it is thus of great concern to assess whether the irrigation with treated wastewater (TW) still improves the soil quality or it could cause degradation to its fertility over the long term.

The objectives of this study are to evaluate the changes in soil fertility and to balance its essential elements in response to 15-year-long treated wastewater irrigation (TW) within the city of Sfax (the second largest city in Tunisia). In this arid region, there are many signs of extremely low groundwater levels which were registered over the last three decades due to the increasing number of wells especially used for irrigation of crops [14]. Treated wastewater in Sfax has thus been used for irrigation since 1989. From that period, the irrigation perimeter has regularly been expanded as to reach the area of 600 ha. New extension of the area irrigated by wastewater is planned for twice over its present surface. This study is a part of a research program which aims at evaluating the impact of wastewater application on both soil and crops properties in the arid region of Sfax. The overall goals are to aid management of crop irrigation by wastewater, to reduce overexploitation of the local groundwater resources and to improve the water recharge of groundwater. Belaid *et al.* [15] have evidenced negative impact on the soil salinity and sodicity, especially in the northern part of the irrigation perimeter covered by a deep permeable fluvisol. Therefore, soil salinization and sodification are mitigated by the amount of exchangeable calcium in irrigated calcisol fields [15]. The present study thus focuses on this type of calcareous soils, which is mostly found in the Southern part of the irrigation perimeter.

2. MATERIALS AND METHODS

2.1. Study Area

The study area is settled at ten kilometres in the West next to the sewage treatment plant (**Figure 1**) near to the town of Sfax (approximately with one million of habitants) in crop fields, which are currently irrigated with treated wastewater whose plant receives domestic as well as industrial effluents from mainly canning factories and textile production. The region has an arid climate with monthly air temperature ranging from 11.3°C to 26.7°C, dry summer and annual rainfalls of 200 mm mostly occurring from October to December. The average annual potential evaporation of 1200 mm, combined with the low rainfall and high temperatures makes irrigation essential for crop production.

The present survey has been carried out in the area of the irrigation perimeter that is covered by a calcisol (according to the FAO World Reference Base for soil resources [16]). This soil presents an isohumic character and shows a homogeneous sandy to sandy loam texture.

As shown in **Table 1**, the selected area produces alternate cycles of crops, in association with permanent harvesting of olives, with successive winter and summer harvest of annual crops (oat, sorghum) sectioned every 10 years by a 3-year-long cropping of alfalfa. This kind of cropping system requires irrigation by open surface furrows distributed every 24 m in-between each row of olive trees. The soil has been submitted to wastewater irrigation for 15 years. In order to assess the effects of the wastewater, a nearby field is taken as a control area which produces only olives and has been preserved from any source of irrigation (**Figure 1**).

2.2. Samples of Preparation and Chemical Analyses

Treated wastewater were sampled at the outlet of the Sfax wastewater treatment plant at different times and conserved at 4°C before characterization. Effluent samples were analyzed for pH and electrical conductivity (ECw) using a pH meter (AFNOR standard method N° NF T 90-008 [17]) and a conductimeter (AFNOR N° NF EN 27888 [17]) respectively. Chemical oxygen demand (COD), suspended solids (SS), biochemical oxygen demand (BOD) and total phosphorus were measured according to standard methods (AFNOR N° NF T 90-018, NF EN 872, NF T 90-103, NF EN 1189 [17]). Cations and anions were measured using chromatography while

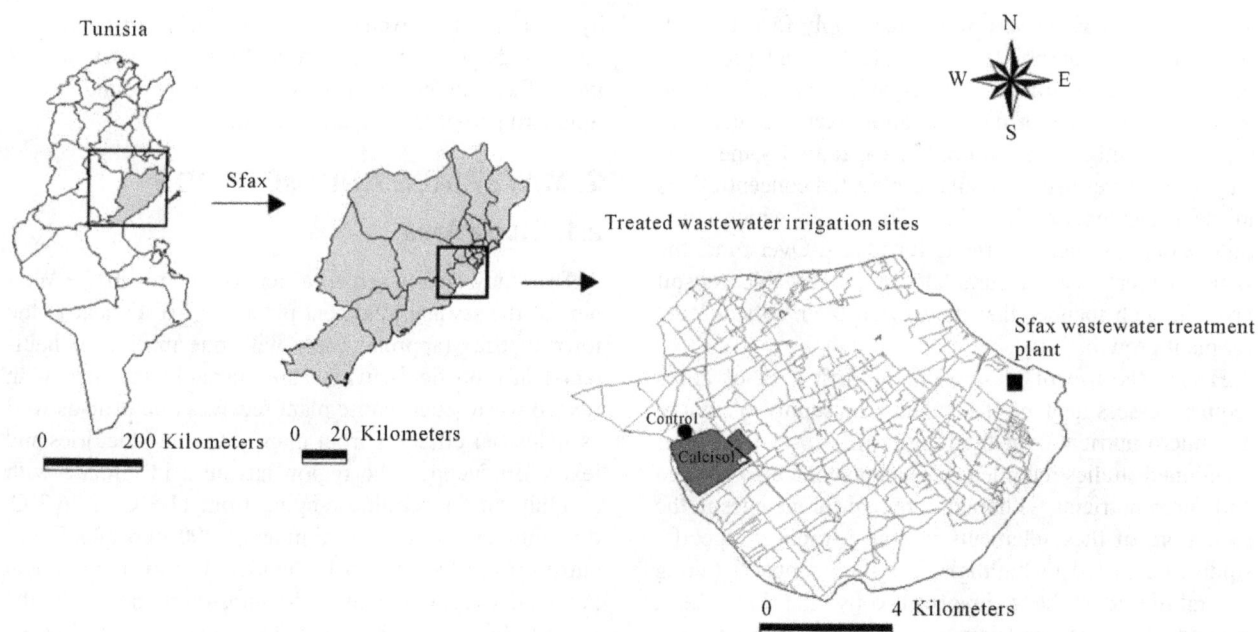

Figure 1. Map of studied area with location of the Sfax water treatment plant and of the calcisol site (TW irrigation perimeter in grey and point for the control site).

Table 1. Main characteristics of the studied calcisol site.

Soil taxonomy	Light texture isohumic calcimagnesic soil according to the Tunisian pedological map	
Soil depth	Moderately deep soil laid over a limestone crust approximately 60 cm deep (the crust was dismantled in large part of the irrigated area)	
Soil texture	Sandy to sandy-loam with calcareous nodules in subsurface	
Soil bulk density (surface layer)	1.4 g/cm^3	
Total CaCO$_3$	5% to 35%	
	Irrigated area	**Control area (not irrigated)**
Cultural system	Associated cultivars (olives trees and forages crops)	Only olives trees
Crop rotation	Winter (oat. ray grass) summer (sorghum) annual (alfalfa)	-
Field area	270 hectares including 90 hectares used for summer crops.	1.5 hectares
TW irrigation rate	1000 mm/yr	-
Irrigation system	Surface irrigation by furrows	-
Irrigation duration	15 years	-
Number of cores sampled	7	1

carbonates and bicarbonates were estimated by titration with HCl of an aliquot of the effluent samples (AFNOR N° NF EN ISO 9963-2 [17]).

Soil sampling was performed in October 2006 after the harvest of summer crops and before the seeding of winter crops. Two soil sampling survey have been carried out. The first one concerned each horizon of soil that has been identified along pedological profiles drilled in the control area and in the irrigated field whereas the second sampling was done on plots covered by summer crops

only by 30 cm thick soil layers using an Edelman-type auger. In order to account for spatial variations of soil texture and depth, 7 replications were performed in the irrigated field (IWC1 to IWC7). The control area was too small to allow such replications (NIC). Only 2 soil layers down to the depth of 60 cm were sampled in the control field as well as in the replication site IWC5 because of the occurrence of a concrete calcareous crust at depth of 60 cm. This crust of sedimentary origin is irregular and has been generally dismantled in the irrigated field in order to help infiltration of irrigation waters, except in the control area, which has never been irrigated.

After air-drying, the soil samples were sieved at 2 mm. Soil pHw and pHKCl were measured in a 2.5 soil to water/KCl 1 M slurry using a glass electrode. Saturation paste extracts of soils were prepared to determine the electrical conductivity of the soil samples (ECs). Soil samples CEC and contents in exchangeable cations were determined at actual soil pH by the cobaltihexamine method [18]. In calcareous soils, Belaid et al., [15] evidenced that the cobaltihexamine method provided more accurate values than the more usual method using 1 M NH_4 acetate solution [19] because of lesser dissolution of carbonates. Concentrations in Ca and Mg were performed by Atomic Absorption Spectrometry (AAS) while Na and K concentrations were determined by Flame Atom Emission Spectrometry (FAES).

NO_3 and NH_4 concentrations were measured in water soluble extracts using ionic chromatography (DIONEX DX-120) after water extraction using 1:5 soil to water ratio for 2 h. Total nitrogen was determined by steam distillation after acid digestion using the Kjeldal procedure. Soil organic carbon (SOC) was determined by the Walkley and Black dichromate oxidation method. Water soluble organic C, which is further referred to as dissolved organic carbon in this paper (DOC), was extracted with deionized water using a 2:1 (water to soil) ratio by shaking at fixed temperature during 3 hours [20]. The extracts were filtered using 0.22 mm filters and DOC concentration was measured using UV-persulfate oxidation on a TOC analyzer (TEKMAR DOHRMANN Phoenix 8000). UV absorption at 254 nm was measured using a Varian CARY 50 Probe UV-visible spectrophotometer. DOC quality; in terms of aromaticity of organic compounds, was determined as specific ultra violet absorbance (SUVA) which was calculated by dividing the absorbance at 254 nm by the DOC.

2.3. Statistical Analysis

One-sample T-test was used for comparing mean values obtained from replicates of measurements at an irrigated site to the values measured at the corresponding non irrigated control one. All measured values correspond to the average composite samples. However, due to the limited area of the control sites, and since the sampling was limited to the central part of them in order to avoid any influence of the neighbouring irrigation then in the absence of sampling replicate in the control area, no variance can be associated with the control values.

Hence, we have assumed that the values of the control site represent exact means to be compared with the variance of the mean values (N = 7) measured at the corresponding irrigated site. Variance is expected to be larger in the irrigated zone than in the corresponding control

one, so that the following sample T-tests can be considered as conservative:

$$T = \frac{\text{mean of TWE irrigated replicated values-control value}}{\text{standard deviation of TWE irrigated replicated values}}$$

A unilateral T-test was used to calculate the parameters, which are clearly increased or decreased after the irrigation by the treated wastewater (ECs, Exchangeable cations, NO_3 and NH_4). Whereas the bilateral T-test was chosen to identify parameters presenting no obvious response to the irrigation (pH, CEC, SOC). The global risk increases with the number of simultaneous tests performed. Therefore, we have also adopted a more severe individual rule than the usual one to minimize the increase of the global risk; the differences are considered significant when $p < 1\%$ instead of $p < 5\%$. In case of p ranging between $1\% < p < 5\%$, we conclude that the differences have to be confirmed. The T-tests were performed using SYSTAT Software version 12

3. RESULTS

3.1. Treated Wastewater Characteristics

During the wastewater survey, the applied treated wastewater (TW) was always remained alkaline with an average basic pH value of 7.7 (**Table 2**). It also always presented a high level of total dissolved solids (TDS) of 3.7 $g \cdot L^{-1}$ and of suspended matter (SS). The level of biochemical oxygen demand (BOD) and chemical oxygen demand (COD) are ranged respectively between 37 and 220 $mg \cdot L^{-1}$ and 123 and 700 $mg \cdot L^{-1}$. The mean electrical conductivity (EC) of effluents reaches 5.7 mS/cm. The sodium absorption ratio (SARw) of the treated wastewater ranges between 9.7 and 15.6. Theses parameters are higher than the usual ranges reported for other Tunisian TW [20] of similar mixed origins (industrial and domestic) than those of the Sfax TW. The Sfax' TW also contains great amounts of nitrate, phosphate and potassium which are crucial nutrients for plant growth and soil fertility.

3.2. TW Irrigation Impacts on the Soil Proprieties

In all sampling points in the irrigated field (**Table 3**) and in the two soil profiles, the pH (pHw and pHKCl) of the soil water extracts remains alkaline at all depths (**Table 4**). There are no significant differences in soil pHw between the wastewater irrigated calcisol and the control calcisol.

By the same way, there are no significant differences in the contents of exchangeable Ca^{2+} between the irri-

Table 2. Mean values of chemical properties of the treated wastewate (TW) generated by the wastewater treatment plant of Sfax from 1984 to 2007.

Parameter	TW of Sfax		Tunisian Standards	Examples of wastewater used for irrigation			
				Treated			Untreated
	Mean	Ranges		Sousse, Tunisia[a]	Sardinia, Italia[b]	Ramtha, Jordan[c]	Kurukshetra, India[d]
pH	7.7	7.1 - 8.7	6.5 - 8.5	7.8	7.73	7.3	7.4
ECw mS·cm^{-1}	5.7	4 - 7.7	7	3.5	1.14	-	1.74
TDS g·L^{-1}	3.7	3.56 - 5.13	-	-	-	0.95	0.9
SS mg·L^{-1}	204	29 - 275	30	32.7	46	-	-
COD mg·L^{-1}	350	123 - 700	90	88	34	-	382
BOD$_5$ mg·L^{-1}	107	37 - 220	30	18.5	-	-	169
Pt mg·L^{-1}	7.5	2.9 - 12.5	-	-	1.64	15.5	-
NO$_3^-$ mg·L^{-1}	21	0.35 - 50	-	-	1.63	29	-
Cl$^-$ mg·L^{-1}	1662	903 - 2580	2000	688	134	-	-
SO$_4^{2-}$ mg·L^{-1}	1022	508 - 1950	-	-	121	-	-
HCO$_3^-$ mg·L^{-1}	630	490 - 732	-	-	-	-	-
Na$^+$ mg·L^{-1}	1137	780 - 2100	-	112	103	-	-
K$^+$ mg·L^{-1}	57	17 - 105	-	333	18	33.3	-
Mg^{2+} mg·L^{-1}	151	129 - 209	-	166	20	-	-
Ca^{2+} mg·L^{-1}	296	103 - 521	-	258	53	-	-
NH$_4^+$ mg·L^{-1}	67	61 - 73	-	-	33	-	-
SAR	12.4	9.7 - 15.6	-	7.7	3	4.6	-

Mean minimum and maximum values of samples characterised since 1984. ECw: Electric conductivity; TDS: Total dissolved solids; SS: Suspended matter; COD: Chemical oxygen demand; BOD: Biochemical oxygen demand; Pt: Total phosphorus; SAR: Sodium absorption ratio: [a]Klays et al., 2010; [b]Cappola et al., 2004; [c]Rusan et al., 2007; [d]Yadav et al., 2002.

gated calcisol and the non-irrigated control calcisol (**Table 3**). This confirms the natural inorganic origin of this element that certainly reflects the presence of Ca-carbonates. On the other hand, contents in exchangeable Mg^{2+}, K$^+$ and Na$^+$ have increased in all examined layers of the irrigated calcisol (**Table 3**). These increases are very significant in the upper soil layer and can be attributed to the relatively high concentrations of the Sfax TW in these cations (**Table 2**). In the deepest layer, the same intensifications may have been enhanced by the wastewater drainage through the ripping of the carbonate crust in the irrigated field. The examination of the electrical conductivity of the saturation paste soil extracts (ECs) approves that 15-year long irrigation period by the Sfax TW results in significant supply of ion into the calcisol (**Table 4**), even in the deepest layers. As a consequence, soil salinity was up to 4 mS·cm^{-1} at all depths, and sometimes exceeded this level.

Analyses of the ammonium and nitrate concentrations

in the soil water extracts and of the soil organic carbon also reveal the impact of the irrigation by the Sfax TW. In the irrigated soil, the contents in extractable NH$_4^+$ remain low and never exceed 0.1 cmol$^+$·Kg^{-1} in the dry soil (**Table 3**). However, as with the other base cations origin- nating from the applied wastewater, the NH$_4^+$ contents of the soil generally increase with depth and are significantly higher in the irrigated soil compared to the non irrigated one (**Table 3**). Similarly, significant contents in extractable nitrate are found in the deepest 60 - 90 cm layer of the irrigated soil confirming drainage by the added TW (**Table 3**). Although much more variable than ammonia, the nitrate contents of the soil water extracts are also systematically higher in the irrigated soil than in the non-irrigated one. Soil organic carbon contents (SOC) follow similar trends with values decreasing with depth in the irrigated soil (**Tables 3** and **4**).

According to the isohumic natureof the studied calcisol, SOC contents are elevated reaching 0.8%. Contents

Table 3. Effects of TW irrigation on some chemical properties (significant differences between irrigated soil compared to control soil showing a p < 0.01).

	Depth (cm)	Control value	Values for irrigated sites			One sample T-test	
			N	Mean	SEM	T	P (%)
pH_w	0 - 30	8.54	7	8.28	0.0764	−3.32	1.581
	30 - 60	8.85	7	8.65	0.0626	−3.17	1.924
ECs mS·cm^{-1}	0 - 30	1.38	7	4.09	0.4608	5.89	0.053*
	30 - 60	0.87	7	4.35	0.4881	7.14	0.018*
NH_4^+ cmol$^+$/kg	0 - 30	0.02	7	0.07	0.0062	8.29	0.008*
	30 - 60	0.03	7	0.06	0.0058	5.98	0.048*
NO_3^- cmol$^-$/kg	0 - 30	0.02	7	0.19	0.0276	6.26	0.038*
	30 - 60	0.00	7	0.09	0.0191	5.00	0.122*
SOC %	0 - 30	0.42	7	0.52	0.1388	1.23	26.311
	30 - 60	0.42	7	0.30	0.0772	−2.68	3.638
DOC mg/kg	0 - 30	17	7	43.92	1.6445	16.36	0.0003*
	30 - 60	-	7	-	-	-	-
K^+ cmol$^+$/kg	0 - 30	5.00	7	10.25	1.0788	4.86	0.1403*
	30 - 60	3.94	7	9.74	1.0028	5.78	0.0583*
Na^+ cmol$^+$/kg	0 - 30	0.32	7	2.59	0.2347	9.69	0.0035*
	30 - 60	0.65	7	2.44	0.2643	6.78	0.0250*
Ca^{2+} cmol$^+$/kg	0 - 30	39.62	7	40.96	2.0748	0.64	54.2487
	30 - 60	49.50	7	43.7	0.9096	−6.32	0.0732*
Mg^{2+} cmol$^+$/kg	0 - 30	1.00	7	6.44	0.2512	21.66	0.0000*
	30 - 60	1.32	7	5.63	0.1172	36.78	0.0000*

N: Number of samples; T: Observed Student statistic (T = Mean-control/SEM); SEM: Mean standard error for measurements in irrigated field; p%: Error risk *significant at p < 1%. SOC: Soil organic matter; DOC: Dissolved organic carbon; ECs: Electric conductivity of soil.

Table 4. Chemical properties of irrigated and not irrigated soil profile.

Soil depth (cm)	pH_W	pH_{KCl}	CEC (cmol$^+$/kg)	S (%)	Nt (%)	SOC (%)	C/N
Irrigated soil profile							
H1 (0 - 10)	8.54	7.79	10.01	83.29	0.0476	0.80	16.80
H2 (10 - 30)	9.15	8.04	9.97	75.67	0.0049	0.44	88.77
H3 (30 - 50)	8.56	8.06	7.81	110.87	0.0245	0.31	12.65
Not irrigated soil profile							
H1 (0 - 20)	8.55	7.9	7.70	79.29	0.0028	0.15	53.57
H2 (20 - 35)	8.85	7.78	10.15	76.55	0.0042	0.10	23.89
H3 (35 - 50)	8.83	7.79	8.12	106.74	0.0245	0.25	10.20

CEC: Cation exchange capacity; S: Saturation in exchangeable cations; Nt: Total nitrogen; SOC: Soil organic carbon.

in organic carbon are not significantly higher in the irrigated soil than in the control one (**Table 3**). However, the same values around 0.4% are found in the two studied soil layers of the non irrigated calcisol while SOC contents are generally higher in the surface layer than in the deeper one in the irrigated soil. This indicates the supply of organic carbon by the irrigation by the treated wastewater.

The examination of the DOC, C/N ratio and SUVA (**Table 5**) highlights the impact of the TW irrigation on soil organic matter quality (SOM). The C/N ratio ranges between 20 and 36 in the irrigated calcisol whereas it remains around 70 in the control site. This indicates an acceleration of the SOM humification in the irrigated field compared to the non irrigated one. The DOC content of the soil water extract, which is related to the fraction of labile carbon, is three times higher in the irrigated calcisol than in the control. By the same way, the fraction of aromatic carbon as estimated by the SUVA ratio is more important in the calcisol that has been submitted to the wastewater irrigation for 15 years.

3.3. Quantitative Supply in Macro and Micro Nutrients by the TW irrigation

Isovolumic mass balance was calculated in order to quantify the supply of micro and macro nutrients by the 15-year-long TW irrigation. Organic carbon, total nitrogen and phosphourous, as well as essential cations pools were expressed in mass per unit of soil area (**Table 6**) using thickness and bulk density of each horizon of soil as employed by Keller and Védy [21]. The percentages of pool change were calculated between the two surficial horizons in reference to the equivalent thickness of the deepest horizon of soil by assuming a conservative element. It was thus assumed that the deepest horizon of soil has not been impacted by the treated wastewater (**Table 4**).

Similarly, the equivalent deep scope of reference was found by using Si and Ti as conservative elements: 31 cm and 40 cm for the irrigated and the control area respectively. A negative pool change reflects a loss of elements from the H1-H2 horizon to the H3 while a positive value reflects a gain. However, because of the risk of error accumulation, a 15% range of uncertainty has to be considered especially for micro nutrients (Mn, Cu, and Zn). Results show no variation in the change of K, Fe and Cu between the control and the irrigated pedons (**Table 6**). Conversely, significant differences appear for other essential elements. Compared to the control pedon, the irrigated pedon shows switch in Ca and Na due to the supply of sodium by the TW. Mn and Zn are less preserved in the irrigated soil profile. Concerning amounts of SOC, P and N (**Table 6**), results indicate that the irrigation by the Sfax TW has mitigated the loss in total nitrogen and increased the pool in organic carbon and in total phosphorus.

4. DISCUSSION

4.1. Impact of Irrigation on the Soil Macro and Micro Nutrients

The quality of the Sfax treated wastewater (TW) has varied since 1984. However, values of parameters indicating the salinity (ECw, Cl) and the sodicity (SAR, Na) remained largely superior to the limits established by the F.A.O. [4] for the reuse of wastewater in agriculture. Aside salts and sodium, the Sfax TW concentrations in dissolved organic carbon, BDO_5 and suspended matter exceed the Tunisian standards for water reuse in irrigation (NT 106.03). Compared to other treated urban effluents from Sardinia, Tunisia and Jordan respectively [6,11,22], the Sfax TW also provide higher concentrations in macro nutrients such as ammonia, nitrate, total phosphorus and

Table 5. Organic pattern of the TW and of the surface soil layers (0 - 30 cm) of irrigated and not irrigated soils.

Samples	Nt (g/100g)	SOC (g/100g)	C/N	DOC (mg/kg)	Abs 254	SUVA
IWC1	0.018	0.53	29.4	43.1	0.80	1.86
IWC2	0.021	0.77	36.6	45.6	0.94	2.07
IWC3	0.017	0.34	20.0	38.8	0.60	1.54
IWC4	0.018	0.35	19.5	40.0	0.62	1.55
IWC5	0.035	0.81	23.0	43.8	0.78	1.80
IWC6	0.019	0.59	31.3	52.1	0.86	1.65
IWC7	0.014	0.28	20.0	44.1	0.71	1.62
NIC	0.006	0.42	70.5	17.0	0.17	1.02

IWC: Irrigated Wastewater Calcisol; NIC = Control: Not Irrigated Calcisol; TN: Total Nitrogen; SOC: Organic Carbon; C/N: Soil C:N Ratios. DOC: Dissolved Organic Carbon; Abs 254: UV absorbance of soil water extract at 254 nm; SUVA: Specific UV Absorption = Abs 254/DOC.

Table 6. Isovolumic masse balance of change between H1-H2 and H3 layers in the irrigated and not irrigated profiles.

Soil depth (cm)	Mg kg/m²	Ca kg/m²	Na kg/m²	K kg/m²	Mn kg/m²	Fe kg/m²	Cu kg/m²	Zn kg/m²	SOC kg/m²	Nt kg/m²	P kg/m²
					Irrigated soil profile						
H1 (0 - 10)	0.7	5.5	0.2	1.2	0.02	1.5	0.68	2.4	1.2	0.07	0.07
H2 (10 - 30)	1.4	10.7	0.4	2.5	0.04	3.1	1.14	4.0	1.2	0.01	0.11
H3 (30 - 50)	1.6	33.6	0.5	2.3	0.03	3.5	1.29	4.3	1.0	0.08	0.07
Change (%)	−17	−69	−23	0	65	−17	−5	−10	49	−35	66
					Not irrigated soil profile (control pedon)						
H1 (0 - 20)	1.0	6.9	0.3	2.1	0.05	2.5	0.96	5.0	0.4	0.01	0.05
H2 (20 - 35)	1.1	9.3	0.3	2.0	0.04	2.9	1.06	3.4	0.2	0.01	0.04
H3 (35 - 50)	0.9	11.1	0.2	1.7	0.03	2.5	0.86	2.7	0.5	0.05	0.04
Change (%)	−16	−45	−6	−8	106	−17	−12	17	−50	−86	−15

potassium (**Table 2**). Nevertheless, concentrations in nitrate, phosphorous and potassium are much lower in the Sfax TW than in untreated wastewater [5].

The quantitative mass balance results obtained in the irrigated and not irrigated fields (**Table 6**) shows the effects of the Sfax's TW properties. Indeed, the 15 year-long irrigation period does not lead to change the pools in Mg, K, Cu and Fe. Conversely, it has increased the pools of SOC (+100%) and of total P (+80%) in the topsoil layer compared to the subsoil level. Likewise, the TW irrigation has led to preservation of 50% of the pool in total N. Consequently, concentrations of available mineral nitrogen (NO_3^- and NH_4^+) have significantly been increased by the irrigation (**Table 3**). The nitrate and ammonium can either be directly brought by the wastewater or indirectly by the turnover of the organic matter and subsequent N mineralization.

In the case study, results suggest simultaneous increases in the organic pool (cf. **Tables 5** and **6** either for SOC and total Nitrogen) and in the rate of organic matter turnover (cf. **Table 4** for nitrate and ammonia and **Table 5** for DOC) indicating enhancement of the C and N mineralisation. Such impacts have already been noticed in incubation experiments [23] for loamy Appalachian forest soils (Typic Hapludult) that have been irrigated with a municipal treated wastewater for two years. The used treated wastewater contained similar nitrate (19.4 mg·L⁻¹) and total P concentration (3.15 mg·L⁻¹) than in the Sfax's TW but much lower concentration in ammonia (1.73 mg·L⁻¹). However, the enhancement of the N mineralization by TW irrigation is not systematic. Ramirez-Fuentes et al. [13] has not recorded any changes in the N turnover during the incubation of various Mexican types of soils irrigated with untreated wastewater. Magesan et al. [12] have even noticed a reduction of the amount of

leached nitrate for soil irrigated with TW. As mentioned by Herpin et al. [7] TW impacts on the nutrients turnover are mostly influenced by the difference between the C/N ratio of the soil and of the wastewater effluent.

In the studied case, the enhancement of the organic C and N mineralization can be explained by the quantitative variations in pools of micro nutrients. Mn and Zn are thus both known to be associated with soil organic matter. The rise of the SOM turnover has certainly intensified the availability of these micro nutrients turnover and facilitates their uptake by plants. Such kind of dependence between the turnover of soil organic matter and Zn translocation to the aerial parts has already been observed in other soil types (Andic soil) [24]. As opposed to the irrigated field, the control area is not used for annual crops production. The cultural differences between the irrigated and the non-irrigated field thus make it difficult to interpret data because the sum of essential nutrients uptaken by plant is certainly much lower in the control area than in the irrigated field.

4.2. Impacts of Irrigation on the Soil Organic Matter Quality

Several parameters have to be considered altogether in order to understand the impact of the TW irrigation on the quality of the soil organic matter (**Table 5**): C/N ratio, total content of dissolved organic carbon (DOC) and specific UV absorption ration (SUVA).

In the irrigated topsoil layer (0 - 30 cm), the soil C/N ratio is low compared to those of non-irrigated soil, implying an enhancement of the SOM biodegradability. This is confirmed by the amount of DOC extracted from the irrigated soil samples. The SUVA was originally designed for estimating amounts of aromatic C in DOC fraction in waters [25,26]. It has been successfully used to compare these overall quantities in humic substances of waters as well [27]. The latter aromatic C fraction can be sought for assessing water contamination by halogen-organic compounds [28]. In soil, aromatic C fraction is considered as being more stable than the labile DOC soil fraction made of proteins or carbohydrates [29]. Thus, a high SUVA value of the soil DOC fraction reveals a more important consumption of the labile C fraction by the soil microbial communities [30]. This not only indicates a higher degree of degradation of the labile organic matter, but also a change in the SOC quality. Korshin *et al.* [31] stated that the absorbance at 254 nm is even increased by the presence of polar functions in aromatic compounds, such as by hydroxyls, carbonyls, carboxylic and ester functions. An elevated SUVA ratio can therefore also outlines the SOM reactivity or its ability to form combinations with the soil mineral fraction.

As mentioned in the previous studies, the input of available micro nutrients and labile fraction of organic C by TW irrigation can over stimulate the microbial activity of the soil [12,13]. In adequate soil conditions, this results in an enhancement of the C mineralization rate, with subsequent decrease of the total SOC content in less than 4 years of TW [7]. Yet, inconsistent results are found over the long term. In Mexican Leptic calcaric soil [10], the irrigation for 50 years with untreated wastewater has depleted the SOC content in rate of 53% whereas in India (various soil types of pH_w ranging between 5.1 to 9.9), Rattan *et al.* [8] recorded an increase of the SOC content in rate of 59% after 20 years of irrigation with untreated effluents.

In the case study, the combination of values of C/N ratio, DOC, SUVA and SOC contents clearly reveals an enhancement of the organic material turnover in the irrigated soil compared to the non irrigated one. However, despite the DOC supply in large amounts by the Sfax's TW (**Table 2**), the irrigation of the studied calcisol has not systematically implied changes in the total content of soil organic carbon (SOC), compared to the non irrigated soil (**Table 3**). Huge variability of SOC contents are indeed noticed in the topsoil layer of the irrigated field, so that the mean values are not statstically different from the reference value of the non irrigated field (**Table 3**). This result suggests the influence of contradictory processes to the organic matter mineralization.

It has to be noticed that the Sfax's TW are particularly rich in salt and sodium. Chow *et al.* [25] have perceived that the soil salinisation and sodification affects the SOM structure as to decrease the DOC content in the leaching soil waters. As shown in **Figure 2**, a significant negative correlation is indeed found between the soil ECs and the DOC in the irrigated soil (R = −0.63, p < 0.05). Romkens and Dolfing [32] explained that free Ca^{2+} cations exchanged by the added Na^+ flocculate more than 50% of

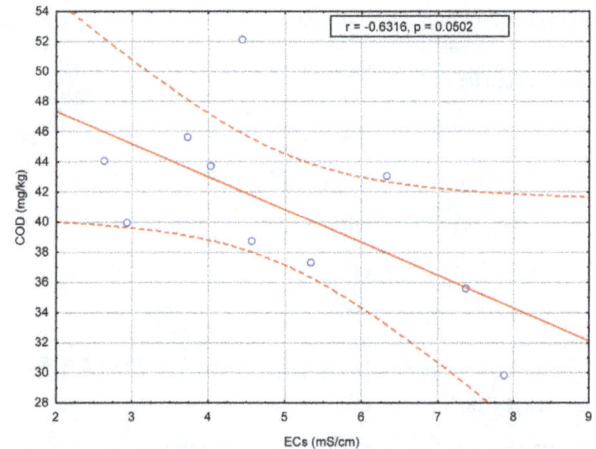

Figure 2. The correlation between dissolved organic carbon (DOC) in the water soil extracts and electric conductivity (ECs) of the surface layers of irrigated soils.

the soil DOC. Rietz and Haynes [33] added that the increase of the soil salinity leads to an inhibition of the soil microbial activity, and thus to significant decrease of the SOM mineralization. Likewise, the latter contents stabilization takes place in the studied calcisol corresponding to its mineralization enhancement by means of TW' application. Subsequently, this procedure tends to explain the variability of resulting contents of SOC in the irrigated field (**Table 5**). Over the long term, salt-inducing SOM stabilization can also lead to a decrease of crop yields [33].

4.3. The Risk of Fertility Loss over the Long Term

Over the long term, salt-inducing SOM stabilization can lead to decrease of crop yields [33]. The relationship between salinity of irrigated soils (ECs) and their content in labile carbon (DOC) thus gives an insight of further risks of soil losses of the studied calcisol fertility afterwards (**Figure 2**). The relationships between nitrate, ammonium contents and the evolution of the quality of the soil organic matter (as seen by the C/N ratio, SUVA and DOC contents in soil water extracts) suggest long term risk of degradation of groundwater quality by the leaching waters.

The irrigation clearly leads to a decrease of the contents in nitrate and ammonium in the water extracts of soils, not only at the soil surface, but also in the deepest soil layer (see **Table 3** and [15]). DOC of the applied effluent, as well as the soil DOC fraction only represent small parts (0.01% up to 0.1%) of the total SOM in soils. It is also the more mobile and the further used C fraction for heterotrophic microbial communities involved in the N turnover. Kim and Burger, [23] showed that nitrogen supply by treated effluents favored nitrification and leaching of nitrate because the amounts of available nitrogen, supplied by the TW, largely exceed the plant demand. As seen in **Tables 4** and **6**, the control calcisol is naturally poor of nitrogen. The presence of active carbonates in this type of soil also usually causes natural inhibition of microbial activity, which is indicated in this present case by a C/N ratio above 70. Moreover, the CEC of the studied calcisol remains low compared to other equivalent to calcareous soils in which CEC reach values up to 20 $cmol^+ \cdot Kg^{-1}$ [15,34]. Due to the low CEC and to the elevated irrigation rate, ammonia cations supplied by the Sfax TW have limited chances of being absorbed by the soil. In such context, nitrate and ammonium contents occurring in the deepest layer are certainly deriving straight from the applied wastewater. The two mineral nitrogen ions are leached down to the root zone and risk to reduce the upper soil layer fertility. As a matter of fact, this can also cause degradation of the soil leaching water and consequently of the free groundwater below roots and transient zones. Apart from risks of degradation of groundwater quality, there are also concerns for the condition of the crop harvest. Indeed the excess of available mineral nitrogen in soil usually lead to the breaking of forage crops during the maturation stage (oat and soghum).

The previous studies have also identified losses in essential nutrients (Ca, Mg, K, Na, Cu, Zn, Mn) and soil acidification as others risks of degradation of the fertility of alkaline calcareous soils being irrigated by domestic effluents for several decades [5,8,10,11]. In the studied case, the 15-year long wastewater irrigation has not affected the pH of the soil. This result can be explained by an important buffer capacity of the examined calcisol. Calcium represents the most abundant exchangeable base cation (**Table 3**). As seen along the two pedological pits, horizons of soil are nearly saturated in essential base cations (**Table 4**). Belaïd *et al.* [15] already mentioned significant linear correlations (p < 0.05) between the CEC and the contents of exchangeable cations added by the TW: Na^+ (R = 0.94), K^+ (R = 0.79) and Mg^{2+} (R = 0.99). The quantitative analysis of pool changes for Ca and Na in the irrigated pedon validates the exchanges of natural Ca^{2+} by the supplied Na^+ (**Table 5**). Exchangeable contents in Mg^{2+}, K^+ and Na^+, have generally increased only in the irrigated soil (**Table 3**). These cations developments are corresponding with its relatively high concentrations in the treated wastewater used for irrigation. Although total amounts in these essential cations have not significantly changed at the scale of the soil profile (**Table 5**), there is a risk over the long term of loss of these elements due to the high rate of irrigation and low soil CEC (**Table 4**). On the other hand, due to the salinity of the Sfax's TW, important amounts of the supplied Na are leached down to the deepest horizon of soil in the irrigated field initiating sodification of groundwater.

5. CONCLUSION

The Sfax treated wastewater (TW) is particularly rich in available organic carbon and mineral nitrogen. Irrigation for 15 years with the Sfax TW has not significantly changed the pH of the studied calcisol and the SOC content of the topsoil layer. The results of the present study also confirm, under field conditions, processes previously identified in laboratory experiments about the impact of wastewater effluents on the quality of the soil organic matter. From a quantitative respect, the 15-year-long irrigation by the Sfax TW has limited the loss of micro nutrients such as Mn and Zn. It has also supplied important amounts in most essential macro nutrients (P and N). The irrigation by the TW has also clearly enhanced the C and N turnover in the studied calcisol which fertility was naturally limited by the elevated content in active carbonates. If these beneficial effects could reduce the cost of mineral fertilization and aid the production of crops, the continuous use of the TW arises some questions concerning soil fertility and groundwater protection over the long term. In the case study, because of the elevated concentrations in sodium and salts of the applied TW, the beneficial activation of the microbial activity and the resulting availability of essential elements could be quickly inhibited. Moreover, due to the elevated irrigation rate and the low CEC of the soil, the TW irrigation has clearly increased the leaching of mineral nutrients such as nitrate and exchangeable K, Mg, and Na. Further irrigation, even with natural water, could stand for an imminent threat for the quality of the free watertable in the Sfax region by increasing concentrations in nitrate, sodium and salt. Apart from this problem, the excessive supply of salts and available nitrogen already constitutes a risk for the future crops production. However, since these speculations only concerned one calcisol filed, the validity of our conclusions need to be verified across a wider study area and for the other types of soils that were irrigated with the Sfax TW.

6. ACKNOWLEDGEMENTS

The authors gratefully acknowledge the staff of the CRDA-Sfax for their cooperation during site selection and soils sampling.

REFERENCES

[1] Bahri, A. (1987) Utilization of treated wastewater and sewage sludge in agriculture in Tunisia. *Desalination*, **67**, 233-244.

[2] Bahri, A. (2002) Water reuse in Tunisia: Stakes and prospects. Actes de l'atelier du PCSI, Montpellier, 1-11.

[3]   Haruvy, N. (1997) Agricultural reuse of wastewater: Nation-wide cost-benefit analysis. *Agriculture Ecosystem Environment*, **66**, 113-119.

[4]   Pescod, M.B. (1992) Wastewater treatment and use in agriculture. Bulletin FAO 47, Rome, 125.

[5]   Yadav, B., Goyal, R.K., Sharma, S.K., Dubey, P.S. and Minhas, R.K. (2002) Post-irrigation impact of domestic sewage effluent on composition of soils, crops and groundwater—A case study. *Environment International*, **28**, 481-486.

[6]   Coppola, A., Santini, A., Botti, P., Vacca, S., Comegna, V. and Severino, G. (2004) Methodological approach for evaluating the response of soil hydrological behavior to irrigation with treated municipal wastewater. *Journal of Hydrology*, **292**, 114-134.

[7]   Herpin, U., Gloaguen, T.V., da Fonseca, A.F., Montes, C.R., Mendonça, F.C., Piveli, R.P., Breulmann, G., Forti, M.C. and Et Melfi, A.J. (2007) Chemical effects on the soil-plant system in a secondary treated wastewater irrigated coffee plantationcA pilot field study in Brazil. *Agriculture Water Management*, **89**, 105-115.

[8]   Rattan, R.K., Datta, S.P., Chhonkar, P.K., Suribabu, K. and Singh, A.K. (2005) Long-term impact of irrigation with sewage effluents on heavy metal content in soils, crops and groundwater—A case study. *Agriculture Ecosystem and Environment*, **109**, 310-322.

[9]   Abbas, S.T., Sarfraz, M., Mehdi, S.M., Hassan, G. and Rehman, O.U. (2007) Trace elements accumulation in soil and rice plants irrigated with the contaminated water. *Soil & Tillage Research*, **94**, 503-509.

[10] Solis, C., Andrade, E., Mireles, A., Reyes-Solis, I.E., Garcia-Calderon, N., Lagunas-Solar, M.C., Pina, R.G. and Flocchini, C.U. (2005) Distribution of heavy metals in plants cultivated with wastewater irrigated soils during different periods of time. *Nuclear Instruments and Methods in Physics Research Section B*, **241**, 351-355.

[11] Rusan, M.J.M., Hinnawi, S. and Rousan, L. (2007) Long term effect of wastewater irrigation of forage crops on soil and plant quality parameters. *Desalination*, **215**, 143-152.

[12] Magesan, G.N., Williamson, J.C., Yeates, G.W. and Loyd-Jones, A.R. (2000) Wastewater C:N ratio effects on soil hydraulic conductivity and potential mechanisms for recovery. *Bioresource Technology*, **71**, 21-27.

[13] Ramirez-Fuentes, E., Lucho-Constantino, C., Escamilla-Silva, E. and Dendooven, L. (2002) Characteristics, and carbon and nitrogen dynamics in soil irrigated with wastewater for different lengths of time. *Bioresource Techology*, **85**, 179-187.

[14] Bouri, S., Abida, H. and Khanfir, H. (2008) Impacts of wastewater irrigation in arid and semi arid regions: Case of Sidi Abid region, Tunisia. *Environmental Geology*, **53**, 1421-1432.

[15] Belaid, N., Neel, C., Kallel, M., Ayoub, T., Ayadi, A. and Baudu, M. (2010) Effects of treated wastewater irrigation on salinity and sodicity of soils: A case study in Sfax (Tunisia). *Journal of Water Science*, **23**, 133-145.

[16] FAO (1998) World Reference Base for Soil Resources, by ISSS-ISRIC-FAO. World Soil Resources Report No. 84. Rome, 88 p.

[17] AFNOR, *et al.* (1997) Water quality, analysis methodes. **3**, 296.

[18] Orsini, L. and Remy, J.C. (1976) Utilisation du chlorure de cobaltihexamine pour la détermination simultanée de la capacité d'échange et des bases échangeables des sols. Sciences du sol, bulletin de l'AFES 4, 269-279.

[19] Metson, A.J. (1956) Methods of chemical analysis for soil survey samples. New Zealand Soil Bureau, 12.

[20] Baker, M.A., Valett, H.M. and Dahm, C.N. (2000) Organic carbon supply and metabolism in a shallow groundwater ecosystem. *Ecology*, **81**, 3111-3148.

[21] Keller, C. and Védy, J.-C. (1994) Distribution of copper and cadmium fractions in two forest soils. *Journal of Environmental Quality*, **23**, 987-999.

[22] Klay, S., Charef, A., Ayed, L., Houman, B. and Rezgui, F. (2010) Effect of irrigation with treated wastewater on geochemical properties (saltiness, C, N and heavy metals) of isohumic soils (Zaouit Sousse perimeter, Oriental Tunisia). *Desalination*, **253**, 180-187.

[23] Kim, D.Y. and Burger, J.A. (1997) Nitrogen transformations and soil processes in a wastewater-irrigated, mature Appalachian hardwood forest. *Forest Ecology and Management*, **90**, 1-11.

[24] Néel, C., Soubrand-Colin, M., Piquet-Pissaloux, A. and Bril, H. (2007) Mobility and availability of Cr, Ni, Cu, Zn and Pb in a basaltic grassland: Comparison of selective extractions with quantitative approaches at different scales. *Applied Geochemistry*, **22**, 724-735.

[25] Chow, A.T., Tanji, K.K. and Gao, S. (2003) Production of dissolved organic carbon (DOC) and trihalomethane (THM) precursor from peat soils. *Water Research*, **37**, 4475-4485.

[26] Xue, S., Zhao, Q.L., Wei, L.L., Wang, L.N. and Liu, Z.G. (2007) Fate of secondary effluent dissolved organic matter during soil-aquifer treatment. *Chinese Science Bulletin*, **52**, 2496-2505.

[27] Traina, S.J., Novak, J. and Smeck, N.E. (1990) An ultraviolet absorbance method of estimating the percent aromatic carbon content of humic acids. *Journal of Environmental Quality*, **19**, 151-153.

[28] Hassouna, M., Theraulaz, F., Lafolie, F. and Massiani, C. (2005) Characterisation and quantitative estimation of the hydrophobic, transphilic and hydrophilic fractions of DOC in soil using direct UV spectroscopy. *Geophysical Research Abstracts*, **7**, 3 p.

[29] Kalbitz, K., Schmerwitz, J., Schwesig, D. and Matzner, E.

(2003) Biodegradation of soil-derived dissolved organic matter as related to its properties. *Geoderma*, **113**, 273-291.

[30] Van Miegroet, H., Boettinger, J.L., Baker, M.A., Nielsen, J., Evans, D. and Stum, A. (2005) Soil carbon distribution and quality in a montane rangeland-forest mosaic in northern Utah. *Forest Ecology Management*, **220**, 284-299.

[31] Korshin, G.V., Li, C.W. and Benjamin, M.M. (1997) Monitoring the properties of natural organic matter through UV spectroscopy: A consistent theory. *Water Research*, **31**, 1787-1795.

[32] Romkens Paul, F.A.M. and Dolfing, J. (1998) Effect of

Ca on the solubility and molecular size distribution of DOC and Cu binding in soil solution samples. *Environmental Science and Technology*, **32**, 363-369.

[33] Rietz, D.N. and Haynes, R.J. (2003) Effects of irrigation-induced salinity and sodicity on soil microbial activity. *Soil Biology and Biochemistry*, **35**, 845-854.

[34] Jalali, M., Merikhpour, H., Kaledhonkar, M.J. and Van Der Zee, S.E.A.T.M. (2008) Effects of wastewater irrigation on soil sodicity and nutrient leaching in calcareous soils. *Agricultural Water Management*, **95**, 143-153.

11

Integrated management of natural resources in the Ecuador Highlands

Víctor Hugo Barrera[1*], Luis Orlando Escudero[1], Jeffrey Alwang[2], Robert Andrade[2]

[1]Instituto Nacional Autonomo de Investigaciones Agropecuarias, Quito, Ecuador;
[*]Corresponding Author
[2]Department of Agricultural and Applied Economics, Virginia Polytechnic Institute and State University, Blacksburg, USA

ABSTRACT

The Andean region of Ecuador is characterized by extreme poverty caused by low agricultural productivity, limited off-farm opportunities, and lack of access to markets. Poverty is related to degradation of natural resources as lagging agricultural productivity leads to incursions into fragile areas and use of erosive farming techniques on steeply sloped hillsides. Food production in fragile areas degrades soil and water resources, contributes to deforestation and loss of biodiversity, and reduces productive potential over time. This article discusses an agricultural development project designed to reduce the long-term downward development spiral in a watershed in Bolivar, Ecuador. The applied research program began with analysis of the state of soil resources, water, and biodiversity in the Chimbo sub-watershed. This information was used to design a plan with the input of local stake-holders to introduce environmentally friendly farming practices, soil and water conservation techniques, and various institutional innovations to promote resource conservation. This adaptive management program has been a solid success. This article describes the project, the challenges it faced, and how the process of adaptive management led to consensus among stakeholders about the appropriateness of sustainable management practices. We show how implementation of enhanced management practices contribute to reduced environmental vulnerability and improved welfare.

Keywords: Component; Natural Capital; Micro-Watershed; Systems Approach; Adaptive and Integrated Watershed Management

1. INTRODUCTION

The South American Andes are rife with environmental problems related to human activities in fragile ecosystems. Andean populations are among the poorest in South America and often depend on rain-fed agriculture. The Andes form the headwaters of many of the great river systems of South America, and runoff and agriculture-related pollution can have negative consequences far from their sources. Humans are encroaching into fragile high plains as population pressures at lower elevations extend the agricultural frontier. Strategies to address these problems include more environmentally benign agricultural technologies in fragile areas, intensified production in less-fragile areas to reduce pressure on more fragile areas, and raising income-earning potential through less land-intensive activities. A key is to alter human behavior. Adaptive management processes show promise as means of altering behavior to attain agreed-upon goals.

A watershed approach to natural resource management has been tried in different settings with varying degrees of success. Watersheds define natural linkages between human populations and their environments [1]. Watershed management is consistent with decentralized governance, which is gaining favor in Andean countries [2]. However, modern watershed management techniques require digitized data that are of limited availability in high mountain areas, and watershed management often requires the cooperation of competing and overlapping levels of local and regional government. A watershed management approach faces many challenges.

Any watershed approach must begin with the notion that watershed-level outcomes are products of individual decisions on fields spread across the catchment's area. These decisions reflect household livelihood strategies of allocating their physical, human, natural, and other assets to earn livings, increase well-being, and manage multiple risks [3]. Individual decisions have compound effects and impacts on aggregate economic and environmental outcomes result from a complex mosaic of economic, social, and physical networks that characterize all watersheds. The driving factor is human decision making. Effective management must identify mechanisms for changing human activities and introduce options to raise incomes while mitigating negative environmental consequences.

Integrated adaptive watershed management is a relatively new concept in Ecuador, but it provides hope that some environmental problems can be addressed through consensus building. The 1970s-era focus of tops-down watershed management has evolved over time and newer concepts recognize the holistic nature of the relationship between land use, agricultural production, natural resource conservation, and reduction of contaminants. It also recognizes that watershed outcomes result from human decisions [4].

Our integrated watershed management program in the Chimbo sub-watershed in Bolivar Province is guided by four concepts: 1) Agricultural intensification can be consistent with sustainable natural resource management [5]; 2) Sustainable agricultural practices can contribute to preservation of bio-diversity [6]; 3) Increased bio-diversity can contribute to household food security by diversifying diets and reducing risks of crop failure [7]; and 4) Even the poorest of the poor are interested in and capable of adopting environmentally friendly technologies [6].

Evidence shows that these arguments are valid in the Ecuadorean highlands [8]. Ecuador's National Autonomous Agricultural Research Institute (INIAP) has engaged farmers in the Chimbo for many years and has found farmers to be receptive to solutions to natural resource problems [8]. Over time, INIAP has created important strategic alliances and generated broad support for integrated adaptive watershed management. INIAP now combines an integrated adaptive management approach with a livelihoods focus, recognizing that any effort to improve environmental conditions must also create economic space (*i.e.* provide sufficient incomes to maintain a family) for conservation actions.

The objectives of this paper are to describe the adaptive watershed management process, obstacles overcome during its implementation, and provide a preliminary assessment of program impacts. We describe the site, present our research methods, and identify specific innovations attributable to the research. We then discuss research findings with respect to returns to management practices and describe how the recommended practices have spread over time. The paper concludes by discussing lessons learned and how the adaptive management process can be applied to other areas.

2. SITE DESCRIPTION

The Chimbo River sub-watershed covers approximately 3.635 km^2 (**Figure 1**); our program focuses on two micro-watersheds: Illangama and Alumbre. The Illangama micro-watershed covers 131 km^2 and extends from

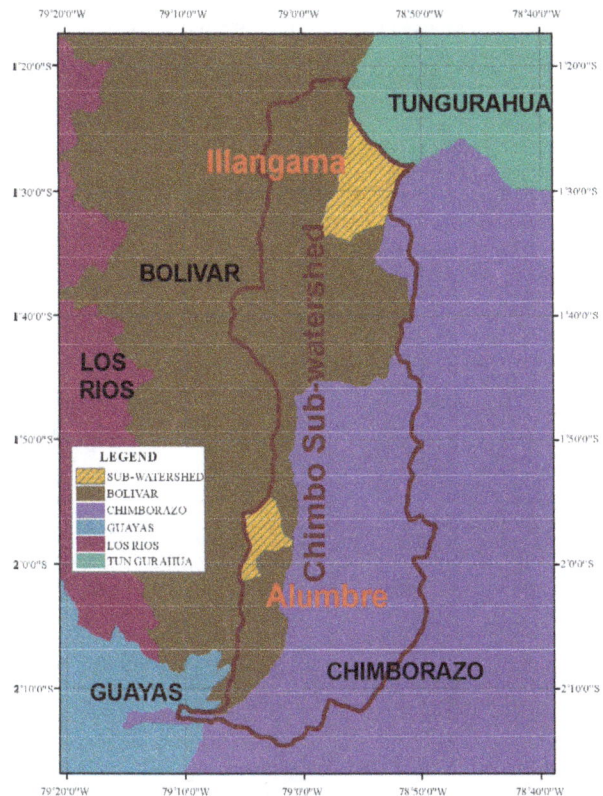

Figure 1. Location of study. The Chimbo River sub-watershed-Ecuador.

a latitude of 1°23'55.30"S through 34'4.80"S and from 78°50'39.38"W to 78°58'29.52"W. The Alumbre covers 65 km^2 and extends from 1°54'29.14"S to 2°1'36.90"S and from 79°0'22.20"W to 79° 6'4.41"W [9]. The Illangama is between 2800 and 4500 masl, with agricultural activity found between 2800 and 3600 masl. The Alumbre area ranges from 2000 to 2800 masl with agriculture throughout [10].

The watershed is characterized by social and economic conditions that threaten environmental sustainability and create long-term risks to human populations [11]. The area is among the poorest in Ecuador [12]. The river system flowing through the watershed provides about 40% of the total flow to the Guayas River, the largest system in Coastal Ecuador. Water quantity and quality has declined in recent years, partly due to upstream erosion, deforestation and expansion of the agricultural frontier into fragile highlands [13]. The highest (páramo) areas are reservoirs of clean water, and incursions into them have major downstream effects [14].

Households in the area depend on agriculture; more than 60% of the economically active population in Bolivar Province is dedicated to agriculture. Agriculture is characterized by small holdings, low productivity and environmental degradation (**Table 1**). Steep slopes, irregular and sudden rainfall, and infrequent use of cover

Table 1. Conditions in Chimbo sub-watershed.

Agro-ecological conditions	Productive activities
Illangama	
• Region: Páramo and Andean mesa	
• Life zones: Subalpine or boreal, montain, low mountain and cool temperate	• Agriculture—potato (*Solanum tuberosum*), pasture, quinoa (*Chenopodium quinoa*), faba (*Vicia faba*), chocho (*Lupinus mutabilis*) and barley (*Hordeum vulgare*);
• Temperature °C: 7 - 13	• Animal production—cattle, swine, sheep and guinea pigs;
• Altitude m: 2800 - 5000	• Tourism, artisan production, commerce, cheese production and sales.
• Annual rainfall: 500 - 1300 mm	
Alumbre	
• Region: Andean mesa and subtropical	• Agriculture—maize (*Zea mays*), beans (*Phaseolus vulgaris*), peas (*Pisum sativum*), blackberry (*Rubus glaucus*), tree tomato (*Ciphomandrea betacea*), vine tomatoes (*Lycopersicum esculentum*);
• Life zone: Low mountain and pre mountain	• Animal production—poultry, swine;
• Temperature °C: 15 - 19	• Agro-industry—medicinal plants, cacao (*Theobroma cacao*), organic coffee (*Coffea arabica*);
• Altitude m: 2000 - 2800	• Tourism, small-scale commerce, artisanal production.
• Annual rainfall: 750 - 1400 mm	

Source: INIAP-SANREM CRSP-SENACYT, 2006.

crops and other means of conserving soils cause severe soil erosion.

3. RESEARCH METHODS AND PROCESS

The program was structured around an adaptive watershed management conceptual framework. This framework begins with the watershed as a geographic entity and recognizes that actors within the watershed make decisions that affect the entire watershed. The adaptive management framework is well-known [15] but has rarely been applied in a developing country context. It begins with an assessment of conditions and identification of problems faced by actors in the watershed. Stakeholders are engaged in goal-setting, and research is designed to address obstacles to achieving goals. Research findings are then used in a participatory process with stakeholders to produce watershed plans. These plans are implemented and outcomes are monitored. Monitoring could lead to changes in plans over time, and the adaptive cycle begins again. We introduce two innovations to this framework: Plans are adapted on a regular basis as the research base and acceptance of it grows, and the land-use plans include consideration of household decision making and how decisions create impact across multiple systems within the watershed.

The household decision process reflects livelihood choices. A livelihood refers to the capabilities, assets (stores, resources, claims, access), and activities required for a means of living [16], or how labor, land, and other assets are distributed among productive and reproductive activities. The decision to adopt a livelihood is based on the household asset base; available alternatives; institutional, policy, and social environments; access to information; and the natural environment. Asset allocation decisions have effects on household wellbeing, the abil-

ity to save and invest, and the natural environment. For example, adoption of a maize technology affects labor and land allocations, income, risk exposure, and may affect erosion, runoff, and future soil quality. All these outcomes were identified as important during implementation of the adaptive management process.

The management program was built on four dimensions: communication, coordination, compromise and cooperation. The project facilitated movement along these dimensions through regular community meetings and a process of participatory research. Interactions helped generate consensus about key problems and solutions most likely to be successful.

Our assessment began with a participatory rural appraisal (PRA) to identify productive activities, assets, and perceptions about environmental conditions. The PRA was followed by a statistically representative household survey that collected information from 286 families. The survey covered household demographics, assets, sources of income, agricultural practices and others. These data were used to categorize households into livelihood typologies and conduct analysis of household decision-making processes. Survey observations were georeferenced, which allowed us to overlay survey information with agro-ecological, soils, infrastructure and other information in a GIS. The GIS was used to create thematic maps for the community engagement process, and to inform and structure research.

Emphasis was placed on identifying alternatives and evaluating them through hands-on research. For example, conversion of lands to permanent pasture or reversion to woodlands was not initially viewed as desirable. An assessment of biodiversity, together with research on alternatives to reduce erosion on productive lands, helped convince stakeholders that a combination of reversion together with adoption of erosion control practices in the most erosion-sensitive areas would help meet objectives about which consensus had been reached. Similar research efforts were undertaken to help find more effective soil fertility management regimes, more environmentally benign pest control methods, etc.

Livelihoods and their diversity: The baseline survey and information from the PRA were used to identify livelihood clusters. A quantitative hierarchical (data based) clustering method [17,18] was combined with expert opinion to create these clusters. The livelihood clusters can be thought of as groupings of households with similar asset bases and different means of combining them to earn incomes. Some clusters were exclusively agriculture, others rely on off-farm incomes, and others on remittances from outside the area [19].

Water quality analysis: Early in the process, stakeholders decided that water quality should be monitored. Monitoring results were used to evaluate impacts of land-use changes on water quality. We measured bio-indicators (macro-invertebrates), physical-chemical compositions, and micro-biological parameters [20]. This monitoring helped engage community members and built ownership of the research. Key macro-invertebrates were identified in exercises with local school children during 2006 and 2007; subsequent monitoring was incorporated into the local curriculum [20]. Monthly chemical analysis begins with samples being extracted by community members and sent to Quito for detailed analysis. Nitrate, phosphorus, total solids, temperature, pH, conductivity, fecal coliform, and total coliform are all measured. Data on rainfall and stream flows are being collected and used to calibrate our watershed models (mainly the Soil and Water Assessment Tool-SWAT).

Biodiversity assessment: The PRA indicated that stakeholders were not aware of biodiversity or its importance. Early on, an assessment activity evaluated the richness and diversity of plant and animal species. The focus was on remaining natural woodlands and areas where water recharge occurs (mainly at upper elevations). The assessment incorporated local knowledge about the value, uses and abundance of native plant and animal species. Stakeholders helped transect the study area, and collect photographic and physical evidence. Evidence was classified and categorized at the National Herbarium in Quito. Strategic transects were also undertaken in remnant woodlands and areas of high vulnerability [21].

Physical and environmental vulnerability: We stratified our on-farm agronomic research (on pilot farms) according to an index of physical vulnerability which included six parameters: slope, vegetative coverage, rainfall frequency and intensity, wind intensity, seasonal variability and soil texture. These indicators were selected following focus group discussions with technicians and local farmers. The index takes a value between 0 and 1,

with 0 signifying no vulnerability and 1 representing areas of extreme vulnerability. The index was especially useful in helping producers understand linkages between farming practices, soil loss, and subsequent off-farm damages. Farmers had their parcels classified and the index values and information on actual land use were incorporated into the GIS. This information was used to identify environmental hot-spots and to inform subsequent land use plans.

Design of environmentally friendly alternatives: The program selected 13 production systems for research on more sustainable practices. All the practices were consistent with livelihood clusters. Illangama systems revolved around a well-established potato-dairy rotation, while in Alumbre maize-beans predominate (**Table 2**). Trials were established on pilot farms to evaluate impacts on income, labor use, environmental degradation, etc. of these practices. Best Management Practices (BMP) were targeted for implementation in high-vulnerability areas (**Table 2**). Farm and watershed-wide plans were created following consensus-building exercises with stakeholders.

4. RESULTS AND DISCUSSION

Four livelihood clusters were identified in the micro-watersheds (**Table 3**). These reflect diversity in asset bases, use of productive resources, and impacts on the environment across the clusters [22].

In Illangama, most family incomes are based on agricultural production and work in agriculture off the farm. In contrast, Alumbre households use a combination of agricultural (own-farm) and diversified off-farm activities (**Table 4**). Households that are more dependent on agriculture and livestock possess and use more natural capital (mainly land) and physical capital (farm equipment and implements). Those that depend more on off-farm incomes have higher levels of human capital (reflected through education of adult members and children's participation in school). They also possess specialized skills, such as carpentry, masonry, etc. Households with greater quantities and qualities of natural and physical capital are most closely linked to agricultural markets and only infrequently participate in non-agricultural income-generating activities. They specialize in agriculture. Households with diversified off-farm incomes sources tend to be wealthiest and have fewer food security challenges.

4.1. Water Quality Analysis

Water quality analysis confirmed perceptions from the PRA that water quality is degraded in both watersheds. In Illangama, only two areas (Culebrillas and Quindigua) had water quality suitable for livestock consumption and

Table 2. Sustainable agricultural production practices evaluated.

Illangama	Alumbre
• Deviation ditches with milín grass (*Phalaris tuberosa*) and native species; • Improved rotations: Natural pasture and improved varieties of potato-barley—faba, and quinoa; • Live barriers with native species (yagual, tilo, romerillo, piquil, chachacoma, aliso, higuerón, tilo); • Chocho associated with improved pasture; • Improved planting and harvest schedule (to manage water and runoff); • Cultivation in belts (wheat, barley) with improved varieties from INIAP; • Improved pastures with forage mixes of annual rye grass, bluegrass, white and red clover, biannual and perennial rye grass.	• Belt/strip cultivation (maize and climbing beans); • Live barriers with native species (nogal, alisos, siete cueros and guarango); • Bench terraces and horticultural production; • Fruit trees on contours to form live barriers (chirimoya, lemon, orange, avocado and blackberry); • Reduced tillage of beans and peas; • Improved pastures with forage mixes of annual rye grass, bluegrass, white and red clover, biannual and perennial rye grass; • Crop rotations (improved varieties of): Hard maize—climbing/bush beans. Evaluation of promising germplasm; • Contour planting and introduction of alfalfa (*Pennisetum sp.*) in strips.

Source: INIAP-SANREM CRSP-SENACYT, 2009.

Table 3. Livelihood clusters in sub-watershed Chimbo River-Ecuador.

Livelihoods	Percent	Households	Members
Diversified households (A)	27	78	432
Engaged in agricultural markets (B)	37	105	576
Rural non-farm economy (C)	17	50	218
Agricultural consumption and wage work (D)	19	53	241
Total	**100**	**286**	**1,467**

Source: Original analysis using household survey (Andrade, 2008).

Table 4. Summary statistics for main variables and livelihoods in Alumbre.

Variables	Livelihood				ANOVA Sig.
	A	B	C	D	
Micro-watershed alumbre %	46	37	98	85	0.00***
Land holding (ha)	3.82	6.79	3.59	3.64	0.00***
Irrigation access %	23	33	6	9	0.00***
Value physical assets $	2008	2348	856	496	0.00***
Distance to closest river (km)	1.12	0.86	2.05	1.58	0.00***
Distance to closest city (km)	7.21	7.58	3.61	5.17	0.00***
Participation in civil societies %	60	55	26	38	0.00***
Family members that migrate %	71	39	54	13	0.00***
Mestizo households %	31	25	64	53	0.00***
Household size	5.54	5.49	4.36	4.55	0.00***
Household head male %	88	90	82	72	0.02**
Secondary education or plus %	65	65	66	45	0.09*
Income per capita annually $	653	785	839	288	0.00***
Expenditures per capita annually $	254	252	252	184	0.03**

Source: Original analysis using household survey (Andrade, 2008). ***Significant at less than 1% level; **Significant at less than 5% level; *Significant at less than 10% level.

no sampling site had quality suitable for human consumption (**Table 5**). Fecal coliform contamination is a severe problem throughout the watershed; in 100% of the samples, we detected fecal coliform (*E. coli*). Highest values were found in Paltabamba and Quindigua and were related to the presence of trout lagoons near the sampling site and human waste. In other areas, livestock grazing in upstream water sources and agricultural runoff are key contributors to water quality degradation. Even at very high elevations in near-pristine environments, water quality is a problem.

In the lower-elevation Alumbre, physical chemical parameters fall within normal levels and are below national limits. Indicators of total solids and turbidity indicate significant siltation from erosion. Measures of fecal and total coliform indicate severe contamination (**Table 5**). Coliforms (*E. coli and others*) were detected in all the samples. The highest concentration of fecal coliform was detected in Chillanes (2240 UFC/100 cc), because it is found at the confluence of two smaller rivers carrying human wastewaters from urban centers. The second highest coliform count was found in Pacay, mainly because of high concentrations of cattle and swine near the river and human wastewaters.

4.2. Biodiversity Assessment

The micro-watersheds are distinguished by major differences in flora biodiversity. We identified around 162 tree and bush species in the area. In Illangama and Alumbre we identified 13 and 32 species, respectively, unique to that micro-watershed. Only 17 families of species were common to both micro-watersheds. Biodiversity is far richer in Alumbre, where the warmer climate is more conducive to species diversification.

Tree and bush species form a significant part of livelihood systems, particularly in Illangama. In Illangama, families have strong interest in species that can be used as animal forage, firewood and charcoal, and varieties with medicinal properties. They also use trees as live barriers in soil conservation structures, to extract dyes, and fibers for artisanal products. Alumbre residents are less aware of the uses of tree species, are unfamiliar with local names, and are mainly interested in trees for the exploitable wood they can produce.

The water quality and the biodiversity assessments were designed to highlight the fragility in high-altitude. The initial assessments showed that farmers' voice concerns about environmental quality, recognize that their productive practices can create environmental damage, and seek alternatives to resource-mining activities. The strongest concern for the environment was voiced by community members in the upper watershed who recognize environmental change such as variable rainfall patterns, less water availability and others. The assessments also strengthened linkages between the research team and community members; the participatory means of conducting them and open sharing of findings built ownership of the adaptive management process.

4.3. Physical and Environmental Vulnerability

In order to prioritize interventions, stakeholders need information on vulnerability and its variability over space. The GIS was combined with watershed modeling to generate a map of vulnerability to runoff. Results showed that about 4000 ha in Illangama and 2000 ha in Alumbre are extremely vulnerable to environmental damage. These areas needed special attention during the planning and management phases.

The vulnerability mapping exercise uncovered evidence of conflicts between ideal and actual land uses: Some of the most environmentally vulnerable lands are currently under intensive crop production. These areas should be reserved for conservation or reforested and managed sustainably. Research thus focused on the physical and economic/social consequences of less intensive

Table 5. Analysis of variance for microbiological indicators of water quality in sub-watershed Chimbo River-Ecuador.

	Illangama			Alumbre	
Collection sites	**UFC** *E. coli* /100 cc	**UFC colif.** total/100 cc	**Collection sites**	**UFC** *E. coli*/100 cc	**UFC colif.** total/100 cc
Culebrillas (3495 m)	550 c	4243 b	Chillanes (2274 m)	2240 a	14,926 a
Quindigua (2930 m)	1075 ab	6793 a	Pacay (2240 m)	840 ab	5746 b
Quindigua (2886 m)	600 bc	4462 b	Guayabal (2193 m)	487 b	4346 b
Paltabamba (2723 m)	1244 a	6512 a			
Mean$_g$	867	5503	**Mean$_g$**	1189	8340
P	0.0136	0.0025	**P**	0.0155	0.0001

Source: INIAP-SANREM CRSP-SENACYT, 2009. Letters indicate statistical significance (P $\leq$ 0.05).

uses on vulnerable lands. Such activities are especially critical in areas of water recharge. Two main challenges constrain efforts at conservation in these areas: Lack of finance to ensure that households can survive during the transition from intensive to extensive production (such as forestry or agro-forestry systems), and low rates of return in extensive production systems. Few own sufficient land resources to sustain a family on forestry production. Part of the problem is institutional; farmers have no means of capturing the off-farm benefits from less intensive land use. The team began negotiations with downstream govern ments to examine if these governments were willing to pay farmers to avoid downstream damages. These negotiations are ongoing, but downstream siltation is increasingly associated with costly flooding and there is strong interest in finding low-cost ways to avoid these damages.

4.4. Design of Environmentally Friendly Farming Alternatives

The research identified several environmentally friendly agricultural production options for farms in the Chimbo (**Table 2**). These alternatives increase productivity, enhance soil retention and improve soil health. They were tested on model farms, where farmers participated in site preparation, cultivation, and evaluation. Field days demonstrated the practices to farmers. Substantial adoption has occurred already and, given the success of the alternatives, we expect more widespread adoption as information becomes more widespread.

The pilot sites were established on farms with average sizes of 7.5 ha. In 2006, production systems included small pine forests, natural pasture, small areas of improved pasture, and potato production together with mashua (*Tropaeolum tuberosum*). At that time, the most vulnerable areas had been devoted to crop production. The research team designed a farm use plan incorporating improved cropping systems and farming practices, pastures and woodlands (**Figure 2**).

We tested and subsequently recommended use of improved potato varieties, faba beans, barley, quinoa and chocho. Conservation agriculture practices such as improved

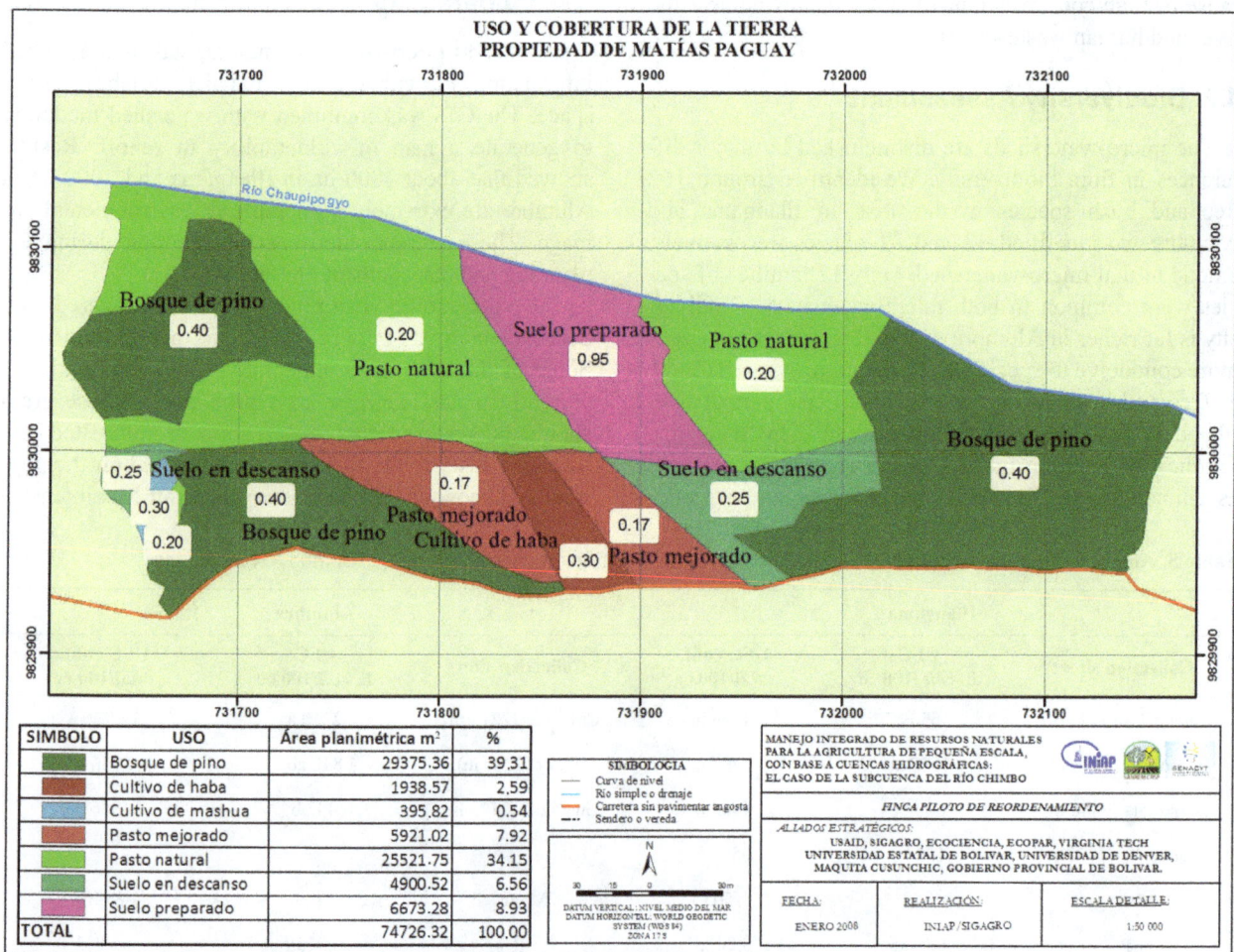

Figure 2. Land use map for model production system in Illangama, 2006.

rotations, reduced tillage, and increased groundcover were included (**Table 2**). We also recommended *in-situ* conservation of native Andean tubers such as native potatoes, oca (*Oxalis tuberosa*), melloco (*Ullucus tuberosus*), mashua and carrot (*Daucus caraota*).

As of 2010, the project had been functioning for 5 years and it was possible to evaluate its impacts. This was done by transecting the sub-watersheds to measure the extent of adoption of the practices and computing changes in farm incomes associated with the practices. **Table 6** shows an assessment of uptake of BMPs in Illangama and Alumbre.

In Illangama, net economic benefits have risen to about $ 1921 per hectare per year, an increase of about

65% compared to 2006. Improvements have resulted from incremental increases in yields of potatoes, faba beans, chocho, barley, quinoa and improved pasture. Soil management has changed dramatically as ground cover is more widespread throughout the year. Part of this change was caused by changes in relative prices; potato and other crop prices have become increasingly variable and farmers are moving toward dairy production with continuous pasture and other more environmentally suitable crops. Potato net profits have, however, grown by as much as 50%, due to improved rotations and reduced pesticide use (a major cost of production). Use of late blight-resistant potato varieties, improved soil fertility and use of better-quality seeds help lower variable

Table 6. Changes in farming practices and uptake of BMPs, 2006-2010.

Micro-watershed	BMP	2006	2010
	Area under potatoes (ha)	1.02 (100)	0.85 (87)
	Area under faba beans (ha)	0.38 (13)	0.45 (48)
	Area under chocho (ha)	0.60 (4)	0.50 (35)
	Area under barley (ha)	0.59 (5)	0.65 (65)
	Area under quinoa (ha)	0	0.15 (40)
	Area under natural grass (ha)	2.13 (48)	1.80 (35)
	Area under improved pasture (ha)	1.51 (71)	2.25 (85)
Illangama	Milk production (l/day)	18	50
	Potato yields (t/ha)	8.35	13.09
	Faba beans yields (t/ha)	0.45	0.95
	Chocho yields (t/ha)	0.60	1.15
	Barley yields (t/ha)	0.73	1.30
	Quinoa yields (t/ha)	N/A	1.20
	Pesticide used in potato production ($/ha)	289	186
	Net benefits ($/year)	1163	1921
	Area under white maize (ha)	1.85 (72)	1.85 (85)
	Area under yellow maize (ha)	1.18 (5)	1.18 (15)
	Area under maize/beans (ha)	4.48 (27)	3.00 (20)
	Area under beans (ha)	2.17 (20)	1.50 (60)
Alumbre	White maize yields (t/ha)	0.44	1.10
	Yellow maize yields (t/ha)	0.41	0.97
	Maize/beans yields (t/ha)	0.57	0.91
	Beans yields (t/ha)	0.40	0.88
	Net benefits ($/year)	898	1629

Source: INIAP-SANREM CRSP-SENACYT, 2006-2010. Illangama: Percentage of farmers with the crop in each year (sample size: 117 in 2006; 250 in 2010). Alumbre: Percentage of farmers with the crop in each year (sample size: 169 in 2006; 80 in 2010).

costs. Milk production per land unit grew by 122% due to improved forages, and better sanitation and feeding practices. Food security has also improved. Diversified grain sources broaden the dietary base, reduce risks from dependence on single crops, and increase energy and protein intake.

The data indicate impressive trends toward more diversified production, with increases in relatively new (to the area) products such as quinoa. Quinoa production has emerged, and the crop provides nutrition for home consumption and high prices in the market. As a result of all these changes, erosion is being reduced and water quality is improving.

In Alumbre, net benefits from agricultural production increased by 81% to $1629 per hectare per year in 2010. This increment was a product of increased yields of white maize, yellow maize and beans, resulting from improved management practices. The main engine was introduction of improved varieties, and more intensive management concentrated in less vulnerable and more productive areas. Planting densities have increased and integrated pest management practices have reduced input costs. Increased agro-diversity and lower profit risks (due to fewer purchased inputs) have also increased food security.

Use of vulnerability maps to guide land use planning has reduced production on most vulnerable lands and improved ecosystem services. Indicators of biodiversity, soil retention and water quality have improved alongside improvements in agricultural profitability. Farmers now concentrate productive activities on the most fertile and least vulnerable lands. Yield improvements and cost reductions allow farmers to earn higher incomes and simultaneously improve environmental conditions. Ability to observe farming practices on the pilot farms has built confidence in the new practices and they have naturally

spread throughout the watershed. Concurrently, the study of biodiversity raised consciousness about the value of native species and led to planting and maintenance of these potential sources of biodiversity. These actions have improved environmental conditions and water availability

Prior to 2006, conservation practices in the area did not exist. Now, various practices are widely found, such as improved crop rotations, strip cultivation, deviation ditches, contour plowing, and use of live barriers. An indigenous innovation has led to the protection of deviation ditches with various local species. These include milín grass and native plants such as Quishuar, Yagual, Chachacoma, Romerillo, Aliso, Pumamaqui, Lupinus, Piquil. Contour cultivation is also widely practiced now in both watersheds, irrigation water management has improved and actions have been taken to protect areas of water recharge. This protection has involved replanting many of the native plants metioned above.

Table 7 summarizes results of the 2010 evaluation. Farmers in the Illangama watershed were more likely to apply all natural resource management methods, except for green fertilizer. Differences over time of use of conservation methods are statistically significant.

4.5. Participatory Planning

Our team identified local stakeholders, institutions and government and non-government partners to engage in participatory planning. Participants identified research themes and designed research activities and collaborated in on-farm trials. The process included meetings, workshops and information exchanges. Stakeholders immediately recognized the need for coordinated cross-sectorial actions and institutional change to increase the value of natural resources. A regular meeting of a project steering group was held; the group identified and promoted the

Table 7. Adoption of improved management practices, 2010.

Method	Alumbre		Illangama		p-value
	% Using	S.E.	% Using	S.E.	
Strip cultivation[*]	3.77	1.23	21.25	4.60	0.0004
Deviation ditches[*]	5.02	1.42	31.25	5.21	0.0000
Contour plowing[*]	6.28	1.57	22.50	4.70	0.0015
Crop rotation[*]	59.41	3.18	92.50	2.96	0.0000
Live barrier[*]	24.69	2.79	63.75	5.41	0.0000
Reduced till[*]	26.36	2.86	76.25	4.79	0.0000
Green fertilizer	7.11	1.67	6.25	2.72	0.7873
No. respondents	239		80		

[*]Denotes statistically significant differences between the watersheds (0.01 level).

idea of integrated adaptive management. This group engaged local and Provincial Governments who are full partners in the process. The Provincial Government created a new unit for environmental management and linkages across government units has facilitated coordinated actions; our technical team has trained the Government's technical team and this strategic alliance has been strengthened over time. The alliance is important because the Provincial Government bears responsibility for creating and enforcing the regulatory and legal regime.

Our research agenda was arrived at after an arduous process of building consensus among stakeholders. Probably the most valuable research output at the start of the process was to help stakeholders understand and appreciate the value of their natural resources. This newfound appreciation of value has strengthened incentives for actions to promote soil retention and health, and to use native species as a contributor to this conservation. Native species of trees and bushes have been widely incorporated into live barriers to reduce water and wind erosion, and as lining biomass for deviation ditches. The team also helped identify a major source of reduced water supply and quality: Incursions into the upper páramo areas. As a result, we have built support for increased intensification at lower elevations and a sense of community-wide disapproval for those who exploit the pristine higher-elevation areas of the Illangama. Social pressure is having an effect.

Social capital has been strengthened in many ways. The participatory planning process is strengthening social networks in the region. In addition, training in biodiversity, natural resource valuation, and natural resource management has built networks of activists in both micro-watersheds. Efforts to understand the potential benefits of higher-valued market chains and obstacles to participation in them have helped identify how networks of producers can have more effects than individual actors. Subsequent efforts to build these networks have also reinforced local social capital.

5. CONCLUSIONS

The participatory land-use planning process led to a functioning watershed planning model. A key component of success was investment in agricultural and other research to increase incomes. Through this research, we are creating the economic space to address longer-term problems associated with natural resource degradation. The participatory process has built confidence among stakeholders who now largely buy into our larger program. Exposure to new technologies to raise incomes has helped build participation.

The nature of the watershed and the variety of stakeholders pushed us toward casting a large net; we made major efforts to involve institutions with any presence in the watersheds. Some of the less-recognized assets of our program: technical expertise, knowledge of successes and failures from elsewhere in Ecuador helped build bridges to institutions that in other cases might be less receptive to innovative ideas. For example, our ability to provide training for units of local governments helped legitimize our presence in the eyes of this important stakeholder.

Our adaptive watershed management project is less a political process and more a process of social learning and empowering community actions. A key barrier to effective local action was the Balkanization that predominated prior to our project: Different actors and stakeholders did not communicate and were even less likely to undertake coordinated actions. By casting a wide net and strongly encouraging participation, the project broke many of the barriers to collective action. Our training and participatory research efforts helped bring down these barriers and created a common consensual set of knowledge about actions. These programs brought down barriers and increased the capability to effect change.

Our most important lesson was the necessity to build consensus and engage stakeholder groups. The effort required to reach this point was substantial and involved tireless exercises in outreach, networking and stakeholder engagement. This process is long and one that may not show immediate results. However, the impact it eventually created was worth the effort. Communities in the watersheds now actively participate.

6. RECOMMENDATIONS

A first recommendation is to continue participatory consensus building. Efforts to engage community members in field experiments, in water quality data collection, and in the biodiversity assessment were especially helpful in gaining local ownership and building credibility for the entire project. We began with small steps and a massive amount of participation, and built in complexity as the process evolved. Wide participation also lowers labor costs associated with project activities.

A second recommendation is to incorporate a multisectorial approach. The primary objective of the project was to create sustainable means of natural resource management through adaptive watershed management. However, we documented the importance of research for more profitable technologies, for increased returns to producers through higher-valued chains (such as dairy production in the Illangama watershed), and for enhanced social and institutional capital. All these actions created space for and consensus about the need to conserve natural resources. Without them, the approach would not have been successful. We are still seeking a means of increasing farmer capture of value from off-farm benefits to on-farm investments (reduced erosion, enhanced water quality, enhanced biodiversity).

A final recommendation is that researchers should take risks. We began the project with skepticism about the adaptive management approach. We were concerned about the ability to generate data and make use of high-tech tools such as GIS and SWAT. We wondered whether the community would accept research results from exotic tools. We found that although the data base is still inadequate and the SWAT results do not reflect reality as well as they might, the results are being used. GIS is probably the most effective tool in our arsenal for presenting results to stakeholders; they understand and can effectively interpret the maps we produce. These tools, while exotic, are not beyond the reach of a modest program in highland Ecuador.

7. ACKNOWLEDGEMENTS

This project was part of the SANREM CRSP, supported by the United States Agency for International Development and the generous support of the American people through Cooperative Agreement No. EPP-A-00-04-00013-00. We thank the people of the communities of the Alumbre and Illangama sub-watersheds for their participation in the project.

REFERENCES

[1]   Doolette, J. and McGrath, W. (1990) Strategic issues in watershed development. In: Doolette, J. and McGrath, W., Eds., *Watershed Development in Asia*, World Bank Technical Paper No. 127, Washington DC.

[2]   Guerra-García, G. and Sample, K. (2007) Policy and poverty in Andean countries. International IDEA, Stockholm.

[3]   Siegel, P. and Alwang, J. (2004) An asset-based approach to social risk management: A conceptual framework. Social Protection Discussion Paper 9926. Social Protection Unit, Human Development Network, World Bank, Washington DC.

[4]   Dourojeanni, A. and Jouravlev, A. (2001) Crisis in governability and water management: Challenges to implementation of agreed-upon recommendations. Economic commission for Latin America and the Caribbean (CEPAL). Series on Natural Resources and Infrastructure, Santiago.

[5]   Scherr, S. and McNeely, J. (2004) Reconciling agriculture and wild biodiversity conservation: Policy and research challenges. In: *Conservation and Sustainable Use of Agricultural Biodiversity: A Sourcebook*, CIP-UPWARD, Lima, 46-55.

[6]   Scherr, S. and Downward, A. (2000) "Spiral? Recent evidence on the relationship between poverty and natural resource degradation. *Food Policy*, **5**, 479-498.

[7]   De Marco, J. and Monteiro Coelho, F. (2004) Services performed by the ecosystem: Forest remnants influence agricultural cultures' pollination and production. *Biodiversity and Conservation*, **13**, 1245-1255.

[8]   Barrera, V., León-Velarde, C., Grijalva, J. and Chamorro, F. (2004) Management of the potato-pasture production system in Andean Ecuador: Technology options. Editorial ABYA, Technical Bulletin INIAP-CIP-PROMSA, Quito, 196 p.

[9]   SIGAGRO (2008) Information on micro-watersheds in Alumbre and Illangama, Ecuador. Geographical Information System, Quito.

[10]   INIAP (2008) Geographical Information System for the Chimbo River, Bolivar-Ecuador. Climate Monitoring. INIAP, Quito.

[11]   Barrera, V., Alwang, J. and Cruz, E. (2008) Integrated management of natural resources for small-scale agriculture in the Chimbo river, Ecuador-lessons learned. INIAP SENACYT Information Bulletin No. 339, Quito.

[12]   INEC-MAG (2002) III National Census of Agriculture: Resultados nacionales, provinciales y cantonales, instituto nacional de estadísticas y censos y ministerio de agricultura y ganadería. Quito.

[13]   GPB (2004) Strategic plan for development, provincial government of Bolivar. Planning office, Guaranda, 224 p.

[14]   Gallardo, G. (2000) Final report for the integrated management of natural resources in Watersheds. Quito, 220 p.

[15]   Salafsky, N., Margoluis, R. and Redford, K. (2001) Adaptive management: A tool for conservation practitiones. Biodiversity Support Program, Washington DC.

[16]   Chambers, R. and Conway, G. (1992) Sustainable rural livelihoods: Practical concepts for the 21st century. IDS Discussion Paper 296. Institute for Development Studies, Brighton.

[17]   Aldenderfer, M. and Blashfield, R. (1984) Cluster analysis; Series: Quantitative applications in the social science. SAGE University Paper, Beverly Hills.

[18]   Ward, H. (1963) Hierarchical grouping to optimize and objective function. *Journal of the American Statistical Association*, **58**, 236-244.

[19]   Alwang, J., Barrera, V., Andrade, R., Hamilton, S. and Norton, G. (2009) Adaptive watershed management in the South American highlands: Learning and teaching on the Fly. In: SWCS, Ed., *The Sciences and Art of Adaptive Management: Innovating for Sustainable Agriculture and Natural Resource Management*. Ankenny, 209-227.

[20]   Calles, J. (2007) Bioindicadores terrestres y acuáticos para las microcuencas de los ríos Illangama y Alumbre, provincia Bolívar. EcoCiencia, Quito, 30 p.

[21]   Cruz, E. (2009) Study of biodiversity in microwater- sheds of Illangama and Alumbre, Ecuador. INIAP, Quito, 26 p.

[22] Andrade, R. (2008) Household assets, Livelihood decisions and well-being in chimbo ecuador. Master of Science Thesis, Department of Agriculture and Applied Economics, Virginia Tec

12

Relationship between cation exchange capacity and the saline phase of Cheliff sol

Djamel Saidi

Biology Department, Faculty of Science, Hassiba Ben Bouali University, Chlef, Algeria

ABSTRACT

T Measurements of the cation exchange capacity (CEC) show significant soil properties, in particular its ability to retain the cations because of their mobility in the soil. Thirteen soil samples rich in electrolytes of the Cheliff plain (Algeria) were analyzed in order to measure their CEC and to draw up the existing relationship between texture, organic matter content and pH. In calcareous soils, the CECe values are always higher than those measured at pH 7. Regression equations using the percentages of organic carbon and clay as independent variables would make it possible to estimate 90% of the variability of the CEC measured in the ammonium acetate buffered at pH 7 and 89% of the variability for that measured at the pH of the soil. These percentages are particularly useful due to the fact that they make it possible to estimate the CEC of the soil according to the pH only starting from the organic matter and texture. The correlations between the salinity indices, the parameters of the saline phase and the physical properties, show that the cobalt-hexamine method makes it possible to characterize the soil of this plain with more precision than the Metson method. It constitutes a means for following-up the chemical quality of the soil. The Metson method makes it possible to approach the reactivity of the soil in relation with the geometry of the components. The measurement of the CEC at pH 7 makes it possible to envisage the water content at the permanent wilting point of the plants. Finally, it is noticed that a sodisation of the adsorbing compound, which consequently generates a reduction in the structural stability and a reduction in the infiltration always leads to the salinity in these soil types.

Keywords: CEC$_{Metson}$; CECe; Salinity Index; Salinity; Sodicity; Physical Properties

1. INTRODUCTION

Currently, it is known that the reaction of a saline soil depends upon the amount and the nature of the clay fraction as well as the salinity level and the nature of the common cations and anions in the soil solution [1-6]. The fine clay fraction essentially ensures the regulation of the physicochemical phenomena. This fraction plays an important role not only in the water retention and the soil structure, but also in the retention and the bioavailability of the nutrient elements that are essential to plants. The surface properties of the soil components can be characterized using two types of data, namely the cation exchange capacity (CEC) and the specific surface of the clay fraction (SS). Generally, the total surface is determined using a strongly adsorbed, polar organic molecule such as ethylene glycol monoethylether (EGME) [7]. The cation exchange capacity of a soil measures the surface electric charge of soil components [8,9]. Both CEC and specific surface of the soil were used as a predicting criteria tool to evaluate the properties of the soil components. The water retention to clays is closely related to the cation exchange capacity and specific surface [6,10] showed that the cation exchange capacity was a good estimate of the water properties in the clay horizons of low organic matter content. The CEC measurement has to be done at a specific soil solution pH in order to avoid the variable charges due to the change in soil pH [9]. In the usual methods, the CEC is measured at pH = 7 (Metson method) for the lightly acid or neutral soils and at pH = 8.2 (Bower method) for the alkaline soils. Furthermore, in the presence of differently soluble salts, these methods seem not to be accurate. Under salinity onditions, it is difficult to make a monoionic absorbent on the soil material [8,11].

The objective of the present investigation is to compare the results obtained using two standardized methods of the cation exchange capacity (CEC) and their importance in the calcareous, saline soils of Cheliff plain area, north Algeria, which have clay loam texture. The relationships between the CEC and the salinity, the sodicity and some physical properties of the soils will be also evaluated.

2. MATERIALS AND METHODS

2.1. Study Area

The study was conducted in the Mina area (lat. 36°10'N, long. 00°30' & 1°20'E) of Algerian lower Cheliff Valley (**Figure 1**). The specific climate is semi-arid where the summer is very hot and the winter is cold. The mean annual rainfall in this area is 350 mm and the mean annual temperature is 18°C with only some major fluctuations through the year. The altitude at this plain is about 70 m above the sea level and the parent material of the soil is alluvium. The studied soils are pedologically young and developed from rich clay calcareous material [12-14]. Soil particles that have a diameter < 2 μm are mainly illites accompanied with a mixture of clay minerals of smectite, kaolinite and of chlorite [15].

2.2. Soils Samples

Soil samples were collected from the upper layer of the Mina fields, Algerian Cheliff plain. Thirteen samples were taken from the surface horizon (0 - 30 cm) of each cultivated soil and analyzed using the standardized methods. The granulometric analysis was carried out without decalcification after the dispersion with sodium hexametaphosphate. The percentage of organic carbon is given according to the Anne method; the pH of the soil is measured in a 1:2.5 of soil to water suspension. Total calcium carbonate is obtained by a volumetric calcimeter of Ber-

nard. The specific surface was measured using ethylene glycol mono ethylether (EGME), according to the protocol developed by [7]. Both cation exchange capacity and the cation exchange extraction were determined in an accredited analysis laboratory of INRA, Arras, France at its usual soil pH and using cobalt hexamine (Cohex) trichloride, $[Co(NH_3)_6]Cl_3$ as an exchange solution [8] and at pH 7 using ammonium acetate as an exchange solution, which was proposed by [16]. These two methods are standardized [17].

2.3. Water Retention

The determination of water retention was related to the fragments size (5 - 10 cm³). The apparent density of the fragments was measured using a petrol method [17,18]. Six water contents were used as metric potential values of −10 kPa (PF = 1), −330 kPa (pF = 2.5), −1000 kPa (pF = 3.0), −3300 kPa (pF = 3.5), −10000 kPa (pF = 4.0), and −15000 kPa (pF = 4.2). Then measurements were carried out using pneumatic devices. This device makes that it is possible to put at balance 30 to 40 bounds on the balance at the same time in only one cell [19]. The water content is measured after 7 days of the setting to the balance at the selected pressure and then, the cell content is ovendried at 105°C for at least 24 hours.

2.4. Structural Stability

The aggregate stability was measured on a diameter of 3 - 5 mm aggregates according to the method proposed by

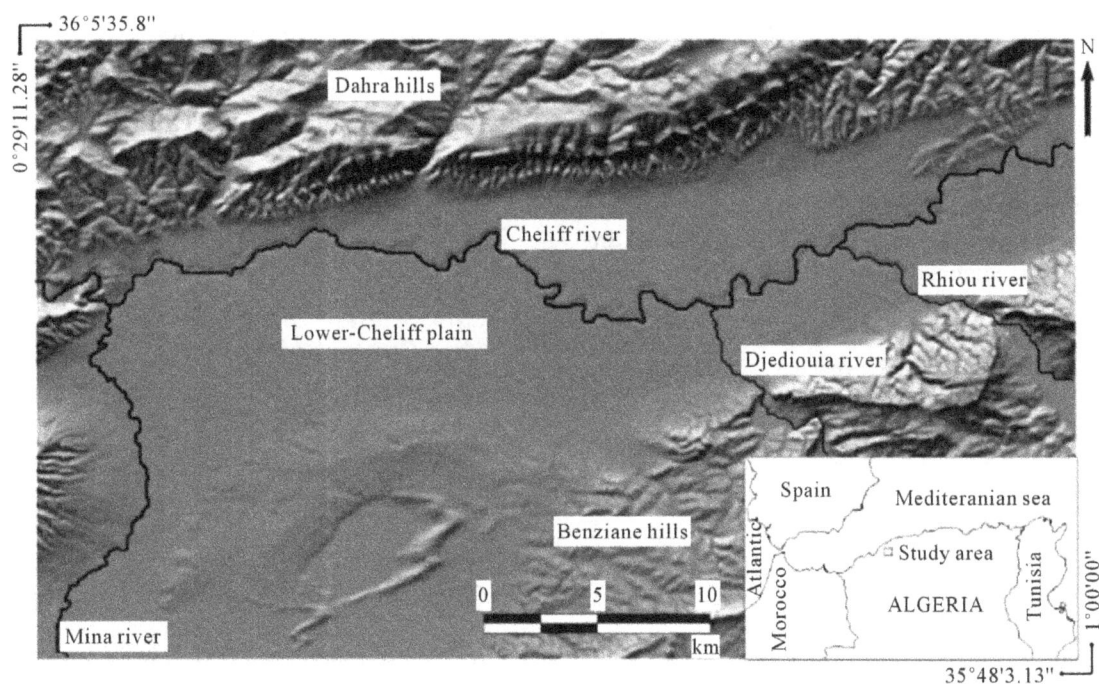

Figure 1. Location of the study area in northern Algeria, showing the lower-Cheliff plain.

Le Bissonnais [20]. Its objective is to give a realistic description of the behaviour of soil materials subjected to the action of the rain and to allow a relative classification of materials with respect to this behaviour. In summary, the method is designed to distinguish between the various mechanisms of breakdown: slaking due to fast wetting, micro-cracking due to slow wetting and mechanical breakdown by stirring of prewetted aggregates. The laser particle-measurement instrument of Mastersizer S (Malvern Ltd Instrument) was used. It is based on the Mie theory of light diffusion to calculate the diameter of the particles. To apply this theory, it is enough to choose the structure type of the sample and nature of liquid phase in the suspension. In this study, it is supposed that the particles are in water suspension. A selected type of lens (300 RF) makes that is possible to measure a diameter between 0.05 and 880 μm. The fraction > 500 μm was obtained by the classic sieving method. The mean weight diameter (MWD), which is the sum of the percentage fractions of the soil remaining on each class multiplied by the mean aperture of the adjacent meshes, is calculated from the fragment-size distribution. The MWD ranges between 0.001 mm and 3.5 mm. An average MWD is calculated to summarize the results of three treatments. This non-destructive technique characterized by using the resulting fragment-size distribution with the laser diffraction instrument has the advantage to ensure the repetitively of measurements acquired in order to control their stability. On the other hand, it allows to measure the disintegration kinetics of the soil structure according to the time of agitation. In this case, the MWD (bis) is the result of calculation after a 5 minutes interval between the first and the second measurement.

2.5. Soil Infiltration Measurement under Simulated Rain

2.5.1. The Rainfall Simulator

The artificial rain was implemented using a sprinkling device. The rain simulator is established according to the model designed by Asseline and Valentin [21]. It consists of a watering system fixed at pyramidic tower of 4 metres in height and protected from the wind action by a removable cover. Sprinkling is ensured by a metering jet (tube no. 6540) assembled on an oscillating arm whose movement is printed by an electric motor. This infiltrometer with sprinkling makes that is possible to simulate rain of controllable intensities on a measurement seat. The range of available intensity varies from 20 to 150 mm/h. The protocol of simulation includes several rain tests of an average intensity of 30 mm/h (± 2 mm·h^{-1}) during one hour and half is 45 mm of precipitation. It corresponds well to that meets in the zone of study for a natural rain event of decennial recurrence. The flux of

water through the sample was measured every 5 min., and the ratio between final infiltration and the rainfall (the final infiltration coefficient) was calculated for the experiment. The treatments were replicated three times for each soil.

2.5.2. Preparation of the Samples

The soils samples of approximately 10 cm distance in-between are distributed on a PVC plate measuring 50 × 50 cm. The soil samples of a volume of 78.5 cm^3 resting on a bed of 1 cm of calibrated and washed sand are put in cylinders made of PVC of a diameter of 5 cm and a 5 cm height. They are transparent in order to be able to control the moistening visually. They are perforated at their bases to be used as holes of evacuation for the infiltration measurement. The soils are initially saturated by the base. They are then exposed to a rainfall simulated by using distilled water until the rate of stabilized infiltration is obtained. The standard conditions of the simulator operation for each episode rain are calibrated and checked. The analysis of each group of data primarily relates to the evaluation release time of the infiltration and thereafter, to the measurement of the water volume infiltrated for each rainy event.

3. RESULTS AND DISCUSSION

3.1. Statistical Characterisation of Sample Population

The general statistics of the physical and chemical characteristics of the soils are given in **Table 1**. They indicate the prevalence of soils that have basic pH and are generally calcareous, with contents of organic matter ranging between 0.5% and 3% and a clay loam texture. The samples are characterised by averages raised for electrical conductivity and the quantity of total soluble cations translating a strong variability of the results in this area. In the calcareous soils, the value of the CECe measured at the soil pH is always higher than that measured at pH 7.

3.2. Importance of CEC Measurements in the Context of the Saline Phase

The cation exchange capacity (CEC) constitutes the privileged characterization tool of the soil surface properties. Two methods were tested in this study. The first one is based on the determination of the CEC at the natural pH of the soil (CECe). The value of the CEC then depends upon the geochemical environment, in particular the pH. The second method determines the CEC at pH 7 which makes to compare materials on the standard bases (CEC$_{Metson}$). The results of **Table 1** show that the CEC values vary from 17 to 25 cmol·kg^{-1} when it is measured at the natural pH of the soil and from

Table 1. General statistics of various physical and chemical characteristics of the soil Cheliff plain samples.

Characteristics	Minimum	Maximum	Mean	Standard Deviation
pH$_{(1/2,5)}$	8.00	8.33	8.17	0.009
SAR[a]	0.8	6.27	2.09	1.64
CE[b] (dS/m)	1.93	41.33	12.48	12.73
CaCO$_3$ (%)	17.53	22.1	19.77	1.42
MO[c] (%)	0.53	3.38	2.37	0.68
Clay (%)	40.80	55.37	46.14	4.35
Silt (%)	35.49	52.15	42.21	4.02
Sand (%)	4.01	18.6	11.65	5.08
ESP[d] (%)	3.24	52.16	14.03	14.36
Bulk Density (g/cm^3)	1.32	1.79	1.66	0.12
FIR[e] (%)	3.51	7.02	4.39	1.01
MWD[f] (mm)	0.26	0.72	0.41	0.15
CECe[g] (*) (cmolc/kg)	17.1	24.6	19.58	1.89
CEC Metson[g] (**) (cmolc/kg)	12.5	19.7	15.13	1.93
Total Cations (*) (cmolc/kg)	19.1	107.69	38.7	28.61
Total Cations (**) (cmolc/kg)	53.6	227.85	85.29	55.86
Specific Area (m^2/g)	310.08	359.79	327.8	15.46

(*): Chlorure de Cobaltihexammine Method; (**): Metson Method; [a]: Sodium Absorptio Ratio SAR = $Na^+/((Ca^{2+} + Mg^{2+})/2)$, [ ]: meq/l^{-1}; [b]: Electric Conductivity of Saturated Paste Extract (CE); [c]: Organic Matter (OM); [d]: Exchange Sodium Percent ESP = $Na^+ \times 100/CEC$, [Na^+]: Sodium adsorbed en cmolc/kg, CEC: Cation Exchange Capacity (cmolc/kg); [e]: Final Infiltration Rate; [f]: Mean Weight Diameter; [g]: Cation Exchange Capacity (cmolc/kg).

12 to 20 cmol·kg^{-1} when it is measured at the pH 7. The difference between the cobalt-hexamine method of [11] and the Metson method [16] takes account of the pH reached during the initial saturation of samples [9]. In the first method, it is the usual pH, that the soil approaches, but in the second one, the soil pH equals to 7. Comparing the values of the two methods in **Figure 2**, it is realized that the CEC values with cobalt-hexamine (CECe) are higher compared to those of the Metson method (CEC7). However, the exchangeable bases of the Metson method are higher compared to those of the cobalt-hexamine method (**Table 1**). According to [22], this difference resides in the fact that during measurement with pH equal to 7, a part of the carbonates is dissolved. The result is that the cations extracted come in particular from the setting in calcium solution. On the other hand, the CEC with the cobalt-hexamine measured at the pH of the soil (CECe). This makes that it is possible to extract the exchangeable cations and those that are present in the solution from the soil without the effect of carbonate dissolution. Moreover, the CEC values of the soil depend, at the same time, upon the mineralogical nature of clays and the electric surface charges developed

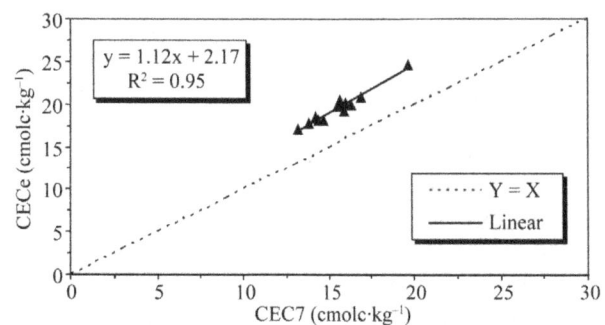

Figure 2. Relationship between the cation exchange capacity measured with the cobalt hexamine trichloride method and with the Metson method.

by the organic matter. The multiple linear regressions (**Table 2**) show that the CEC of the clays and the organic matter are regarded as additives [23,24]. An average value of CECe of 41.26 cmol·kg^{-1} is obtained for clays of Cheliff plain. Clay content explains 65% of the CEC variability of the soil measured with the cobalt-hexamine. On the other hand, when one uses the Metson method, the average CEC value of clay is 39.94 cmol·kg^{-1} and clay content explains only 60% of the variability. The

Table 2. Linear regression between the CEC of the soil and their physical and chemical characteristics.

ΔpH	Linear regression	r^2	Estimate standard error	Standard Error on r^2
$\Delta pH = (pH_{sol} - 8)$	CEC cobaltihexammine $CEC_{sol} = 41.26 \times 10^{-2} A + 281.24 \times 10^{-2} C - 1942.23 \times 10^{-2} \Delta pH$ Avec $r^2 = 0.89$ (dont A = 0.65; C = 0.004 et ΔpH = 0.23)	0.89	0.9591	0.1430
$\Delta pH = (pH_{sol} - 7)$	CEC Metson $CEC_7 = 39.94 \times 10^{-2} A + 295.46 \times 10^{-2} C - 628.88 \times 10^{-2} \Delta pH$ Avec $r^2 = 0.90$ (dont A = 0.60; C = 0.15 et ΔpH = 0.15).	0.90	0.9335	0.1390

A = clay%; C = organic carbon%.

organic matter explains a large part of variability for Metson method than cobalt-hexamine method. Differences between the CEC values of the organic matter are obtained. These last differences (on the basis of 58% of carbon) are evaluated, with 1710 mmol(+)·kg^{-1} for the Metson method and 1630 mmol(+)·kg^{-1} for that with the cobalt-hexamine. However, as the organic matter content is weak, its weight is not as significant as that of clay. This explains 0.4% and 15% of the variability of the cobalt-hexamine CECe and the CEC Metson, respectively.

It appears that the value of the CEC results from the cumulated properties of the clay and the organic matter. The fact that there is a measurement difference between both methods according to their implementation (pH of the soil, CECe; and pH 7, CEC7), makes the possibility to coarsely characterize the state of reactive surfaces and their contribution to the pH of the sample. In this direction, for the Metson method, the used (ΔpH) corresponds to the difference between the pH of the soil and the pH of the extraction solution. With regard to the cobalt-hexamine method, the used ΔpH corresponds to the difference between the real pH and the minimum pH found in the studied soils. The regression equations indeed use a differentiation factor of the (ΔpH) as independent variable to justify 23% and 15% of the CECe and CEC7 variability, respectively. This CEC variation due to (ΔpH) is attributed to the action of protonation/deprotonation. Shortly, to evaluate the CEC by these two methods, one can realize that the independent variables explain 89% and 90% of the CEC variability. The CECe that carried out with the pH close to the soil is less influenced by the organic matter, since its contribution is, in this type of soil, often unimportant. It is justified only by 0.4% of variability due to organic matter and would be dependent on the clay which presents variable loads of edge according to the pH. The CEC at the standard pH 7 (Metson method), is justified by 15% of variability due to the dissociation of the phenol groupings of the organic components at pH 7. According to regression equations, these various ratios are useful because they make it possible to estimate the CEC of the soil at any pH only starting from the organic matter and of texture. These values agree with the bibliographical data [23].

3.3. Salinity Index Determination

According to the various CEC results obtained and the exchangeable cations, which were extracted using ammonium acetate and the cobalt hexamine (Cohex) trichloride, two salinity indices could be defined. These indices will be applied to both methods used in the determination of CEC and exchangeable cations.

$$\text{Salinity index 1 (SI}_1) = \Sigma\text{cations/CEC} \qquad (1)$$

$$\text{Salinity index 2 (cmolc·kg}^{-1}) \text{ (SI}_2) = \Sigma\text{cations} - \text{CEC} \qquad (2)$$

SI_1 and SI_2 are related to the method with the cobalt-hexamine, whereas for the Metson method SI_3 and SI_4 are used. The electrical conductivity of the saturated paste (ECe), the sodium of adsorption ratio (SAR) and the exchangeable sodium percentage (ESP) were measured and estimated from the equations that were modelled by the linear relations 3 and 4 as follow:

$$Y = aSI_1 + b \qquad (3)$$

$$Y = cSI_2 + d \qquad (4)$$

SI is the index of the salinity, the b and d are constants, a and c are the slopes of the regression.

The statistical analysis that relates to manpower of 13 samples (**Table 3**) shows that SI_1, SI_2 and SI_3 explain 87% of the electrical conductivity (EC) variance. On the other hand, SI_4 explains only 80% of this variance. It is noticed that when the sum of the exchangeable cations extracted with the cobalt-hexamine is equal to the cation exchange capacity, indices SI_1 and SI_2 take the values of 1 and 0, respectively, whereas the salinity, which is expressed by electrical conductivity, is higher than 4 dS·m^{-1}. The presence of cations in excess compared to the CEC confirms the presence of salts, including magnesium and calcium salts. In addition, the sum of the exchangeable cations was never equal to the cation exchange capacity extracted with ammonium acetate. A ratio of 6/1 for index SI_3 and a difference of 70 cmolc/kg for index SI_4 make that it is possible to classify the soil samples in the category of the strongly saline soils because of their electrical conductivity has values higher than 12.00 dS·m^{-1}. This clearly shows that the quantity of cations extracted with pH 7 is much higher than that extracted with

Table 3. Linear regression between the parameters of salinity & sodicity and the salinity indices. The data in boldface characters correspond to the regressions obtained to the measurement made with the pH close to the soil (cobalt-hexamine).

Parameters	Linear regression	R^2	Estimate standard error	Standard error on R^2
CE	**CE=15.26 IS$_1$ – 10.23**	**0.87**	**4.79**	**0.0118**
	CE= 0.80 IS$_2$ + 4.90	**0.87**	**4.84**	**0.0120**
	CE=3.31 IS$_3$ – 6.31	0.87	4.85	0.0121
	CE= 0.20 IS$_4$ – 1.85	0.80	5.94	0.0182
SAR	**SAR= 1.94 IS$_1$ – 0.79**	**0.85**	**0.67**	**0.0139**
	SAR = 0.10 IS$_2$ + 1.23	**0.84**	**0.68**	**0.0143**
	SAR = 0.36 IS$_3$ + 0.07	0.61	1.07	0.0356
	SAR= 0.02 IS$_4$ + 0.67	0.47	1.24	0.0477
ESPeq	**ESP = 2.61 IS$_1$ – 0.94**	**0.85**	**0.91**	**0.0140**
	ESP= 0.14 IS$_2$ + 1.65	**0.84**	**0.91**	**0.0143**
	ESP=0.48 IS$_3$ + 0.21	0.61	1.44	0.0357
	ESP=0.03 IS$_4$ + 1.03	0.48	1.67	0.0477
ESP	**ESP = 15.69 IS$_1$ – 9.32**	**0.72**	**7.91**	**0.0253**
	ESP = 0.82 IS$_2$ + 6.27	**0.72**	**8.01**	**0.0259**
	ESP = 2.66 IS$_3$ – 1.04	0.44	11.25	0.0511
	ESP = 0.14 IS$_4$ + 4.66	0.30	12.51	0.0632

ESP$_{eq}$: ESP estimated by the equation of USSL Staff [25].

the pH close to the soil. So, the cobalt-hexamine method is an indicator about the state of the soil salinity (**Table 3**). For the Metson method, the electrical conductivity (EC) expresses at the same time the electrolytic load of the solution, which comes from the phenomenon of dissolution of calcite at pH 7. The salinity index is also dependent on the SAR (formula 7) of the soil solution. These indices explain 60% to 85% of the sodicity variance of Mina plain soils. As it is shown in **Table 4**, the values of the determining coefficient are higher for the SAR and ESP when the explanatory variables use the data of the cobalt-hexamine method. By using the Metson method, the relation between the salinity index and the parameters of sodicity is low. The typical error and the standard error r^2 are justifying it, which are appreciably higher than the others. The correlation matrix (**Table 4**) shows the various connections which can exist between salinity and sodicity. In the case of the salinity of the saturated paste extract, it corresponds to many soluble salts.

Figure 3 indicates the relation between the salinity index 2, which represents the quantity of cations in soil solution, and the total soluble cations of the extract of saturated soil paste. The equation of the straight regression line shows that the slope value is lower than 1,

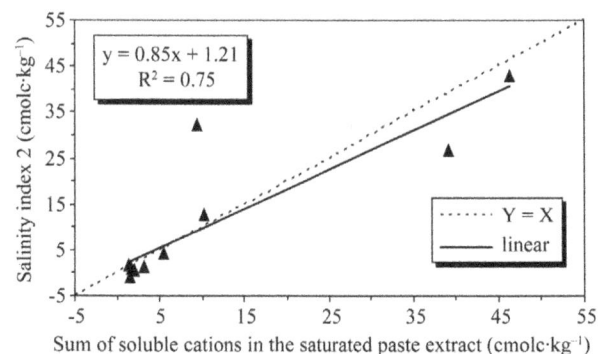

Figure 3. Relationship between the salinity index 2 estimated by the cobalt hexamine trichloride method and the total soluble cations from the saturated paste extract.

which explains why the sum of total soluble salts extracted from saturated soil paste is higher than those extracted with the cobalt-hexamine. It can be concluded from the cobalt-hexamine extraction that the dilution ratio (1/10) used with water was not sufficient to extract all the cations that are present in the saline soils of Cheliff plain.

To characterize the saline soils of Cheliff plain, the CEC effective and the exchangeable cations extracted at the pH of the soil used as a criterion of evaluation must

Table 4. Matrix correlation between the salinity indexes and the parameter of salinity & sodicity (the signifycant correlations at p = 0.005 are underlined).

	IS_1	IS_2	IS_3	IS_4	CE	SCE	ESP	ESPeq	SCS	SAR
IS_1	**1.00**									
IS_2	0.99	**1.00**								
IS_3	0.87	0.86	**1.00**							
IS_4	0.80	0.79	0.99	**1.00**						
CE	0.93	0.93	0.93	0.89	**1.00**					
SCE	0.99	0.99	0.84	0.78	0.94	**1.00**				
ESP	0.85	0.85	0.66	0.55	0.83	0.84	**1.00**			
ESPeq	0.92	0.92	0.78	0.69	0.92	0.91	0.97	**1.00**		
SCS	0.87	0.87	0.99	0.97	0.94	0.85	0.70	0.80	**1.00**	
SAR	0.92	0.92	0.78	0.69	0.92	0.91	0.97	1.00	0.81	**1.00**

ESP$_{eq}$: ESP estimated by the equation of USSL Staff (1954); SEC: Sum of exchangeable cations; SSC: Sum of soluble cations; r_{th} = 0.5529 et 0.6835 pour p= 0.950 et 0.990 et n = 13.

be adapted well compared to those extracted with standard pH 7. Through the total cations and the CEC, it can be regrouped the soils of Mina in Cheliff plain. It is also necessary to define the parameters of the saline phase as follows:

$$ESP = (100 * Exchange\ Na)/cationic\ exchange\ capacity \quad (5)$$

$$ESP = (100 * Echangeable\ Na)/\Sigma(Exchangeable\ Ca + Mg + K + Na) \quad (6)$$

$$SAR = [Na^+]/([Ca^{2+} + Mg^{2+}]/2)1/2 \quad (7)$$

$$ESPeq = 1.475 * SAR/(1 + 0.0147 * SAR)\ [25].$$

- The none-saline soils regroup which are the soils whose electrical conductivity is lower than 4 dS/m and the quantity of the total cations is almost identical to the CEC. Therefore, without excess cations in the soil solution, ESP will be evaluated on the basis of CEC (Formula 5).
- The saline soils regroup which are the soils whose electrical conductivity is higher than 4 dS/m and the quantity of the total cations is higher than the CEC. In this case, the exchangeable cations are obtained by the difference between total cations and soluble cations of the saturated soil paste extract. ESP will be evaluated on the basis of sum of the exchangeable bases (Formula 6).
- The saline soils have an evolution dominated by the presence of strong quantities of salts more soluble than gypsum. In general, the salinity is measured by the electric conductivity of the saturated soil paste extract (ECe). The richness of the soil exchange complex by sodium ion and the degradation susceptibility of soil physical properties are characterized by

the exchangeable sodium percentage (ESP) and sodium adsorption ratio (SAR) when they exceed the values of 15% and 10%, respectively. They are usually used to envision the evolution of the exchange complex composition and to define the term of sodicity. It is noticed, according to the matrix of correlation (**Table 5**) that the SAR and ESP increase with increasing the electrical conductivity with a correlation coefficient higher than 0.90. This indicates that the salinity of the soils is accompanied by a sodization of the soil exchange complex. In addition, both salinity indices (SI_1 and SI_2) make that it is possible to well characterize the saline phases (sodicity & salinity) and could be used to estimate the level of salinization and alkalization.

3.4. Influences of Salinity Index on Soil Physical Properties

3.4.1. Water Retention

Measurements of water retention were taken on the natural water samples of various salinity levels. The results had a range of potentials from 10 kPa (pF = 1.0) to 15000 kPa (pF = 4.2), *i.e.* that represented a range of moisture content going from the field capacity to the point of permanent witting of the plants. With water potential bottoms (pF = 4.2), the water retention appears to be very strongly correlated to the cation exchange capacity when it is measured at the standard pH 7 (CEC7) and to the specific surface (SS) area (**Table 5**). On the other hand, the relationship obtained with CEC effective measured using cobalt-hexamine method is not close to the coefficient of correlation (r = 0.68).

It also appears that the standard CEC (Metson method)

Table 5. Matrix of linear correlation between the salinity indexes and the water retention properties (the significant correlations at p = 0.005 are underlined).

	pF1	pF2.5	pF3	pF3.5	pF4	pF4.2	1/Da	CECe	CECm	SS
pF1	**1.00**									
pF2.5	0.85	**1.00**								
pF3	0.79	0.95	**1.00**							
pF3.5	0.65	0.83	0.93	**1.00**						
pF4	0.71	0.84	0.93	0.97	**1.00**					
pF4.2	0.34	0.44	0.55	0.50	0.53	**1.00**				
1/Da	0.80	0.98	0.90	0.76	0.76	0.38	**1.00**			
CECco	0.15	0.20	0.40	0.58	0.53	0.68	0.09	**1.00**		
CECm	0.38	0.44	0.59	0.70	0.68	0.80	0.33	0.97	**1.00**	
SS	0.52	0.69	0.69	0.56	0.59	0.86	0.66	0.43	0.65	**1.00**
IS1	0.15	0.31	0.23	0.15	0.00	−0.29	0.33	−0.11	−0.08	−0.07
IS2	0.17	0.33	0.25	0.17	0.03	−0.29	0.34	−0.09	−0.06	−0.07
IS3	0.31	0.54	0.42	0.24	0.13	−0.03	0.59	−0.16	−0.05	0.23
IS4	0.39	0.62	0.51	0.32	0.22	0.10	0.67	−0.09	0.04	0.35

ESP_{eq}: ESP estimated by the equation of USSL Staff (1954); 1/Da: The inverse of the bulk soil density.

and specific surface area (SS) could be used as a suitable criterion for the soil water retention. These results agree with those of [6] and [24]. For high water potentials in the vicinity of the field capacity (pF between 1 and 3), the correlation coefficients obtained between the CEC and the water content are weak. On the other hand, the best relations are obtained with the inverse of the apparent soil density (1/Da). The apparent soil density varies from 1.3 to 1.8 for the studied soil samples. The apparent density of the soil is also an indicator of the water retention for the field capacity because the variation of water content is primarily related to the evolution of structural porosity and thus to the structure of the soil. These results are in accordance with these of [26], which showed that the water retention at pF 2.5 was well correlated to the inverse of the apparent soil density (1/Da).

3.4.2. The Structural Stability and the Hydrodynamic Behaviour of the Soils

As it is shown in **Table 6**, the salinity index calculated using the results of cobalt-hexamine measurement is well correlated to the inverse of the mean weight diameter (1/MWD) and to the final infiltration coefficient (FIC). It is also noticed that the relation is better between both SI1 and SI2 indices and the structural disintegration under an agitation during 5 minutes (MWDbis). The structural stability and the water infiltration of the soils remain the most adapted means to evaluate the sensitivity of the soils

to the mechanisms of disintegration. The linear correlation coefficients estimated between these tests (1/MWD and FIC) and the parameters of the saline phase (**Table 6**) indicate rather strong relations between the electrical conductivity of saturated paste extract (ECe), the sum of the exchangeable cations (SEC), the exchangeable sodium percentage (ESP) and the sodium absorption ratio (SAR). The sum of the soluble cations (SSC) is not significantly dependent on any of the physical parameters. So, the effect of the soil solution concentration is less apparent. Both the inverse of the mean weight diameter (1/MWD) and the final infiltration coefficient of (FIC) have correlation coefficients slightly higher with sodicity than with salinity. It is noticed that the variables of sodicity (ESP, SAR and ESPeq) are better than the variables of salinity (ECe and SI) with respect to the disintegration under an agitation during 5 minutes (I/MWDbis). This means that the exchangeable sodium content controls the physicochemical mechanism of dispersion during disintegration, which involves the production of fine particles. This influence is explained by the role of sodium in the dispersion of soil particles [4,27] and the reduction of water infiltration. These two effects strongly contribute to closing the surface when the samples are subjected to the simulated rains of 30 mm/h [28,29].

4. CONCLUSION

The present study shows that it is possible to estimate

Table 6. Matrix of linear correlations between the salinity indexes, the salinity and sodicity parameters and the physical properties (the significant correlations at p = 0.005 are underlined).

	IS1	IS2	IS3	IS4	CE	SCS	SCE	ESP	SAR	ESP$_{eq}$	1/MWD	1/MWD bis	CIF
IS1	**1.00**												
IS2	0.99	**1.00**											
IS3	0.87	0.86	**1.00**										
IS4	0.80	0.79	0.99	**1.00**									
CE	0.93	0.93	0.93	0.89	**1.00**								
SCS	0.87	0.87	0.99	0.97	0.94	**1.00**							
SCE	0.98	0.98	0.90	0.84	0.98	0.91	**1.00**						
ESP	0.93	0.92	0.84	0.76	0.94	0.85	0.96	**1.00**					
SAR	0.92	0.92	0.78	0.69	0.92	0.81	0.94	0.98	**1.00**				
ESP$_{eq}$	0.92	0.92	0.78	0.69	0.92	0.80	0.95	0.98	1.00	**1.00**			
1/MW	0.72	0.72	0.53	0.45	0.61	0.52	0.68	0.70	0.67	0.67	**1.00**		
1/MWD bis	0.73	0.73	0.47	0.37	0.57	0.45	0.67	0.67	0.66	0.67	0.98	**1.00**	
CIF	−0.58	−0.59	−0.51	−0.47	−0.62	−0.48	−0.62	−0.67	−0.64	−0.65	−0.77	−0.73	**1.00**

ESP$_{eq}$: ESP estimated by the equation of USSL Staff (1954); SSC: Sum of soluble cations; SEC: Sum of exchangeable cations; r_{th} = 0.5529 et 0.6835 pour p = 0.950 et 0.990 et n = 13.

the cation exchange capacity of the Cheliff soils from the clay content, the organic matter content and the pH of the soil. It also reveals that measurements of the CEC and the total cations carried out at the pH of the soil with the cobalt-hexamine cation make that it is possible to approach the chemical properties of the soils, such as sodicity. On the other hand, the measurement of the CEC at pH 7 presents the disadvantage of dissolving a part of carbonates and thus it is not adapted to measure the exchangeable cations and CEC in calcareous soils. However, the measurement of the CEC carried out at the standard pH 7 seems an indicator of the hydraulic soil properties at water potential bottoms, in relation to specific surface area. The inverse of the apparent density and the specific surface seems good indicators to estimate the properties of water retention at the field capacity and the permanent witting point, respectively. The influence of salinity is relatively less significant than the sodicity and the cation concentration of soil solution is not synonymous with bad stability. In general, It is clear that the relation of the saline phase and the physical properties results is not on the agenda. A detailed study is necessary. So, in this direction, we started a study on the effects of exchangeable sodium and cation concentration of the solution on the physical properties of clay materials of the Cheliff plain. We can say shortly that our study is a contribution to evaluate the quality of the estimate and the development of the pedotransfer functions to predict the behaviours of the Cheliff plains' saline soils.

REFERENCES

[1] McNeal, B.L. and Coleman, N.T. (1966) Effect of solution composition on soil hydraulic conductivity. *Soil Science American Proceeding*, **20**, 308-312.

[2] El-Swaify, S.A. (1973) Structural change in tropical soils due to anions in irrigation water. *Soil Science*, **115**, 64-72.

[3] Cass, A. and Sumner, M.E. (1982) Soil pore structural stability and irrigation water quality. I. Empirical sodium stability model. *Soil Science Society of American Journal*, **46**, 503-506.

[4] Halitim, A., Robert, M., Tessier, D. and Prost, R. (1984) Influence de cations échangeables (Na$^+$, Ca^{2+}, Mg^{2+}) et de la concentration saline sur le comportement physique (rétention en eau et conductivité hydraulique) de la montmorillonite. *Agronomie*, **4**, 451-459.

[5] Daoud, Y. and Robert, M. (1992) Influence of particle size and clay organisation on hydraulic conductivity and moisture retention of clays from saline soils. *Applied Clay Science*, **6**, 293-299.

[6] Tessier, D., Biggore, F. and Bruand, H. (1999) La capacité d'échange: Outil de prévision des propriétés physiques des sols. *Compte Rendu d'Académie des Sciences*,

85, 37-46.

[7]   Heilman, M.O., Carter, O.L. and Gonzalez, C.L. (1965) Ethylene glycol mono ethyl ether for determining surface area of silicate minerals. *Soil Science*, **100**, 356-360.

[8]   Ciesielski, H. and Sterckemann, T. (1997) Determination of exchange capacity and exchangeable cations in soils by means of cobalt hexamine trichloride. Effects of experimental conditions. *Agronomie*, **17**, 1-7.

[9]   Charlet, I. and Schlegel, M.L. (1999) La capacité d'échange des sols. Structures et charges à l'interface eau/particule. *Compte Rendu d'Académie d'Agriculture*, **85**, 7-24.

[10] Bruand, A. and Zimmer, D. (1992) Relation entre la capacité d'échange cationique et le volume poral dans les sols argileux: Incidences sur la morphologie de la phase argileuse à l'échelle des assemblages élémentaires. *Compte Rendu d'Académie des Sciences*, **315**, 223-229.

[11] Orsini, L. and Remy, J.C. (1976) Utilisation du chlorure de cobaltihexammine pour la détermination simultanée de la capacité d'échange et des bases échangeables des sols. *Bulletin de l'AFES Science du Sol*, **4**, 269-275.

[12] Boulaine, J. (1957) Etude des sols des plaines du chélif. Thèse d'Etat, l'Université d'Alger.

[13] Saidi, D. (1985) Etude agropédologique de la plaine de la Mina (Relizane) et évaluation des propriétés physiques des sols. Thèse Ing., INA., Alger.

[14] Daoud, Y. (1983) Contribution à l'étude de la dynamique des sels dans un sol irrigué du périmètre du Haut Cheliff (Algérie). Thèse de Docteur Ingénieur de l'ENSA de Rennes.

[15] Saidi, D. (2005) Influence de la phase saline sur les propriétés physiques des matériaux argileux du Bas Cheliff. Thèse de Doctorat d'Etat, INA, Alger.

[16] Metson, A.J. (1956) Methods of chemical analysis for soil survey samples. New Zealand Soil Bureau Bulletin No. 12.

[17] Monnier, G., Stengel, P. and Fies, J.C. (1973) Une méthode de mesure de la densité apparente de petits agglomérats terreux. Application à l'analyse des systèmes de porosité du sol. *Annals Agronomique*, **25**, 533-545.

[18] AFNOR (1996) Qualité des sols. Recueils de normes Française, AFNOR, Paris.

[19] Tessier, D. and Berrier, J. (1979) Utilisation de la microscopie électronique à balayage dans l'étude des sols. Observations des sols humides soumis à différents pF. *Sciences du Sol*, **1**, 67-82.

[20] Le Bissonnais, Y. (1996) Aggregate stability and assessment of soil crustability and erodibility. I. Theory and methodology. *European Journal of Soil Science*, **47**, 425-437.

[21] Asseline, J. and Valentin, C. (1978) Le simulateur de pluies de l'ORSTOM. *Cahier Hydrologique de l'ORSTOM*, **4**, 321-347.

[22] Julien, J.L. and Turpin, A. (1999) Surfaces réactives et raisonnement de quelques propriétés chimiques des sols acides. *Compte Rendu d'Académie d'Agriculture*, **85**, 25-35.

[23] Curtin, O. and Rostad, H.P.W. (1997) Cation exchange and buffer potential of Saskatchewan soils estimated from texture, organic matter and pH. *Canadian Journal of Soil Science*, **77**, 621-626.

[24] Bigorre, F., Tessier, D. and Pedro, G. (1999) Contribution des argiles et des matières organiques à la rétention de l'eau dans les sols. Signification et rôle fondamental de la capacité d'échange en cations. *Compte Rendu d'Académie des Sciences*, **330**, 245-250.

[25] US Salinity Laboratory Staff (1954) Diagnosis and improvement of saline and alkali soils. USDA Handbook.

[26] Bruand, A., Duval, O., Gaillard, H., Darthout, R. and Jamagne, M. (1996) Variabilité des propriétés de rétention en eau des sols. Importance de la densité apparente. *Etude et Gestion des Sols*, **3**, 27-40.

[27] Shainberg, I. and Letey, J. (1984) Response of soil to sodic and saline conditions. *Hilgardia*, **52**, 1-57.

[28] Le Bissonnais, Y. (1988) Aggregate stability and assessment of soil crustability and erodibility. I. Theory and methodology. *European Journal of Soil Science*, **47**, 425-437.

[29] Yousaf, M., Ali, O.M. and Rhoades, J.D. (1987) Clay dispersion and hydraulic conductivity of some salt-affected arid land soil. *Soil Science Society of America Journal*, **51**, 905-907.

13

Analysis of cotton water productivity in Fergana Valley of Central Asia

J. Mohan Reddy[1*], Shukhrat Muhammedjanov[2], Kahramon Jumaboev[1], Davron Eshmuratov[1]

[1]International Water Management Institute, Tashkent, Uzbekistan; *Corresponding Author
[2]Scientific Information Center of the Interstate Committee for Water Coordination, Tashkent, Uzbekistan

ABSTRACT

Cotton water productivity was studied in Fergana Valley of Central Asia during the years of 2009, 2010 and 2011. Data was collected from 18 demonstration fields (13 in Uzbekistan, 5 in Tajikistan). The demonstration field farmers implemented several improved agronomic and irrigation water management practices. The average values of crop yield, estimated crop consumptive use (ET_a) and total water applied (TWA) for the demonstration sites were, respectively, 3700 kg/ha, 6360 m^3/ha, and 8120 m^3/ha. The range of values for TWA and ET_a were, respectively, 5000 m^3/ha to 12,000 m^3/ha and 4500 m^3/ha to 8000 m^3/ha. A quadratic relationship was found between TWA and ET_a. The average yield of the adjacent fields was 3300 kg/ha, whereas the average yield of cotton in Fergana Valley as a whole was 2900 kg/ha, indicating 28% and 14% increase in crop yield, respectively, from, demonstration fields and adjacent fields. There was no significant difference in crop yields between the wet years (2009 and 2010) and the dry year (2011), which is explained by the quadratic relationship between TWA and ET_a. The water productivity values ranged from 0.35 kg/m^3 to 0.89 kg/m^3, indicating a significant potential for improving water productivity through agronomic and irrigation management interventions. The ratio of average ET_a divided by average TWA gave an average application efficiency of 78% (some fields under-irrigated and some fields over-irrigated), the remaining 22% of water applied leaving the field. Since more than 60% of the water used for irrigation in Tajikistan and Uzbekistan is pumped from, even if all this 22% of water returns to the stream, substantial energy savings would accrue from improving the average application efficiency at field level. The range of values for TWA indicates the inequity in water distribution/accessibility. Addressing this inequity would also increase water productivity at field and project level.

Keywords: Furrow Irrigation of Cotton; Irrigation in Fergana Valley; Water Productivity

1. INTRODUCTION

After independence from the former Soviet Union (in 1991), the operation and maintenance of irrigation and drainage systems was neglected due to lack of adequate financial resources. This exacerbated the pre-existing problem of waterlogging and salinity of irrigated lands. In Central Asia as a whole, more than 5.97 million ha of irrigated area out of the total irrigated area of 8 million hectares requires artificial drainage. There were significant investments in drainage in the region until 1990s. However, with the collapse of the Soviet Union, drainage systems are no longer properly maintained and the area under waterlogging and salinity has been steadily increasing: 35% increase in waterlogged area and 62% increase in area under moderate to high salinity [1].

Furthermore, the State/Collective farms disintegrated, with nobody to claim the ownership of irrigation and drainage infrastructure. Land was distributed to local people, irrespective of their prior background in agriculture. In Kyrgyzstan, Kazakhstan and Tajikistan, farmers own their land, whereas in Turkmenistan and Uzbekistan farmers lease their land from the government. Disintegration of large farms has increased the number of farmers the majority of whom have inadequate knowledge/ skills of irrigated agriculture. There was insufficient on-farm irrigation infrastructure to distribute water to individual farmers. During the Soviet era, every State/Collective farm had professional agronomists and irrigation specialists for providing advisory services for irrigated agriculture. However, with the collapse of the system, some of this expertise was lost. Without adequate irrigation infrastructure and organizational support for water distribution below the tertiary canal level, irrigated agriculture became chaotic-head-end/tail-end problems, inequity and unreliability in water supply, lack of advisory services on agricultural practices, lack of appropriate farm machinery for operation on small farms, etc.

After year 2000, through Agricultural Reform Acts, Water Users Associations (WUAs) have been formed. This process is not complete in Tajikistan and Turkmenistan. The Government agencies provide bulk water supply to WUAs, and then it is the responsibility of WUAs to supply this water equitably to individual farmers. Yet, there are problems of equity and unreliability of water supply within WUAs hindering improved water management at plot level. This situation combined with waterlogging and salinity has resulted in significant reductions in crop yields.

With a view to increase crop yields from irrigated agriculture, the Swiss Agency for Development and Cooperation (SDC) financed a project for improving water productivity at plot level (WPI-PL). The project had two objectives. The first objective of the project was to develop and evaluate an effective mechanism called "Innovation Cycle" for dissemination of knowledge on improving water productivity to farmers in the Fergana Valley of Central Asia on an experimental basis. This objective was accomplished successfully during the three year period of the project. A separate paper is being prepared on the structure and functioning of the developed Innovation Cycle. The second objective of the project was to evaluate the effect of the Innovation Cycle on improving water productivity of agricultural crops in Fergana Valley. To this end, data on irrigated agricultural production of several major crops such as cotton and wheat, and other crops such as potato, maize, sunflower, watermelons, cucumbers, onions, etc., were collected from several demonstration sites in the countries of Kyrgyzstan, Tajikistan and Uzbekistan. In the past, some general studies were undertaken on water productivity of major crops (cotton and rice) in Syr Darya basin [2,3].

However, no water productivity studies were conducted at field level after implementing the Agricultural Reform Acts of early 2000's. In addition, no data are available on water productivity of agricultural crops under improved agronomic and irrigation water management practices because no effective mechanisms for dissemination of irrigated agriculture knowledge exist in Central Asia today. This paper discusses cotton water productivity from demonstration fields in Tajikistan and Uzbekistan that received irrigated agriculture advisory services from the WPI-PL project.

2. DESCRIPTION OF SITE

To assess cotton water productivity at field level, dem-

onstration sites were selected in Fergana Valley of Central Asia. Fergana Valley is located in the Southeastern part of Central Asia region and the Eastern part of Aral Sea basin, and its territory is shared by three countries—Kyrgyzstan, Tajikistan and Uzbekistan. The Fergana Valley forms the upper and mid-reach of the Syr Darya Basin. Syr Darya is formed from the confluence of Naryn and Karadarya rivers. The average temperature in the Valley is 13.1°C, ranging from −8°C to 3°C in January and from 17°C to 36°C in July. Annual precipitation varies from 109 mm to 502 mm, whereas evaporation ranges from 1133 mm to 1294 mm throughout Fergana Valley. Fergana Valley is home for 11,342,000 people over an area of 124,200 km^2.

Data on water productivity were collected from a total of 23 demonstration sites—13 sites in Uzbekistan, 5 sites in Kyrgyzstan, and 5 sites in Tajikistan. The main criterion used in the selection of demonstration sites was that the farmer must be a "progressive farmer", *i.e.* a farmer with background in irrigated agriculture and was willing to experiment with innovative agronomic and irrigation practices. All the selected demonstration site farmers in Kyrgyzstan did not grow cotton. Therefore, data only from the remaining 18 sites was used to calculate water productivity of cotton. The location of these 18 sites is presented in **Figure 1**. At each of these 18 demonstration farms, an adjacent farm was also selected for comparison purposes.

For all the 18 demonstration fields, information on soil texture, soil-moisture content at field capacity, and depth of watertable from ground surface was collected (**Table 1**). Soil salinity is not an issue at most of the demonstration sites. All the fields practiced furrow irrigation, with runoff from the downstream-end of the fields. The fields are sloping with undulations. No data was collected on the degree of undulations in each field. Flow measurement structures were installed at all the demonstration sites to measure the amount of irrigation water applied to the fields and the amount of runoff from the fields. Information on the irrigation norms (based upon hydromodule zoning) was also provided to relevant WUAs. In Uzbekistan, cotton crop is mandated to be grown in order to meet the annual production quota that is determined by the government. In order to facilitate the meeting of total national quota, a target yield level is set for each field based upon the soil-texture, soil fertility, condition of watertable, level of soil salinity and salinity of water used for irrigation. In addition, farmers are provided with credit facilities for acquiring the necessary agricultural inputs. Farmers are required to produce cotton yields that are at least equal to the target level set by the government. If any farmer fails to meet the production target, his/her land lease will be re-negotiated. Furthermore, the farmers are expected to sell cotton only to the government. Cot-

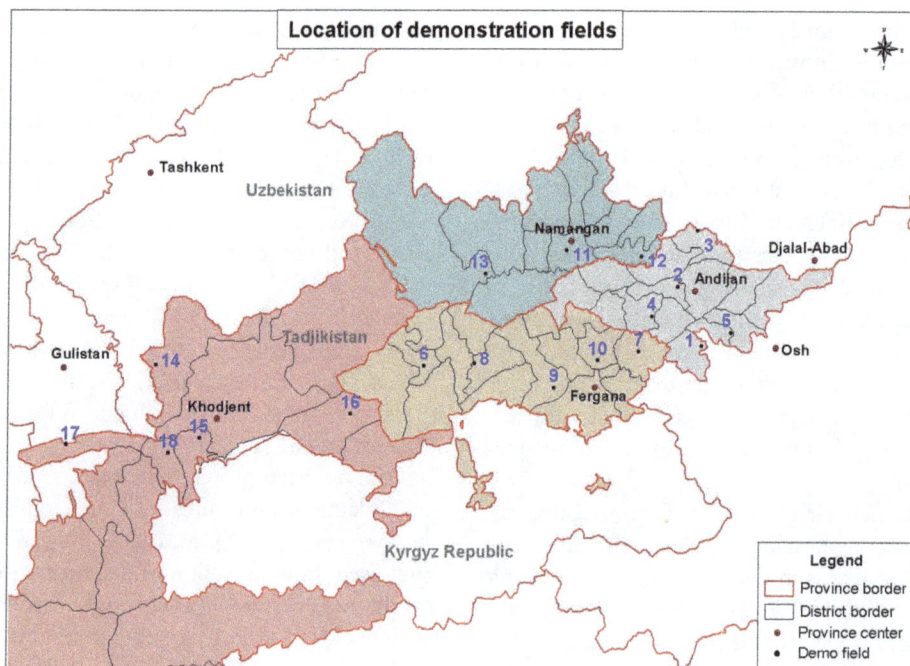

Figure 1. Location of demonstration sites in Tajikistan and Uzbekistan.

Table 1. Field capacity and depth to watertable of demonstration sites in Tajikistan and Uzbekistan.

Province	Site №	Field capacity, mm/m	Depth to Watertable, cm
Andijon	1	177	>300
	2	181	150
	3	184	>300
	4	167	200
	5	189	>300
Fergana	6	192	160
	7	173	160
	8	166	150
	9	139	140
	10	125	250
Namangan	11	150	>300
	12	184	150
	13	192	>300
Sogd	14	167	>300
	15	192	>300
	16	167	>300
	17	125	>300
	18	192	>300

ton is not a mandated crop in Tajikistan; therefore, no government financed credit facilities are provided to farmers. Though the farmers in Tajikistan have the incentive to produce high yields of cotton, lack of credit may be a constraint for increasing agricultural production per unit area. In addition, farmers' income is also vulnerable to the world market prices.

All the demonstration site farmers received information on a set of innovative agronomic and irrigation practices to improve water productivity at field level (**Table 2**). These innovative practices included: land preparation, agro-ameliorative certification of farms, proper sizing of irrigation schemes, mixing of mineral fertilizers with organic fertilizers (manure), application of liquid mineral fertilizers through irrigation water in furrows, adoption of volumetric water delivery method, irrigation scheduleing, measurement of irrigation flow using Sokolok method, short furrow irrigation, alternate furrow irrigation, installation of plastic films at the head of furrows, runoff recovery, cutback irrigation, water rotation, interrow cultivation, and leaching of salts. As shown in **Table 2**, most of these recommendations were implemented by several demonstration field farmers. Almost all the farmers used alternate furrow irrigation, short furrow irrigation, good pest control measures, inter-row cultivation, and re-use of runoff water from fields.

In order to calculate the net benefits accrued to the demonstration farm farmers, the following information was collected: type and kilogram of seed farmer applied per hectare, amount and cost of fertilizer and pesticides used per hectare, cost of equipment for tillage and culti-

Table 2. Technologies used at demonstration sites.

Site №	Agro technical activities and land preparation for irrigation season	Leveling of irrigated land	Agro-ameliorative certification of farms	Selecting technological irrigation scheme	Application of mineral fertilizers (soil fertility improvement) for cotton and other crops in conditions of Fergana Valley (to irrigate through pits filled with organic fertilizers)	Fertilizer irrigation through application of liquid mineral fertilizers with irrigation water (fertigation)	Crop pest control	Water accounting (construction of water measuring devices, implementation of water accounting) is organized on inlets and outlets	Adoption of volumetric water delivery method (payment per m^3 of water received). Preparation of necessary documents (contract between WUA and farmer, water accounting log, act, etc)	Implementation of crop irrigation regime	Irrigation considering soil moisture	Mechanism of efficient irrigation water use in small farms (Sokolok)	Defining needs and problems of farmers hindering land and water productivity improvement	Short furrows	Every second furrow irrigation	Variable jet furrow irrigation	Night irrigation	Installation of plastic films in the furrow heads	Use of plastic bottle heads and siphons	Decrease of technological discharges from the fields (use of drainage water or secondary use of water on downstream fields)	Water rotation	Inter-row cultivation	Leaching of saline soils with different types of salinity
2009																							
1	1			1	1	1	1			1		1		1	1		1	1	1		1	1	1
2	1			1	1	1	1			1				1			1	1				1	1
3	1			1	1	1	1			1				1	1		1	1	1		1	1	1
4	1			1	1	1	1			1				1			1	1				1	1
5	1			1	1	1	1			1				1	1		1	1					
6	1	1			1		1	1		1				1	1		1	1			1		1
7	1	1		1	1	1	1			1				1	1		1	1			1	1	1
8	1			1	1		1	1		1				1							1		1
9	1	1	1		1	1	1	1		1	1			1	1		1	1	1		1	1	1
10	1	1		1	1		1	1		1	1			1	1		1	1	1		1		1
11	1			1		1				1				1	1		1	1			1	1	1
12	1			1		1				1				1	1		1	1			1		1
13	1			1	1	1	1							1	1		1	1			1		
14	1	1		1	1		1	1	1	1				1		1	1	1	1		1	1	1
15	1	1		1	1		1	1	1	1				1		1	1	1	1		1	1	1
16	1	1		1	1		1	1	1	1				1		1	1	1	1		1	1	1
17	1	1		1	1		1	1	1	1				1		1	1	1	1		1	1	1
18	1	1		1	1		1	1	1	1				1		1	1	1	1		1	1	1
2010																							
1	1			1	1	1	1			1		1		1	1		1	1	1		1	1	1
2	1			1	1	1	1			1				1	1		1	1				1	
3	1			1	1	1	1			1	1			1	1		1	1	1	1	1	1	1
4	1			1	1	1	1			1				1			1	1				1	1
5	1			1	1	1	1			1	1			1	1		1	1	1	1			
6	1	1			1		1	1		1				1	1		1	1			1		1
7	1	1		1	1	1	1			1				1	1	1	1	1			1	1	1

Continued

8	1	1		1	1	1	1	1		1			1	1			1	1		1			1
9	1	1	1		1	1	1	1		1	1		1	1	1		1	1		1	1	1	
10	1	1		1	1		1	1		1	1		1	1	1		1	1		1		1	
11	1	1	1		1		1	1					1		1		1	1		1		1	
12	1	1	1		1	1	1	1					1	1	1		1	1		1			
14	1	1		1	1	1	1		1	1			1	1	1	1	1	1		1	1	1	
15	1	1		1	1	1	1		1	1		1	1	1	1	1	1	1		1	1	1	
16	1	1		1	1	1	1		1	1		1	1	1	1	1	1	1		1	1	1	
17	1	1		1	1	1	1		1	1		1	1	1	1	1	1	1		1	1	1	
18	1	1		1	1	1	1		1	1		1	1	1	1	1	1	1		1	1	1	
										2011													
1	1			1	1	1	1	1		1		1	1	1			1	1		1	1	1	
2	1			1	1	1	1	1		1	1		1	1	1		1	1		1			
3	1			1	1	1	1	1		1	1		1	1	1		1	1	1	1	1	1	
5	1			1	1	1	1	1		1	1		1	1	1		1	1	1	1	1		
7	1	1		1	1	1	1	1		1			1	1	1	1	1	1		1	1	1	
8	1	1		1	1	1	1	1		1			1	1			1	1		1			1
9	1	1	1		1	1	1	1		1	1		1	1	1		1	1		1		1	
10	1	1		1	1	1		1		1	1		1	1	1		1	1		1		1	
11	1	1	1		1	1	1			1			1		1		1	1		1		1	
14	1	1		1	1	1	1		1	1	1	1	1	1	1	1	1	1		1	1	1	
15	1	1		1	1	1	1		1	1	1	1	1	1	1	1	1	1		1	1	1	
17	1	1		1	1	1	1		1	1	1	1	1	1	1	1	1	1		1	1	1	
18	1	1		1	1	1	1		1	1		1	1	1	1	1	1	1		1	1	1	

vation, cost of labor, amount of irrigation water applied per hectare, cost of transportation, fixed costs for agricultural production, and finally yield of major crops. In addition, climatic data from the nearest weather station for each of the 18 sites was gathered for calculating reference evapotranspiration of cotton crop at the given locations.

3. METHODOLOGY TO CALCULATE WATER PRODUCTIVITY

There are several definitions of water productivity (WP). The most commonly used definition [4] is given as the ratio of the crop yield, Y_{crop} (kg/ha), divided by the consumptive use of water by the crop, ET_a (m^3/ha), i.e.

$$WP = Y_{crop}/ET_a \qquad (1)$$

in which Y_{crop} = measured crop yield under natural and irrigated conditions, kg/ha; and ET_a = estimated/measured seasonal evapotranspiration or crop water use, m^3/ha. The above definition is independent of the source of water made available for ET_a, and assumes that any water losses that occur at field level, in the form of runoff and deep percolation, are recaptured and re-used somewhere else in the basin, ignoring or discounting some "co-benefits" such as improved water quality (particularly under Central Asian conditions where salinity is a major issue), increased crop production, increased reliability in water supply, decreased energy demands and carbon emissions, and reduced or delayed infrastructure investments [5] that accrue from improved application efficiency at field level. The source of water for ET_a may be a combination of one or more of the following: rainfall, groundwater, residual soil-moisture from previous season or irrigation water.

Sometimes, we are interested in the incremental change

in crop yields due to the addition of irrigation water to fields. Therefore, another productivity term called irrigation water productivity [6] is defined as follows:

$$WP_I = \left(Y_{crop} - Y_D\right)\big/V_I \qquad (2)$$

in which WP_I = irrigation water productivity of crop, kg/m^3; Y_D = crop yield under dryland conditions (rainfall, residual initial soil-moisture content from previous season, groundwater contribution) without any irrigation, kg/ha; V_I = cumulative volume of irrigation water applied during the crop growing season, m^3/ha. To calculate WP_I, information on crop production Y_D at different levels of natural water supply must be available.

The water productivity definitions provided above (**Eqs.1** and **2**) do not provide any indication of inefficiency of water application at field level. Sometimes, farmers apply 50% to 100% more water than the amount of water required by the crop; yet, the actual water use by crop (ET_a) only goes up slightly compared to its water use under normal conditions. In order to capture the inefficiency of water use by farmers, the following definition of water productivity is proposed here:

$$WP_G = Y_{crop}\big/V_{all} \qquad (3)$$

in which WP_G = gross water productivity, kg/m^3; and V_{all} = volume of water applied to a field from all sources (rainfall, residual soil-moisture, groundwater, and irrigation water), m^3/ha, and is calculated as follows:

$$V_{all} = V_{irri} + V_{GW} + V_{imc} + V_{rainfall} \qquad (4)$$

in which V_{irri} = volume of irrigation water applied to a field, m^3/ha; V_{GW} = volume of groundwater contribution to crop root zone, m^3/ha; V_{imc} = volume of initial soil-moisture content at the time of planting, m^3/ha; and $V_{rainfall}$ = volume of rainfall received on the field during the crop growing season, m^3/ha.

In order to calculate WP using **Eq.1**, the seasonal crop water use (ET_a) must be estimated for the given location. The most accurate methods of measuring ET_a are lysimeters, neutron probes, and gravimetric methods. However, since the 18 demonstration sites were scattered over a large area, and since no lysimeters and neutron probes were available, the ET_a was estimated using a standard soil-moisture balance equation [6,7]:

$$ET_a = R + I + F - Rf - \Delta S \qquad (5)$$

in which ET_a = seasonal crop evapotranspiration or water use, mm; R = rainfall during the growing season, mm; I = irrigation amount applied during the growing season, mm; F = net soil-moisture flux (taken positive into the rootzone) at the bottom of the crop rootzone, mm; Rf = runoff from the soil-surface, mm; and ΔS = change in soil-moisture content (taken as positive when the soil-moisture content increases over the season) within the

crop rootzone during the crop growth season, mm. All the quantities on the right-hand-side of **Eq.5** must be carefully estimated in order to estimate seasonal crop water use. In using **Eq.5**, the most difficult variable to estimate is F, the net soil-moisture flux from/to the crop rootzone. In the absence of a high watertable, the net soil-moisture flux is always negative, and is basically due to deep percolation from irrigation and/or rainfall amount added to the crop rootzone. The change in the rootzone soil-moisture content is typically estimated using gravimetric sampling or neutron moisture meter. If a lysimeter is used, all the quantities on the right-hand-side of **Eq.6** are measured in order to compute the crop consumptive use on a daily, weekly, or seasonal basis. In the absence of lysimeters and soil-moisture sensing devices, a different method can be used to estimate the seasonal consumptive water use (ET_a) of a given crop. It is given as follows:

$$ET_a = \Sigma ET_a(i), \quad i = 1 \text{ to } N \qquad (6)$$

in which $ET_a(i)$ = estimated crop evapotranspiration on day i, mm/day; and N= number of days in the growth period for the given crop. $ET_a(i)$ is estimated as follows:

$$ET_a(i) = K_c(i)K_s(i)ET_r(i) \qquad (7)$$

in which $ET_r(i)$ = estimated evapotranspiration of a reference crop on day i, typically estimated using the available climatic data, mm/day; $K_c(i)$ = crop coefficient values as a function of different growth periods of the given crop; and $K_s(i)$ = soil-moisture stress coefficient on day i which is related to the maximum available soil-moisture content and the actual soil-moisture content in the rootzone on day i. In the literature, three different types of relationships are provided between K_s and the soil-moisture content in the rootzone. The following relationship is used in this paper:

$$K_s(i) = \ln\left(1 + PAW(i)\right)\big/\ln(101) \qquad (8)$$

in which PAW(i) = percent available water within the crop rootzone on day i, and is calculated using

$$PAW(i) = 100\left(\theta_a - \theta_{wp}\right)\big/\left(\theta_{fc} - \theta_{wp}\right) \qquad (9)$$

in which θ_{fc} = soil-moisture content at field capacity, mm; θ_{wp} = soil-moisture content at wilting point, mm; and θ_a = actual soil-moisture content, mm. Soil-moisture content on a volume-basis is calculated using the following equation:

$$\theta = \varphi\gamma_{bd}z_r(t) \qquad (10)$$

in which θ = soil-moisture content on a volume-basis; φ = soil-moisture content on a weight-basis; γ_{bd} = soil bulk density; and $z_r(t)$ = crop rooting depth, mm. Since it is tedious to measure soil-moisture content, θ_a, on a daily basis, frequently a soil-moisture balance equation is used to estimate θ_a on a daily basis

$$\theta_a(i+1) = \theta_a(i) + I(i) + R(i) - ET_a(i) + GW(i) \quad (11)$$

in which i = index of day; I(i) = irrigation amount on day i, mm; R(i) = rainfall amount on day i, mm; and GW(i) = groundwater contribution on day i, mm. To use **Eq.11**, information on the soil-moisture content on the day of planting (i = 1), the dates and amounts of irrigation water applied (minus runoff from field), the dates and amounts of rainfall, and groundwater contribution to the crop rootzone on a daily basis must be known. Groundwater contribution, in mm/day, to the crop rootzone depends upon the soil texture, the evaporative demand of the atmosphere, and the depth of the watertable from the ground surface. **Figure 2** shows the dependence of groundwater contribution to crop rootzone as a function of soil-texture and depth of watertable from the ground surface [8]. Information from this graph was used to estimate groundwater contribution to crop rootzone on a daily basis.

In **Eq.7**, the ET_r was estimated using the Penman-Monteith equation [9] along with the local climatic data available from the weather stations operated by the Meteorological Departments of Uzbekistan and Tajikistan. The daily climatic data on sunshine hours, solar radiation, minimum and maximum temperatures and wind speed were available from the weather stations. The reference crop considered was short grass. Since there were no locally calibrated crop coefficients for cotton, the crop coefficients for cotton were obtained from FAO-56 report [9]. Since crop coefficients depend upon the climatic conditions and growth characteristics of given crop vari-

ety, use of general crop coefficients provided in FAO-56 might introduce some error in the estimation of ET_a, and this error was considered acceptable for the current large-scale study.

4. RESULTS AND DISCUSSION

Yield data for the 13 demonstration sites in Uzbekistan and the 5 demonstration sites in Tajikistan was obtained from the farmers. Crop yield data along with the cost of production, and net profits are presented in Table 3 for the years 2009, 2010 and 2011. During the 2010 and 2011 irrigation seasons, all the demonstration farmers did not grow cotton on their demonstration fields. Hence, the number of cotton fields was less than 18. Yields of cotton from the demonstration sites ranged from 2000 kg/ha to 5500 kg/ha (**Figure 3**). This difference in yields was due to a combination of factors such as quality of advisory services received, the quality and quantity of seed used, the availability and quality of inputs received or applied by farmers, crop variety, and the irrigated cotton production knowledge-base of the farmers. Two things are obvious from **Figure 3**. First, the average yield of cotton from the demonstration sites in Tajikistan was lower (less than 3000 kg/ha) than the average yield of cotton from demonstration sites in Uzbekistan (about 3500 kg/ha). This difference in yields may be partly explained by the availability of credit for purchasing agricultural inputs plus application of land use practices which are also supported and monitored by the State. In Uzbekistan, since cotton is one of the two crops that is mandated by the State, the State provides the nec-

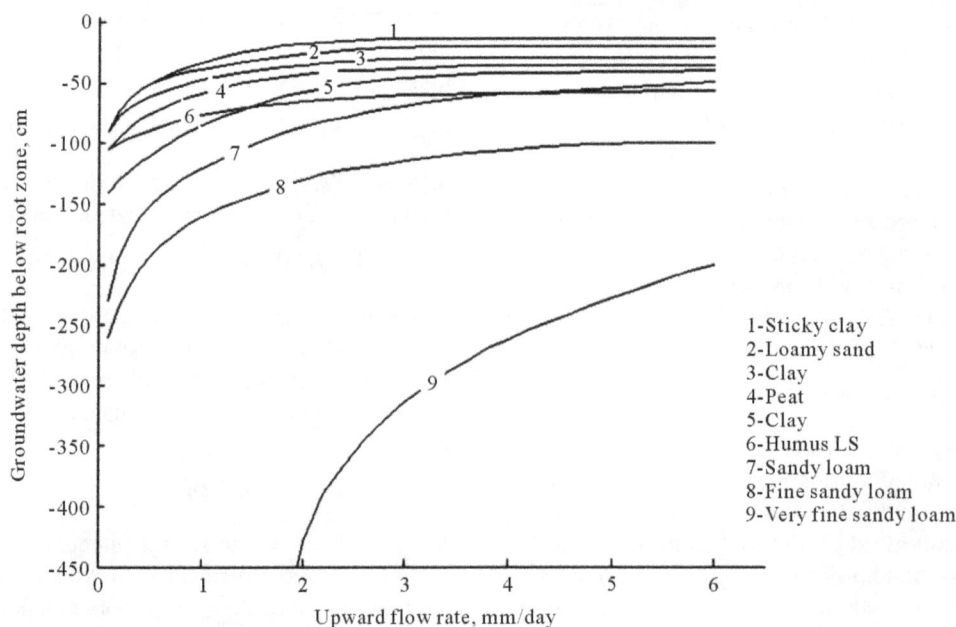

Figure 2. Groundwater contribution to crop rootzone.

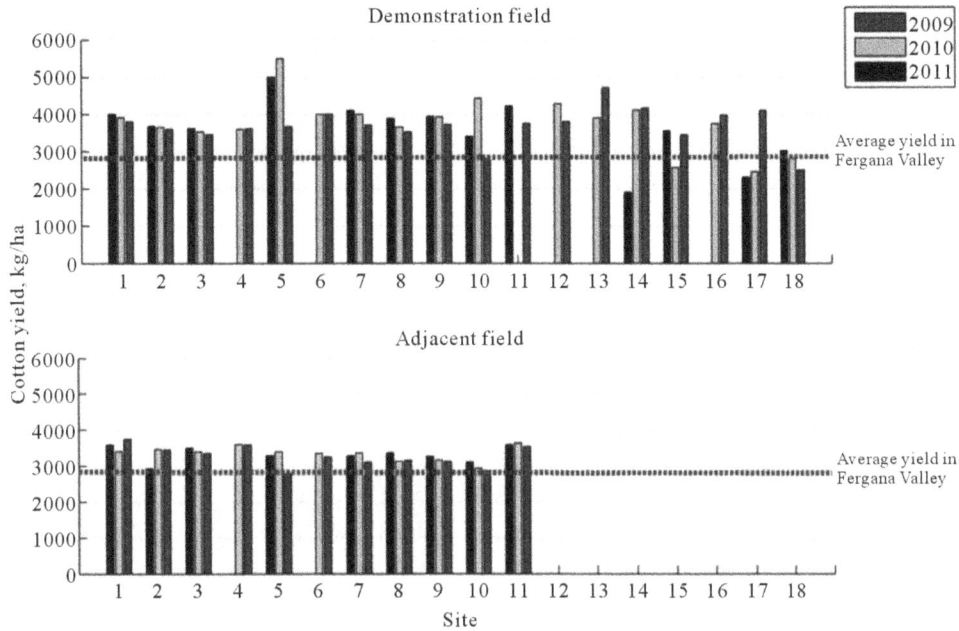

Figure 3. Cotton yields for the demonstration sites and the adjacent fields.

essary credit to farmers for purchasing all the necessary inputs to grow cotton, and the district-level government officials prod the farmers to apply irrigation water on-time according to the irrigation norms (though outdated), and to apply appropriate plant protection measures. Though data on the type and quantity of each input was collected, no information was collected on the quality and the timing of the inputs. In the case of Tajikistan, no such credit is available to farmers, and hence no such monitoring of inputs including water is done by the State. Secondly, as mentioned elsewhere, 2009 and 2010 (more rainfall and more water was available for irrigation during the vegetation period) were considered as wet years whereas 2011 was a dry year (less rainfall and less amount of irrigation water was available during the vegetation period). Yet, on the average, there was no significant difference in the yield of cotton between the wet and dry years. The farmers used irrigation water efficiently by under-irrigating the crop during the dry year. Under-irrigation was practiced not by choice, but by default!

Cotton yields from the adjacent fields are also shown in **Figure 3**. No data was available from the adjacent fields in Tajikistan. As expected, the average yields from the adjacent fields were lower (around 3300 kg/ha) than the average yields (3700 kg/ha) obtained from the demonstration sites in Uzbekistan. The demonstration field farmers implemented a variety of "innovations" or improved agronomic and irrigation practices, as shown in **Table 2**, whereas the adjacent farmers were using one or more of the innovative practices implemented by the demonstration farmers. The average cotton yields in

Fergana Valley were about 2900 kg/ha, indicating that the average crop yields from the demonstration fields were 28% higher than the average crop yields in the area. From the above it is evident that there is a substantial opportunity to increase crop yields, and thus water productivity, in Fergana Valley through a combination of agronomic and irrigation water management intervenetions (**Table 2**). Since there was so much variability in the quality of inputs used at various demonstration sites (including crop varieties), it was not possible to identify the most important factors for increase in crop yields. Also, the yields of adjacent fields increased by 14% (above the average for Fergana Valley) suggesting that, with time, more farmers would adapt these interventions to raise the average yield of cotton in Fergana Valley. Some of the demonstration field farmers were using some additional innovative agronomic practices such as irradiation of seed, plastic mulching, and passing irrigation water through a magnetic field. All of these practices contributed to decent increases in crop yields. However, detailed field investigations are still underway to confirm and document the benefits of using these technologies so that they can be disseminated through the Innovation Cycle.

To estimate water productivity of cotton, the consumptive use of cotton crop was estimated using a simulation model (using **Eqs.7-11** in an Excel Spreadsheet) on all the 18 sites. The total amount of water supplied from all the sources-initial soil-moisture content, groundwater contribution, irrigation, and rainfall was also calculated for all the 18 demonstration sites. Information on daily rainfall amounts (in millimeters) and daily weather conditions was obtained from the nearest weather station for each of the 18 demonstration sites. Then, the Penman-Monteith equation, as described in [9] was used to compute the daily evapotranspiration of a reference crop (short grass), ET_{ref}, for each of the 18 sites. Based upon the depth of the watertable and the soil-texture, daily groundwater contributions to the crop rootzone were estimated using **Figure 2**. Information on dates and amounts of rainfall, daily groundwater contributions, daily ET_{ref}, and dates and amount of irrigation water infiltrated into the crop rootzone for each site was used to calculate soil-moisture balance (**Eq.11**) in the rootzone. In the simulation, the following assumptions were made:

1) The soil-moisture content in the crop rootzone was assumed to be close to field capacity at the beginning of the season.

2) The maximum rooting depth of cotton was assumed to be 1.6 m. The active rooting depth at the beginning of the season was assumed to be 0.15 m, and the rooting depth was assumed to increase to its maximum rooting depth linearly by the end of vegetative period.

3) In situations where there was a high watertable, the maximum rooting depth was set equal to the highest level of the watertable which typically occurred during the second half of the crop growth season.

4) If the calculated soil-moisture content on any given day was higher than the field capacity soil-moisture content for that soil, due to irrigation or rainfall, the soil-moisture content was set equal to the field capacity soil-moisture content for that soil.

These simulated values of daily soil-moisture content

were used to calculate the daily soil-moisture stress coefficient, K_s, using **Eq.8**, which was then used to estimate the daily actual evapotranspiration, ET_a, of cotton. The daily K_c values were obtained by linear interpolation of the values suggested by [9].

The seasonal amount of irrigation water applied, the rainfall amounts received, the groundwater contributions to crop rootzone, and the simulated total consumptive water use of cotton crop were calculated (**Figure 4**) for all the sites. It is clear from **Figure 4** that the seasonal consumptive water use of cotton crop, ET_a, varied from 4500 m^3/ha to 8000 m^3/ha, depending upon the total amount of water supplied (TWA) from all sources, the timing of irrigations and rainfall amounts, and the local climatic conditions. The TWA to fields varied from 5000 m^3/ha to 12,000 m^3/ha. In general, the lowest total water applied and the lowest estimated ET_a occurred in 2011 because it was a dry year! On the average, the TWA values were higher in Tajikistan than in Uzbekistan. This may be partly due to the tighter monitoring that is exercised on following the irrigation norms in Uzbekistan. Based upon the data in **Figure 4**, a quadratic relationship ($R^2 = 0.70$) was found between TWA and ET_a, with ET_a values flattening at higher values of TWA (**Figure 5**). This relationship between TWA and ET_a is not a new finding but confirms the existing knowledge [10]. The ET_a value reaches an upper limit under a given set of climatic conditions; hence, at higher values of TWA, a large decrease in TWA results in a small decrease in ET_a, and thus a small decrease in crop yields. This probably explains why there was no significant difference in the

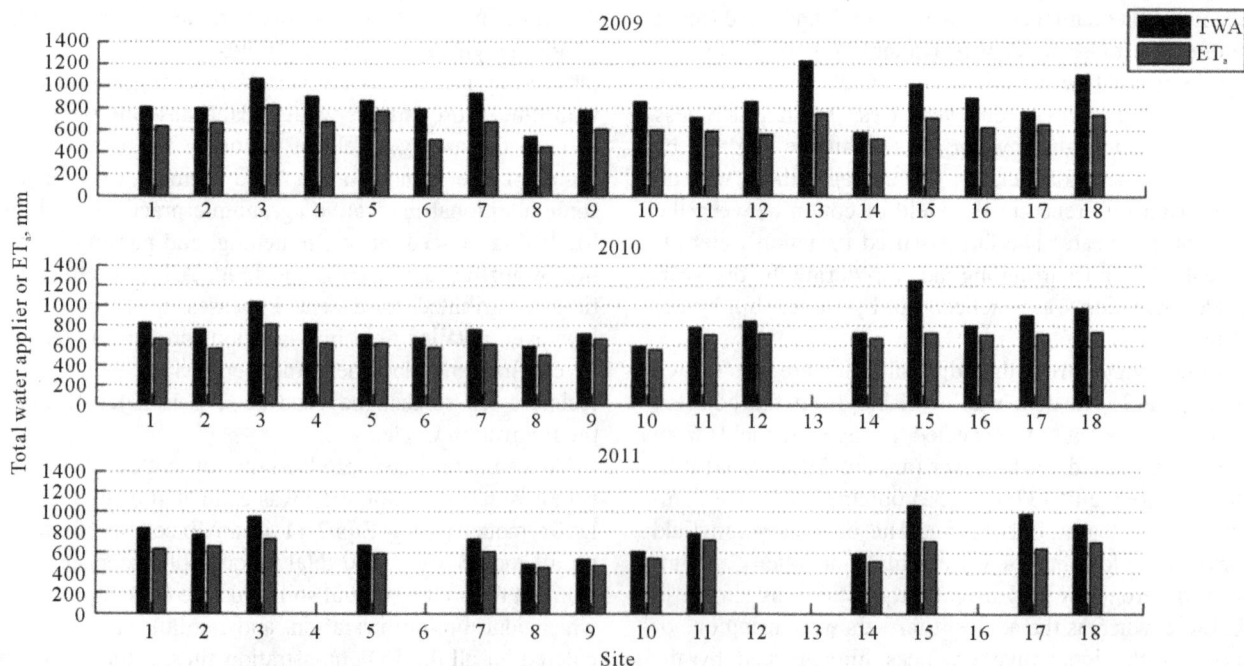

Figure 4. Total water applied and simulated ET_a values for the demonstration sites.

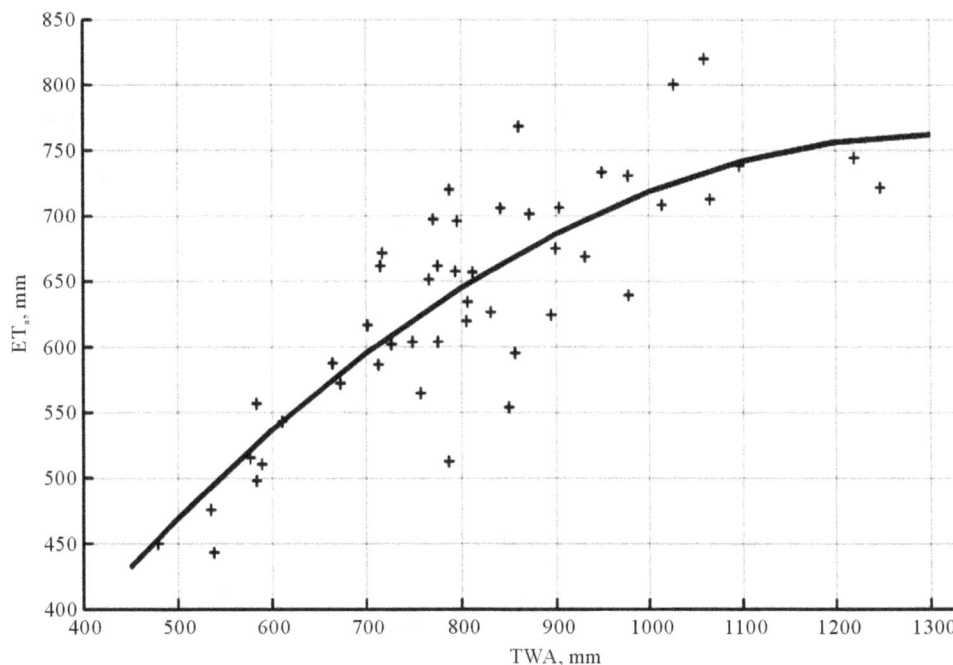

Figure 5. Relationship between total water applied and simulated ET_a values for the demonstration sites.

average yield of cotton (**Figure 3**) between 2011 (a dry year) and 2009 and 2010 (wet years).

A regression analysis was also performed between crop yield data (**Figure 3**) and the simulated ET_a data (**Figure 4**) from the demonstration fields. No definite correlation was found between the two variables. This is no surprise because, though water is the most important input for crop production, crop yields at any given location also depend upon a variety of other factors such as crop variety, seed quality and seeding rate per hectare, soil fertility and fertilizer management, plant protection measures used, climatic conditions and degree of unevenness of land surface, etc. The collected data was not sufficient to do a multiple regression analysis between the input variables and crop yields in order to identify the most important input variables for increasing crop yields. Some or all of these inputs, in addition to water, contributed to achieve higher crop yields.

Water productivity (WP) values (**Figure 6**) were calculated based upon crop yield (**Figure 3**) and ET_a values (**Figure 4**). The WP values ranged from 0.35 kg/m³ to 0.89 kg/m³, with an average value of 0.58 kg/m³. The average TWA value was 28% higher than the ET_a value, indicating that, on the average, the field irrigation systems were operating at 78% application efficiency. The remaining 22% was lost from the fields. However, from a basin perspective, if all of this 22% water was used by some other farmers somewhere else in the project area or returned to the same stream, then this water was not really lost, implying that no real water savings would

accrue by improved application efficiency at field level. However, since more than 60% of the area irrigated in Tajikistan and Uzbekistan receives pumped water, there would be considerable savings in energy used for pumping irrigation water if the application efficiency is further improved. In addition, there would be a proportionate decrease in the amount of salts returning to the stream. Improved application efficiency is achieved by decreasing surface runoff and/or deep percolation water (from fields) through improved layout of irrigation systems and proper irrigation scheduling [11]. However, improved application efficiency comes at a cost! Therefore, one has to weigh the costs and benefits (reduced energy costs, reduction in salinity of downstream areas and the resulting increases in crop yields, improved reliability of water supply to downstream areas) of improving application efficiency at field level. In this research no WP_I values were calculated because no information was available on Y_D values for the demonstration sites.

Sometimes the average values do not tell the whole story. Therefore, we need to look at the range of values for these variables. For example, the WP values ranged from 0.35 kg/m³ to 0.89 kg/m³ which suggest that there is a significant potential for the farmers that are at the lower-end to improve their water productivity through improved water management and/or agronomic practices, depending upon their situation. Similarly, the range of values for ET_a (4500 m³/ha to 8000 m³/ha) suggest that some fields are under-irrigated. Yet, the decrease in yield from the under-irrigated, *i.e.* deficit irrigated, fields was

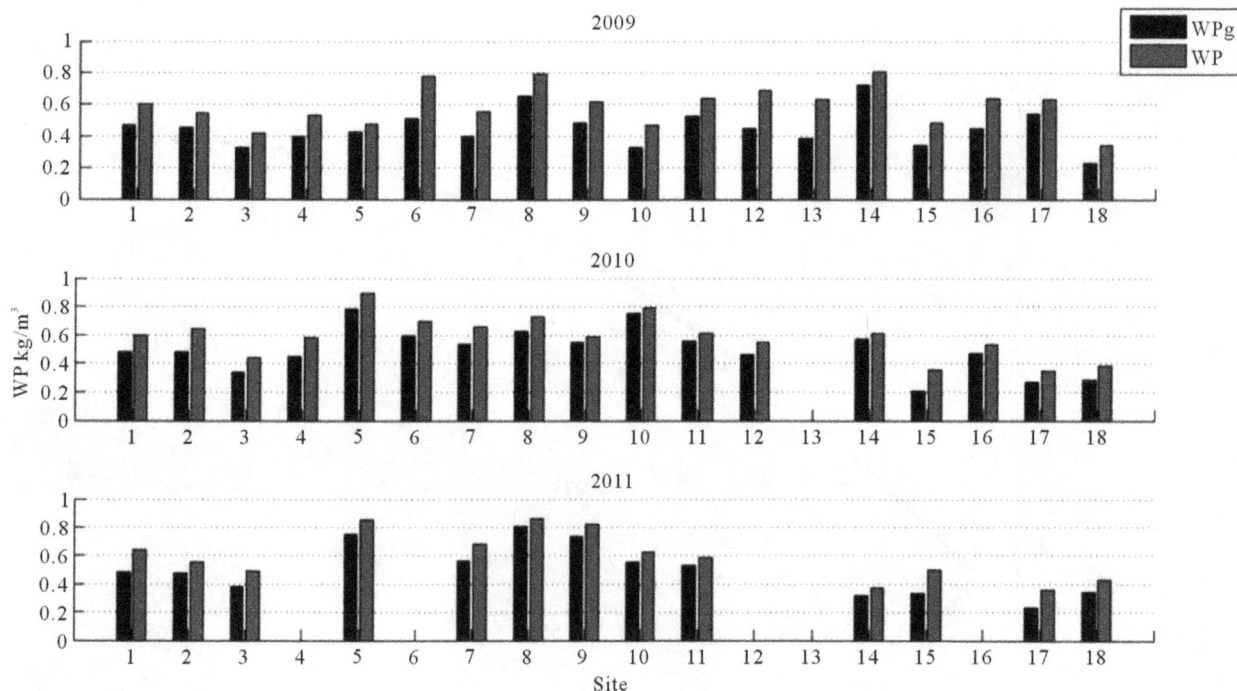

Figure 6. Water productivity and gross water productivity of cotton.

not significantly different from the fields that did not experience water stress, resulting in higher WP. Considering the range of values for TWA and ET_a indicates that the application efficiency can be significantly lower and higher than the average application efficiency of 78% in this case. Significantly higher than 78% application efficiency cannot be achieved without some level of under-irrigation, and consequently some reduction in crop yields. Conversely, significantly lower than 78% application efficiency would result in waterlogging (though it is equally likely that the inefficiency is due to high run-off than deep percolation) and reduced crop yields. This shows the potential for increasing WP by addressing the issue of inequity in water distribution which is a major problem in irrigation projects [12]. Thus there are two avenues for improving water productivity in irrigation projects-through improved technical and agronomic practices at field level, and by improving equity and reliability in water supply to farmers.

Finally, the net profit from crop production was calculated as the difference between gross returns from crop production and the cost of production (**Table 3**). The net profits ranged from $173 to $1911, depending upon the quality of cotton lint, irrigation and financial management skills of the farmer, and the market price for cotton. The net profits were higher in years 2010 and 2011 compared to year 2009 because the market price for cotton was higher during 2010 and 2011. The average net profit was higher in Tajikistan compared to Uzbekistan because the farmers in Tajikistan sold their cotton in

open market compared to the farmers in Uzbekistan where the cotton was sold to the government at the price fixed by the government. Comparing the data on net profits (**Table 3**) with the data on WP (**Figure 6**), it is clear that a high value of WP does not necessarily mean high net profit to the farmer. Because of the earlier mentioned co-benefits of "efficient" irrigation [5], the irrigation system managers would be more interested in improving water productivity through efficient irrigation (deficit or under-irrigation) practices, whereas the farmers are more interested in increasing the net profit per unit area. For farmers, in general, more water means more yields. These objectives are conflicting with each other. From their experience in Tunisia [13], a combination of water pricing and subsidies for improved technologies are required to reconcile this divergence of interests of farmers and irrigation system managers.

5. SUMMARY AND CONCLUSIONS

A total of 18 demonstration fields were selected in Fergana Valley (13 in Uzbekistan and 5 in Tajikistan) where the farmers were provided with a set of agronomic and irrigation management interventions to improve yields and water productivity of cotton. The average yields of cotton, for the years 2009, 2010 and 2011, from the demonstration fields and the adjacent fields were, respectively, 28% and 14% higher than the average yields for Fergana Valley. The total water applied (TWA) and the ET_a values were calculated for all the 18 demon-

Table 3. Cost of production, gross returns, and net returns from the demonstration sites for years 2009-2011.

Site №	Country	Province	2009					2010					2011				
			Area, ha	Total expenses $/ha	Yield kg/ha	Gross revenue $/ha	Net profit $/ha	Area, ha	Total expenses $/ha	Yield kg/ha	Gross revenue $/ha	Net profit $/ha	Area, ha	Total expenses $/ha	Yield kg/ha	Gross revenue $/ha	Net profit $/ha
1	Uz	Andijan	6	834	3800	1393	560						6	1472	4000	2347	875
2	Uz	Andijan	14	821	3600	1212	392						6	1079	3600	1532	453
3	Uz	Andijan	6.5	1096	3460	1269	173						11.5	1178	3620	1572	395
4	Uz	Andijan	7	855	3610	1221	365	2	948.2	3600	1310	362					
5	Uz	Andijan	16.4	733	3660	1637	904	3	1017	5500	2879	1862	3	1379	5000	2559	1179
6	Uz	Fergana	4.8	857	4000	1360	503	4.5	577.1	4000	1593	1015					
7	Uz	Fergana	2	1288	3700	1607	320	2	731.7	4000	1246	514	1	1415	4100	1768	353
8	Uz	Fergana	4	648	3520	1412	763	4	982.9	3650	1934	951	0.15	1050	3887	1623	573
9	Uz	Fergana	2.5	838	3720	1116	278	4	622.8	3925	1557	934	3	756	3930	1631	876
10	Uz	Fergana	2.2	830	2800	1020	190	4.2	573	4420	1691	1118	2.5	1122	3380	1458	336
11	Uz	Namangan	10	684	3750	1294	610	6	1026.4	4280	1315	288	10	834	4210	2689	1855
12	Uz	Namangan	2	735	3800	1405	670	2	506.6	3900	1545	1038					
13	Uz	Namangan	2	857	4700	1589	732										
14	Taj	Soght	2.4	873	4160	1725	852	2.4	1004.4	4100	2596	1592	2.4	560	1900	1120	560
15	Taj	Soght	4.2	763	3440	3422	2659	4.2	884.1	2580	1976	1092	4.2	877	3548	2765	1888
16	Taj	Soght	1	1067	3970	1824	757	1	912	4150	2623	1711					
17	Taj	Soght	1.6	944	4100	1699	755	1.3	749.2	3200	2023	1273	1.3	651	2308	1360	710
18	Taj	Soght	2	966	2500	2877	1911	2	1112.9	2800	1949	836	2	1115	3020	2955	1839

stration sites for the irrigation seasons of 2009, 2010 and 2011, and these values for TWA and ET_a ranged, respectively, from 5000 m^3/ha to 12,000 m^3/ha and 4500 m^3/ha to 8000 m^3/ha, suggesting a quadratic relationship between TWA and ET_a, with ET_a values flattening off at higher values of TWA. During the three irrigation seasons, the calculated WP values ranged from 0.38 kg/m^3 to 0.89 kg/m^3, indicating that the farmers with a WP value less than the average WP of 0.58 kg/m^3 have a high potential to increase crop yields (and thus WP) through improved irrigation and agronomic practices (including selection of appropriate crop variety). Cotton yields in year 2011 (dry year) were not significantly different from the yields achieved during 2009 and 2010 (wet years), which is basically explained by the quadratic relationship between TWA and ET_a.

On the average, the TWA values were 28% higher than the ET_a, suggesting an average application efficiency of 78%. The remaining 22% of the water is lost from individual fields, but may or may not be lost from the basin. This needs to be carefully evaluated for each project. Considering the fact that more than 60% of the water used for irrigation in Tajikistan and Uzbekistan is pumped from rivers and collector drains, even if all this 22% of the water returns to the stream without any degradation in water quality, considerable energy savings would accrue from improved water management at field level. Since salinity of return flows is also a major issue in Central Asia, improved efficiency at field level would alleviate the problems of salinity in lower reaches of the river basins. An average application efficiency of 78% suggests that there were some fields that were under-irrigated (yield losses due to water stress), and some fields that were over-irrigated (yield losses due to leaching of fertilizers and temporary waterlogging conditions). Addressing the issue of inequity and reliability in water supply, through improved water management, would also increase crop yields and water productivity from project areas. In general, there is significant potential for increasing water productivity in Central Asia through a combination of improved agronomic and irrigation practices at field level, and improved equity and reliability in water delivery to fields.

6. ACKNOWLEDGEMENTS

The funding for this research was provided by the Swiss Agency for Development and Cooperation (SDC)-Tashkent office, Uzbekistan, as part of the Water Productivity Improvement at Plot Level (WPI-PL) project. Their financial support is highly appreciated.

REFERENCES

[1]   Abdullaev, I. and Molden, D. (2004) Spatial and temporal variability of water productivity in the Syr Darya basin, Central Asia. *Water Resources Research*, **40**, 1-6.

[2]   Allen, R., Pereira, L.S., Raes, D. and Smith, M. (2000) Crop Evapotranspiration. FAO Irrigation and Drainage Paper 56, Rome.

[3]   Evett, S.R. (2002) Water and energy balances at soil-plant-atmosphere interfaces. In: Warrick, A.A., Ed., *The Soil Physics Companion*, CRC Press LLC, Boca Raton.

[4]   Food and Agriculture Organization (1979) Yield Response to Water. FAO Irrigation and Drainage Paper 33, Rome.

[5]   Food and Agriculture Organization (1984) Crop water requirements. FAO Irrigation and Drainage Paper 24, Rome.

[6]   Gleick, P.H., Christian-Smith, J. and Cooley, H. (2011) Water use efficiency and productivity: Rethinking the basin approach. *Water International*, **36**, 784-798.

[7]   Ibragimov, N., Evett, S.R., Esanbekov, Y., Kamilov, B.S., Mirzaev, L. and Lamers, J.P.A. (2007) Water use efficiency of irrigated cotton in Uzbekistan under drip irrigation and furrow irrigation. *Agricultural Water Management*, **90**, 112-120.

[8]   Kranz, W., Eisenhauer, D. and Retka, M. (1992) Water and energy conservation using irrigation scheduling with center-pivot irrigation systems. *Agricultural Water Management*, **22**, 325-334.

[9]   Luquet, D., Vidal, A., Smith, M. and Dauzat, J. (2005) "More crop per drop": How to make it acceptable for farmers? *Agricultural Water Management*, **76**, 108-119.

[10] Molden, D.J., Murray-Rust, H., Sakthivadivel, R. and Makin, I. (2003) A water productivity framework for understanding and action. In: Kijne, J.W., Barker, R. and Molden, D.J., Eds., *Water Productivity in Agriculture: Limits and Opportunities for Improvement*, CABI Publishing, CAB International, Wallingford.

[11] Murray-Rust, H., Abdullaev, I., Ul-Hassan, M. and Horinkova, V. (2003) Water productivity in the Syr Darya River Basin. *Research Report* 67, IWMI, Colombo.

[12] Pereira, L.S., Dukhovny, V.A. and Horst, M.G., Eds. (2005) Irrigation management for combating desertification in the Aral Sea Basin: Assessment and tools. Vita Color Press, Tashkent.

[13] Reddy, J.M., Matyakubov, B., Jumaboev, K. and Eshmuratov, D. (2012) Analysis of equity and adequacy in water distribution within water users associations of Fergana Valley in Uzbekistan. *Irrigation and Drainage Systems*, unpublished.

14

Global sensitivity analysis for choosing the main soil parameters of a crop model to be determined

Hubert Varella[1*], Samuel Buis[2], Marie Launay[3], Martine Guérif[2]

[1]Syngenta Crop Protection Münchwilen AG, Stein/AG, Switzerland; [*]Corresponding Author
[2]INRA, UMR 1114 INRA-UAPV EMMAH, Avignon, France
[3]INRA, Unite Agroclim, Avignon, France

ABSTRACT

The use of a crop model like STICS for appropriate management decision support requires a good knowledge of all the parameters of the model. Among them, the soil parameters are difficult to know at each point of interest and costly techniques may be used to measure them. It is therefore important to know which soil parameters need to be determined. It can be stated that those which affect significantly the output variable deserve an accurate determination while those which slightly affect the model output variable do not. This paper demonstrates how a global sensitivity analysis method based on variance decomposition can be applied on soil parameters in order to divide them in the two categories. The Extended FAST method applied to the crop model STICS and a set of 13 soil parameters first allows to calculate the part of variance explained by each soil parameter (giving global sensitivity indices of the soil parameters) and the coefficient of variation of the output variables (measuring the effect of the parameter uncertainty on each variable). These metrics are therefore used for deciding on the importance of the parameter value measurement. Different output variables (Leaf Area Index and chlorophyll content) are evaluated at different stages of interest while others (crop yield, grain protein content, soil mineral nitrogen) are evaluated at harvest. The analysis is applied on two different annual crops (wheat and sugar beet), two contrasted weather and two types of soil depth. When the uncertainty of the output generated by the soil parameters is large (coefficient of variation > 1/3), only the parameters having a significant global sensitivity indices (higher than 10%) are retained. The results show that the number of soil parameters which de-serve an accurate determination can be significantly reduced by the use of this relevant method for appropriate management decision support.

Keywords: Global Sensitivity Analysis; Uncertainty Analysis; Soil Parameters; Crop Model STICS; Management Decision Support; Agro-Environmental Variables

1. INTRODUCTION

Dynamic crop models are very useful to predict the behavior of crops in their environment and are widely used in a lot of agro-environmental work such as crop monitoring, yield prediction or decision making for cultural practices [1-3]. These models usually have many parameters and their spatial application for agro-environmental predictions is difficult without a good knowledge of these parameters [4-6].

The crop model parameters can be divided in three groups: those related to the agricultural techniques, those related to the genotype of the crop and those related to the soil properties. Generally, agricultural techniques are quite easy to know as they are those used by the farmer. Crop parameters can be determined from literature, or estimated from experimental work or calibrated on a large database [7-9]. The knowledge of the soil parameters is an important issue because the spatial variability of the crop model simulations depends for a large part on the soil parameter values [10] and predictions obtained with the model are not reliable when inaccurate parameter values are used. This knowledge may be especially difficult to acquire because parameter values can greatly vary in space [11,12]. The use of existing soil maps and associated pedotransfer functions can be considered where accurate soil map are available [13]; but in many cases, the spatial accuracy of the map is too limited for accurate applications such as for example precision agriculture [14]. In those cases, these parameters should be determined in another way. Measurements can be made directly with soil sampling analysis at different locations of the study area or indirectly by using electrical geophysical measurements [15,16]. Whatever the technique of measurement used, it is submitted to practical limitations and to time and financial constraints. Another way of gathering quite accurate values on soil parameters consists in estimating them through an inverse modeling approach using a crop model and observations on the crop state variables [11,17,18]. However, the soil parameters may not have the same contribution to the performance of the crop model and do not require the same precision of determination for a given objective: some of them deserve an accurate determination while the others can be fixed at nominal values [19,20]. Considering this aspect, the practical limitations of soil parameter measurements, as well as time and financial constraints should be reduced by considering only a subset of the crop model soil parameters depending on the objective and configuration of the study.

The combination of uncertainty analysis and sensitivity analysis techniques should help in identifying these key parameters. The objective of sensitivity analysis is to study how the variation of selected outputs of a model can be apportioned to different sources of variation [21]. In particular, sensitivity analysis methods can be used to rank uncertain input factors with respect to their effects on the model output variables by calculating quantitative or qualitative indices. Nevertheless, the fact that some factors are detected as important for a given output variable on the basis of sensitivity analysis results is not sufficient to decide that the uncertainties on these factors should be reduced. Indeed, if the variation of the considered output variable induced by the uncertainties on the factors is low, the results of sensitivity analysis on this output variable should not be taken into account. The description and quantification of these variations is the objective of uncertainty analysis.

Some authors [19,22-26] used uncertainty analysis techniques to quantify the uncertainties of a selection of crop models output variables generated by uncertainties on some selections of input parameters. Others authors [27-32] used global sensitivity analysis to evaluate the contribution of the parameters to the variance of the model output variables. In this study we propose a helpful combination of these techniques to identify soil parameters that need particular accuracy for simulating a set of given output variables of interest in spite of the financial and practical interests of such a study.

A variance-based sensitivity analysis method is used in order to rank the soil parameters relatively to their importance on some selected outputs of the crop model STICS (Simulateur multidisciplinaire pour les Cultures Standard) [33] and to select those which deserve an ac-

curate determination by considering also the coefficient of variation of the outputs, that is the variation of the outputs compared to their magnitude. We considered 13 soil parameters and their effects on 5 dynamic output variables of the STICS crop model, at different phenological stages, which are involved in decision making for crop management. Two different crops (winter wheat and sugar beet) growing on different seasons are considered in order to illustrate the impact of soil properties on crop growth. Each crop considered is simulated under different pedological conditions and weather.

2. METHODS

2.1. The STICS Model

The STICS model [33,34] is a nonlinear dynamic crop model simulating various crops. For a given crop, STICS takes into account the weather, the type of soil and the cropping techniques used, and simulates the carbon, water and nitrogen balances of the crop-soil system on a daily time-scale. In this study, winter wheat and sugar beet crops are simulated. The crop is essentially characterized by its above-ground biomass carbon and nitrogen, and leaf area index. The soil is considered as a series of layers where the transfer of water and nitrate is described by a reservoir-type analogy. The main outputs are agronomic variables (yield, grain protein content for wheat) as well as environmental variables (water and nitrate leaching).

The STICS model includes more than 200 parameters. The global sensitivity analysis described in this study only concerns the soil parameters. The values of the parameters related to the crop have been determined either from literature, from experimental works conducted on specific processes included in the model (e.g. mineralization rate, critical nitrogen dilution curve etc.) or from a calibration based on a large experimental database [35]. Cropping techniques and soil parameters ranges are described in Section 2.5.

2.2. The Soil Parameters

Among the available options for simulating the soil system, the simplest was chosen in this study, by considering only the transfers in the microporosity and ignoring those in the macroporosity, the cracks, pebbles, and processes like capillary rise and nitrification. We then considered the soil as a succession of two horizontal layers, the top layer having a thickness fixed at 30 cm.

These different hypotheses made on the soil description lead to consider a set of 13 soil parameters, defined in **Table 1**. They refer to permanent characteristics and initial conditions. Among the permanent characteristics, clay and organic nitrogen content of the top layer are

Table 1. Definition of the 13 soil parameters and their ranges of variation.

Parameter	Definition	Range	Unit	Label
argi	Clay content of the 1st layer	14 - 37	%	*ar*
Norg	Organic nitrogen content of the 1st layer	0.049 - 0.131	%	*N*
calc	Limestone content of the 1st layer	0 - 28	%	*c*
albedo	Albedo of the bare dry soil	0.13 - 0.31	-	*al*
q_0	Threshold of daily evapotranspiration	7.5 - 14.5	mm	*q*
ruisolnu	Fraction of drip rainfall on a bare soil	0 - 0.065	-	*r*
epc(2)	Thickness of the 2nd layer	0 - 70 or 50 - 130[*]	cm	*e*
DA(1)	Bulk density (1st layer)	1.22 - 1.42	-	*D1*
DA(2)	Bulk density (2nd layer)	1.39 - 1.59	-	*D2*
HCC(1)	Water content at field capacity (1st layer)	14 - 30	$g·g^{-1}$	*H1*
HCC(2)	Water content at field capacity (2nd layer)	14 - 30	$g·g^{-1}$	*H2*
Hinit	Initial water content	4 - 29	% of weight	*h*
NO3init	Initial mineral nitrogen content	4 - 21.5 or 12 - 55[**]	$kg·N·ha^{-1}$	*n*

[*]The first range determine a shallow soil and the second determine a deep soil; [**]The first range concern the wheat (cultivated after beet) and the second concern the beet (cultivated after a bare soil).

involved mainly in organic matter decomposition processes. Water content at field capacity of both layers affects the water (and nitrogen) movements and storage in the soil reservoir and the thickness of the second layer defines the volume of the reservoir. The initial conditions correspond to the water and nitrogen content, *Hinit* and *NO3init*, at the beginning of the simulation, in this case the sowing date.

2.3. Model Output

In this study, the STICS output variables of interest are:

1) The amount of nitrogen absorbed by the plant (*QN*) and the leaf area index (*LAI*) at two (for sugar beet) or three (for wheat) different key stages during the season, which are important variables for making a diagnosis on crop growth,

2) The yield, and the mineral nitrogen content in the soil at harvest (for both crops) plus the grain protein content (for wheat), which are of particular interest for decision making, especially for monitoring nitrogen fertilization.

The different stages of interest and the corresponding variables are displayed for each crop on **Table 2**. For the wheat, the three key stages concern the maximum leaf growth rate-beginning of stream elongation-(*AMF*), the maximum leaf area- or booting-(*LAX*) and the flowering (*FLO*). For the sugar beet, the two key stages concern the maximum leaf growth rate (*AMF*) and the maximum leaf

area (*Summer*).

2.4. Sensitivity and Uncertainty Analysis

Among the available methods of sensitivity analysis, variance-based methods are well adapted for non-linear models that need less than 1 minute for a simulation [36]. These methods are widely used in different contexts [27,28,30-32]. Their principle is to evaluate the contribution of the given uncertain factors to the variance of the model output variables selected. We will describe in this section the sensitivity indices that can be computed with these methods and the EFAST variance-based method we have used in this study to compute these indices. Uncertainty analysis is performed here by computing the coefficient of variation of the output variables considered from the simulations realized for the sensitivity analysis.

2.4.1. Sensitivity Indices and Coefficient of Variation

We note further *Y* an output variable of STICS. *Y* will represent in turn *LAI* and *QN* computed at the different phenological stages and the variables computed at harvest. The total variance of *Y*, *V(Y)*, is partitioned as follows [37]:

$$V(Y) = \sum_{i=1}^{13} V_i + \sum_{1 \le i < j \le 13} V_{ij} + \cdots + V_{1,2,\cdots,13} \quad (1)$$

where *V(Y)* is the total variance of the output variable *Y* induced by the 13 soil parameters *θ*, $V_i = V\left[E\left(Y \mid q_i\right)\right]$

Table 2. Definition of the variables and the stages of interest.

Crop simulated	Variable of interest	Stage of interest	Signification of the stage
Wheat	LAI and QN	AMF	Stage of maximum leaf growth rate (beginning of steam elongation)
		LAX	Stage of maximum leaf area (booting)
		FLO	Flowering
	Yld, Prot and Nit	Harvest	Harvest
Sugar beet	LAI and QN	AMF	Stage of maximum leaf growth rate
		Summer	Day where maximum leaf area is achieved in most cases
	Yld and Nit	Harvest	Harvest

measures the main effect of the parameter θ_i, $i = 1, \cdots,$ 13, and the other terms measure the interaction effects. Decomposition (1) is used to derive two types of sensitivity indices defined by:

$$S_i = \frac{V_i}{V(Y)} \qquad (2)$$

$$ST_i = \frac{V(Y) - V_{-i}}{V(Y)} \qquad (3)$$

where V_{-i} is the sum of all the variance terms that do not include the index i.

S_i is the first-order sensitivity index of the ith parameter. It computes the fraction of Y variance explained by the uncertainty of parameter θ_i and represents the main effect of this parameter on the output variable Y. ST_i is the total sensitivity index of the ith parameter and is the sum of all effects (first and higher order) involving the parameter θ_i. S_i and ST_i are both in the range $(0, 1)$, low values indicating negligible effects, and values close to 1 huge effects. ST_i takes into account both S_i and the interactions between the ith parameter and the 12 other parameters, interactions which can therefore be assessed by the difference between ST_i and S_i. The interaction terms of a set of parameters represent the fraction of $V(Y)$ induced by these parameters but that is not explained by the sum of their main effects. The two sensitivity indices S_i and ST_i are equal if the effect of the ith parameter on the model output is independent of the values of the other parameters: in this case, there is no interaction between this parameter and the others and the model is said to be additive with respect to θ_i. Selecting the parameters that have a negligible effect and that can thus be fixed to nominal values is called factor fixing in the literature [38]. Total effects must be considered in this case. Indeed, a factor that has a small main effect but a medium to high total effect cannot be considered as negligible: its effect depends on the value of other uncertain factors and can be important in some cases.

The coefficient of variation of the output variable Y can be calculated by:

$$CV(Y) = \frac{\sqrt{V(Y)}}{\bar{Y}} = \frac{\sigma(Y)}{\bar{Y}} \qquad (4)$$

where, $\sigma(Y)$ is the standard deviation of the output variable Y and $\bar{Y}$ is the mean of Y induced by the 13 soil parameters θ.

2.4.2. Extended FAST

Sobol's method and Fourier Amplitude Sensitivity Test (FAST) are two of the most widely used methods to compute S_i and ST_i [11]. We have chosen here to use the extended FAST (EFAST) method, which has been proved, in several studies [30,39,40], to be more efficient in terms of number of model evaluations than Sobol's method. The main difficulty in evaluating the first-order and total sensitivity indices is that they require the computation of high dimensional integrals. The EFAST algorithm performs a judicious deterministic sampling to explore the parameter space which makes it possible to reduce these integrals to one-dimensional ones using Fourier decompositions. The reader interested in a detailed description of EFAST can refer to [40].

We have implemented the EFAST method in the Matlab® software, as well as a specific tool for computing and easily handling numerous STICS simulations. The uncertainties considered for the soil parameters are described in the next section. A preliminary study of the convergence of the sensitivity indices allowed us to set the number of simulations per parameter to 5000, leading to a total number of model runs of $13 \times 5000 = 65,000$ to compute the main and total effects for all output variables and parameters considered here. One run of the STICS model takes about 1s with a Pentium 4, 2.9 GHz processor.

2.5. Data

In this study, we have considered two crops: winter wheat and sugar beet. This allows to illustrate the differ-

ence of sensitivities of different crops to the soil properties. For the same reason, each crop is simulated for two
different weathers and two different types of soil depth.
The weather data were obtained from the meteorological
station of Roupy (49.48°N, 3.11°E). The first set of data
is chosen to characterize a dry weather (1975-1976) and
the second set is chosen to characterize a wet weather
(1990-1991). **Table 3** shows the amount of rainfall calculated for each season and weather data set. The wheat
crop simulated in this study is sown on October 30th
while the sugar beet crop is sown on March 30th. The
amount of fertilizer provided on wheat varies between
200 kg (shallow soil) and 240 kg (deep soil), while the
amount provided on sugar beet varies between 150 kg
(shallow soil) and 200 kg (deep soil).

The range of parameter values considered in this study
correspond to the soil description of the precision agriculture experimental site in northern France near Laon,
Picardie (Chambry 49.35°N, 3.37°E) [41]. In this study,
the uncertainties of these 13 soil parameters are observed
in the literature (for parameters related to albedo, evapotranspiration or drip rainfall) or measured in the experimental site (for the other parameters), and their
ranges of variation are displayed on **Table 1**. Concerning
the parameter $NO3init$ two ranges of variation are considered, depending on the crop cultivated just before the
one considered: in this study, the wheat is cultivated after
sugar beet and the sugar beet is cultivated after a bare
soil. The different previous crops used determine the
quantity of nitrogen $NO3init$ at the beginning of the corresponding crop season. The two different types of soil
depth are defined by their ranges of variation (**Table 1**)
and correspond to a shallow soil and a deep soil. The
uncertainties considered in the global sensitivity analysis
for the soil parameters are assumed independent and follow uniform distributions. The ranges of variation of the
distributions are given in **Table 1**.

3. RESULTS AND DISCUSSION

Only the main results of the study are presented here
for the sake of clarity. These results concern: 1) wheat
crop simulated with dry, then wet weather and a shallow
soil and 2) sugar beet crop simulated with dry, then wet
weather and a deep soil.

3.1. Global Sensitivity Analysis

Figure 1 shows the sensitivity indices calculated for
the 13 soil parameters and for each output variable of the
wheat crop simulated with a dry weather and a shallow
soil. For the early stage the initial water content is dominant because in the considered weather, the rainfall is
light in autumn when the wheat is sown (see **Table 3**): at
the stage AMF (**Figure 1(a)**), $Hinit$ is the only parameter

Table 3. Amount of rainfall (in mm) calculated for each season
and weather.

	Spring	Summer	Autumn	Winter
Dry weather	343.4	167.8	222.4	218.8
Wet weather	361.4	247.9	239.4	316.4

contributing (for more than 90%) to the output variance
for both variables LAI and QN. In the later stages, the
effects of parameters evolve with the soil volume explored by the roots (first layer, then second one), up to
the flowering stage where the root development is
maximum: at the stage LAX (**Figure 1(b)**) and FLO
(**Figure 1(c)**), the effect of $Hinit$ disappears and that of
$HCC(1)$ and $epc(2)$ increase, with a dominant position of
$epc(2)$. At $Harvest$ (**Figure 1(d)**), the variables are much
sensitive to $epc(2)$ followed by $HCC(1)$, $HCC(2)$ and
$Hinit$ for the three variables and $albedo$ for the variable
protein of the grain. In those conditions of dry weather
and shallow soil, the parameters related to water availability ($epc(2)$, $HCC(1)$, $HCC(2)$ and $Hinit$) are the main
parameters contributing to the variance of the outputs.
Those concerned by the turnover of organic nitrogen in
the soil are not concerned, because the water stress is the
dominant limiting factor and also because the mineralization processes are reduced by dry conditions.

When considering wet conditions (**Figure 2**), the water stress is not so much a limiting factor: maximum LAI
is equal to 3.61 in average, whereas it is equal to 2.57 in
dry conditions (see **Table 4**). The roots grow more rapidly at the beginning of the season and the size of the soil
reservoir (via the parameter $epc(2)$) is important since
the AMF stage: the depth of roots is equal to 55.84 in
average (3 months after the sowing), whereas it is equal
to 45.62 in dry conditions (see **Table 4**). Moreover, in
these conditions, the mineralization of the soil organic
matter is increased and the effects of the concerned parameters $argi$ and $Norg$ do so: the cumulative mineral
nitrogen arising from humus is equal to 23.95 in average
(at stage LAX), whereas it is equal to 18.09 in dry conditions (see **Table 4**). This does not seem to influence the
effects of the different parameters on LAI at stage AMF
since they are very similar to these obtained with the dry
weather. On the contrary, the sensitivities of the variable
QN to the different parameters are very different from the
ones obtained with a dry weather: there is no contribution of $Hinit$ but $epc(2)$, $HCC(1)$ and parameters involved in the mineralization process ($argi$, $Norg$ and
$NO3init$) contribute to the output variance.

This is also the case for both LAI and QN on later
stages, with an increasing dominancy of $epc(2)$. At $Har-$
$vest$ (**Figure 2(d)**), the variables are sensitive to the parameters $epc(2)$, $HCC(1)$ and $HCC(2)$ with still a slight
sensitivity to $argi$ and $Norg$ for the soil mineral nitrogen

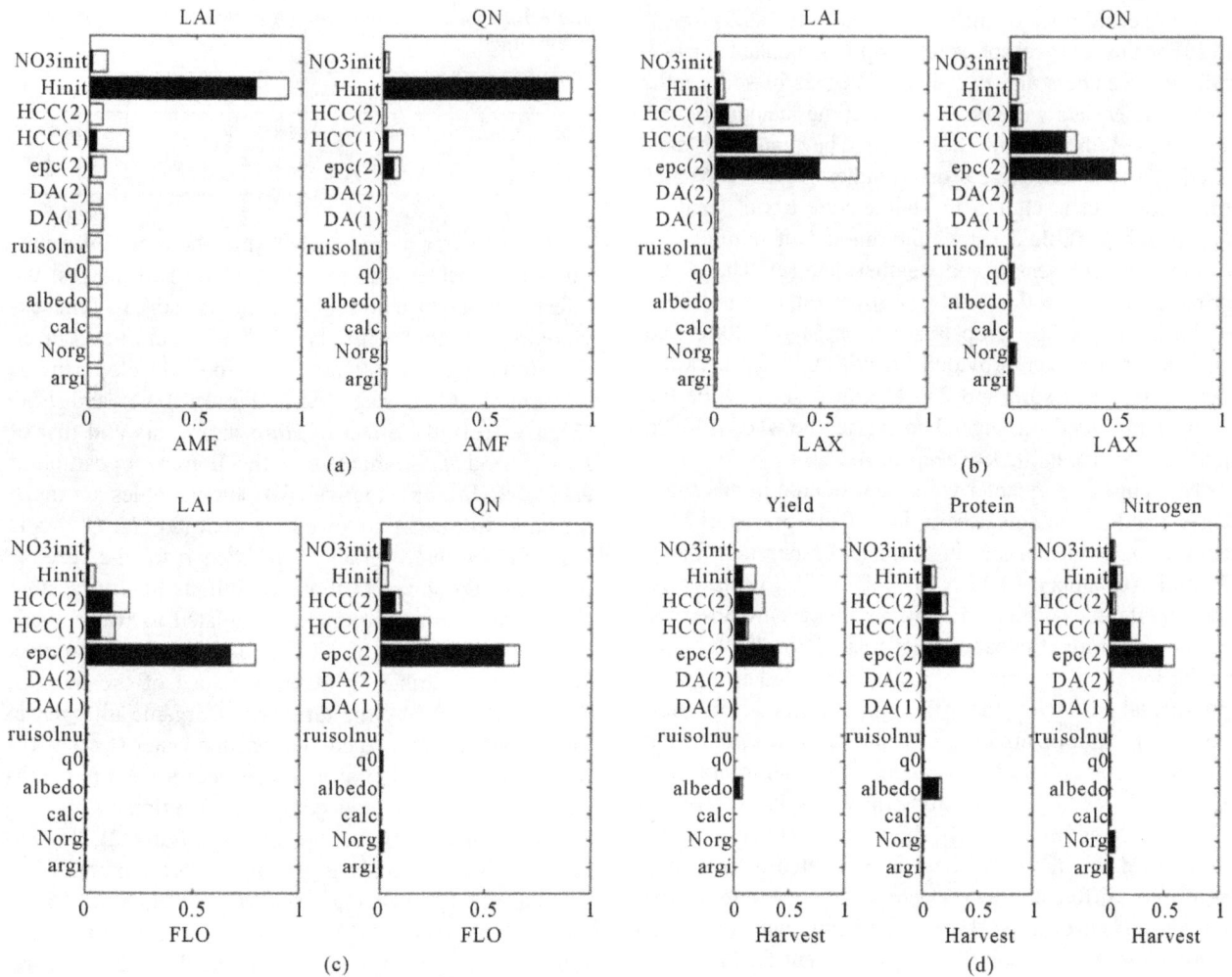

Figure 1. Sensitivity indices of the 13 soil parameters for each model output of the wheat crop simulated with a dry weather and a shallow soil. The outputs (a) correspond to *LAI* and *QN* at stage *AMF*; (b) correspond to *LAI* and *QN* at stage *LAX*; (c) correspond to *LAI* and *QN* at stage *FLO* and (d) correspond to *Yld*, *Prot* and *Nit* at *Harvest*. First-order indices are in black and interactions in white.

content. The main difference between these results and those presented in **Figure 1** lies in the sensitivity to parameters involved in the mineralization process (especially *argi* and *Norg*).

When considering wet conditions (**Figure 2**), the water stress is not so much a limiting factor: maximum *LAI* is equal to 3.61 in average, whereas it is equal to 2.57 in dry conditions (see **Table 4**). The roots grow more rapidly at the beginning of the season and the size of the soil reservoir (via the parameter *epc*(2)) is important since the *AMF* stage: the depth of roots is equal to 55.84 in average (3 months after the sowing), whereas it is equal to 45.62 in dry conditions (see **Table 4**). Moreover, in these conditions, the mineralization of the soil organic matter is increased and the effects of the concerned parameters *argi* and *Norg* do so: the cumulative mineral nitrogen arising from humus is equal to 23.95 in average (at stage *LAX*), whereas it is equal to 18.09 in dry condi-

tions (see **Table 4**). This does not seem to influence the effects of the different parameters on *LAI* at stage *AMF* since they are very similar to these obtained with the dry weather. On the contrary, the sensitivities of the variable *QN* to the different parameters are very different from the ones obtained with a dry weather: there is no contribution of *Hinit* but *epc*(2), *HCC*(1) and parameters involved in the mineralization process (*argi*, *Norg* and *NO3init*) significantly contributes to the variance of this variable. This is also the case for both *LAI* and *QN* on later stages, with an increasing dominancy of *epc*(2). At *Harvest* (**Figure 2(d)**), the variables are sensitive to the parameters *epc*(2), *HCC*(1) and *HCC*(2) with still a slight sensitivity to *argi* and *Norg* for the soil mineral nitrogen content. The main difference between these results and those presented in **Figure 1** lies in the sensitivity to parameters involved in the mineralization process (especially *argi* and *Norg*).

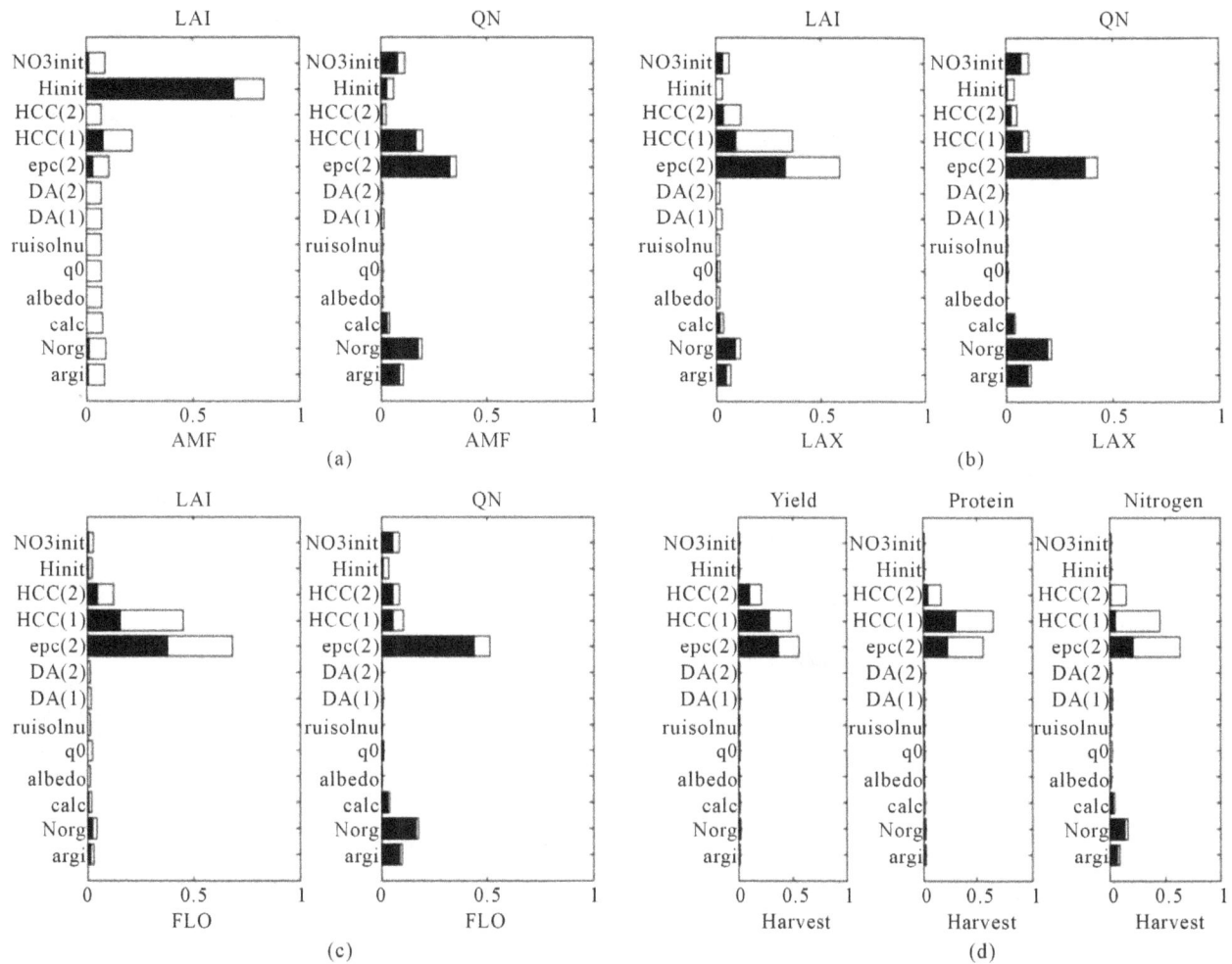

Figure 2. Sensitivity indices of the 13 soil parameters for each model output of the wheat crop simulated with a wet weather and a shallow soil. The outputs (a) correspond to *LAI* and *QN* at stage *AMF*; (b) correspond to *LAI* and *QN* at stage *LAX*; (c) correspond to *LAI* and *QN* at stage *FLO* and (d) correspond to *Yld*, *Prot* and *Nit* at *Harvest*. First-order indices are in black and interactions in white.

Table 4. Ranges of some output variables uncertainties generated by the uncertainties on the soil parameters. The output concerns the value of maximum *LAI*, the cumulative mineral nitrogen arising from humus *Qminh* (calculated at the stage *LAX* or *Summer*) and the depth of roots *Zrac* (calculated 3 months after the sowing date).

Configuration of simulation[*]	Maximum *LAI*			*Qminh*			*Zrac*		
	min	mean	max	min	mean	max	min	mean	max
W C1–	0.78	2.57	3.73	7.19	18.09	40.51	30.1	45.62	56.52
W C2–	2.51	3.61	5.08	9.8	23.95	48.85	30.1	55.84	69.91
SB C1+	0	1.42	4.61	0	24.91	83.38	12.06	77.58	129.61
SB C2+	0.19	4	6.06	19.19	50.19	121.45	71.08	85.55	102.58

[*]Wheat crop, shallow soil and dry weather (*W* C1–) or wet weather (*W* C2–); sugar beet crop, deep soil and dry weather (*SB* C1 +) or wet weather (*SB* C2 +).

Figure 3 shows the sensitivity indices calculated for the 13 soil parameters and for each output variable of the sugar beet crop simulated with a deep soil and a dry weather. In this case, the crop grows mainly in summer where it experiences a severe water stress, leading to a

value of maximum *LAI* equal to 1.42 in average (see **Table 4**). The depth of the second layer (parameter *epc*(2)) does not have any importance here. This is also the case for wheat crop with a deep soil (results not shown here). Indeed, as the root growth is no longer limited by the

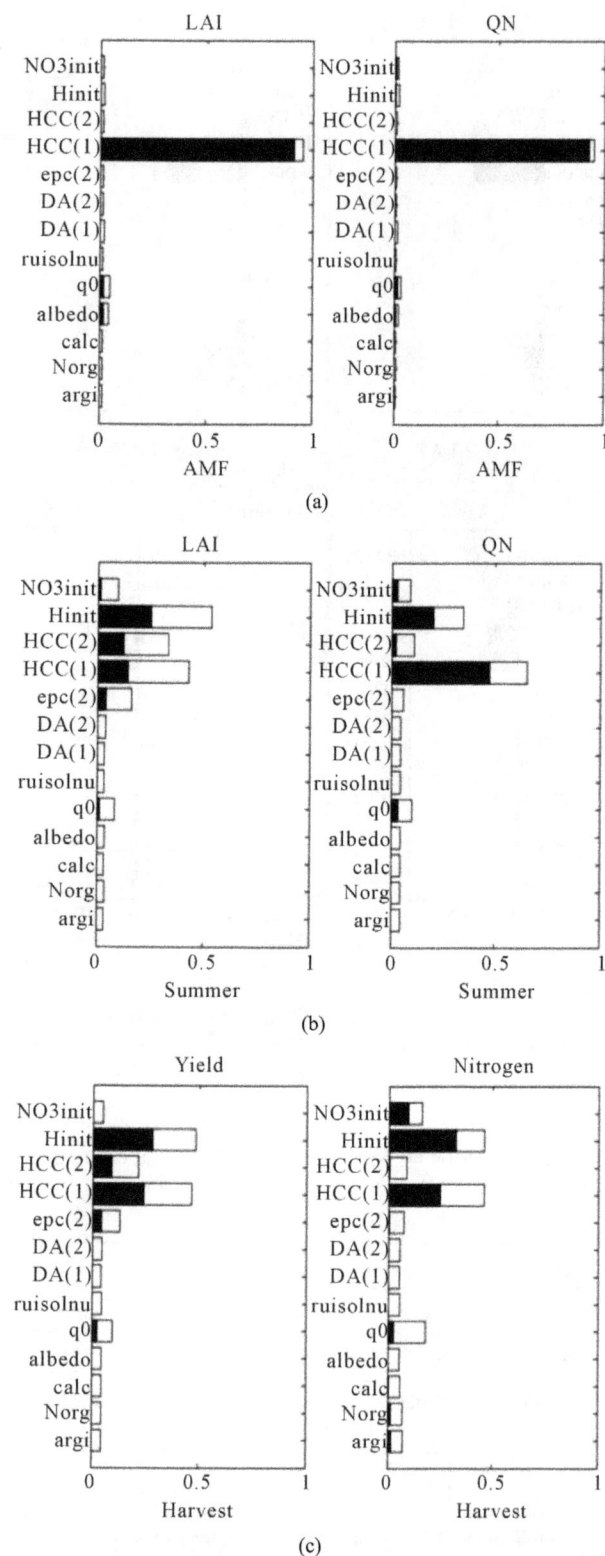

Figure 3. Sensitivity indices of the 13 soil parameters for each model output of the sugar beet crop simulated with a dry weather and a deep soil. The outputs (a) correspond to *LAI* and *QN* at stage *AMF*; (b) correspond to *LAI* and *QN* at *Summer* and (c) correspond to *Yld* and *Nit* at *Harvest*. First-order indices are in black and interactions in white.

thickness of soil (the depth of roots is equal to 77.58 in average), the output variables are no longer sensitive to the parameter $epc(2)$ when the soil is deep. Moreover, the outputs are not at all sensitive to the initial water content *Hinit* because the amount of rainfall is quite important in spring, when the sugar beet is sown (see **Table 3**). The soil water reserve is therefore the main limiting factor and it depends only on $HCC(1)$ for the early stage *AMF*: it contributes for 95% of the total output variance of *LAI* and *QN*. For the *Summer* stage (**Figure 3(b)**), which correspond to the maximum of water stress index, *LAI* is mainly sensitive to parameters linked to water availability of both soil layers ($HCC(1)$ and $HCC(2)$) with an increase of the sensitivity to *Hinit*. *QN* is more sensitive to characteristics of the top layer ($HCC(1)$ and *Hinit*) where is concentrated the organic nitrogen, as it influences the fate of available nitrogen coming from mineralization. The same tendencies are noticed for the output variables at *Harvest*, the yield being more linked to *LAI* and soil mineral nitrogen to *QN*. Many interactions are visible between all these parameters. It is also noticeable that, as in the case of wheat, the output variables have very low sensitivity to the parameters concerned with nitrogen turnover in the soil, due to the dry weather and limited mineralization. The main differences of these results with respect to those presented for the wheat crop (**Figures 1** and **2**) is: 1) that $HCC(1)$ contributes a lot to the variance of the output variables during all the crop season; 2) that *Hinit* has no contribution to the variance of the output variables at the beginning of the sugar beet season and 3) that $epc(2)$ does not affect the output variables when the soil is deep.

When considering wet conditions (**Figure 4**), the sugar beet crop growth is less affected by the water stress: maximum *LAI* is equal to 4 in average, whereas is equal to 1.42 in dry conditions (see **Table 4**). The soil water reserve of the second layer is not a limiting factor in deep soil and wet conditions for both stages *AMF* and *Summer* because the soil reservoir has a large size and the water stress is low. Thus, *LAI* and *QN* are only sensitive to the soil water reserve of the first layer which only depends on $HCC(1)$ (it does not depend on *Hinit* because of the high amount of rainfall in spring). Nevertheless, the soil water reserve of the second layer becomes a limiting factor at the end of the sugar beet crop season, when the roots are deep, involving a significant sensitivity of the output *Yld* to the parameters $HCC(1)$, $HCC(2)$ and $epc(2)$. Moreover, the mineralization of the soil organic matter slightly increases in wet conditions and so do the effects of the concerned parameters on *QN* at *Summer* and *Nit* at yield: the cumulative mineral nitrogen arising from humus is equal to 50.19 in average, whereas is equal to 24.91 in dry conditions (see **Table 4**). The main difference between these results and those presented in **Figure 3**, lies

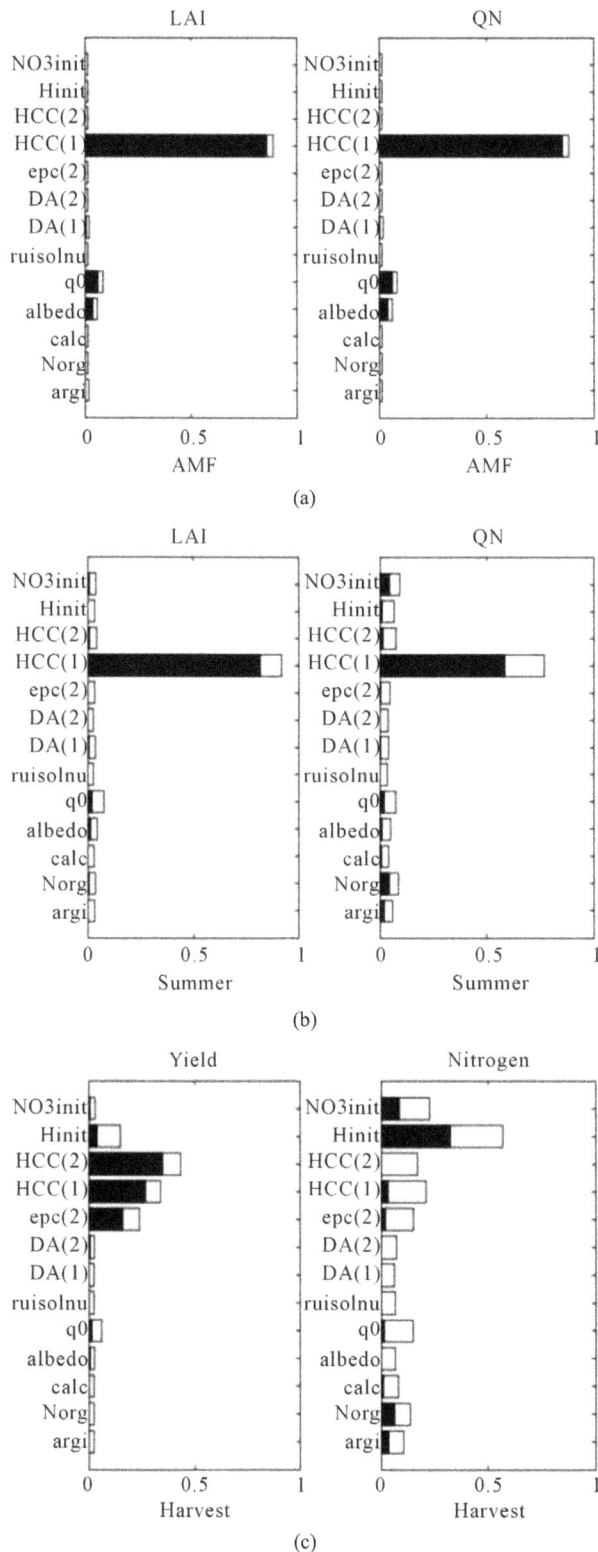

Figure 4. Sensitivity indices of the 13 soil parameters for each model output of the sugar beet crop simulated with a wet weather and a deep soil. The outputs (a) correspond to *LAI* and *QN* at stage *AMF*; (b) correspond to *LAI* and *QN* at *Summer* and (c) correspond to *Yld* and *Nit* at *Harvest*. First-order indices are in black and interactions in white.

in the lower sensitivity of the soil water reserve parameters of the second layer at the two first stages of interest.

3.2. Total Effect and Coefficient of Variation

For each configuration of simulation presented above, **Figure 5** shows the coefficient of variation *CV* of each output variable and the corresponding total effect *ST* of each parameter. The horizontal dashed line is situated at an arbitrary minimum value *ST* = 10% and the vertical dashed line is situated at another arbitrary minimum value *CV* = 1/3. The threshold of 10% for *ST* has been proposed by [30] for screening the significant sensitivity values. When wheat crop is simulated with a dry weather and a shallow soil (see **Figure 5(a)**), three output variables have a coefficient of variation higher than 1/3: *Prot* (*CV* = 0.37), *Yld* (*CV* = 0.54) and *LAI* at the stage *FLO* (*CV* = 0.62). For these outputs, only 5 soil parameters have a ST higher than 10%: *epc*(2), *HCC*(1), *HCC*(2), *Hinit* and *albedo*. This means that for simulating correctly these output variables of the wheat crop when the weather is dry and the soil depth is shallow, only *epc*(2), *HCC*(1), *HCC*(2), *Hinit* and *albedo* have to be determined accurately and the other parameters can be fixed at a nominal value (assuming the arbitrary thresholds *ST* = 10% and *CV* = 1/3). When wheat crop is simulated with a wet weather and a shallow soil (**Figure 5(b)**), only the variable *Nit* has a coefficient of variation slightly higher than 1/3 (*CV* = 0.38). The corresponding parameters having a *ST* higher than 10% are *epc*(2), *HCC*(1), *HCC*(2) and *Norg*, meaning that these parameters are important to be determined accurately for simulating correctly the wheat crop in this case. The first main difference between the results presented in **Figures 5(a)** and **(b)** is that only one output variable has a *CV* higher than 1/3 when the weather is wet, instead of three when the weather is dry. The second main difference is that the parameters *albedo* and *Hinit*, which contribute for a significant part of the output variance when the weather is dry, are replaced by the parameter *Norg*, which is involved in the mineralization process and contribute for a significant part of the variance of the outputs when the weather is wet.

Considering the results presented in **Figures 5(a)** and **(b)**, the parameter *Hinit*, which contributes for a large part to the variance of the output variables *LAI* and *QN* at the stage *AMF* of the wheat crop (see Section 3.1), does not need in fact an accurate determination for simulating correctly these output variables. Its *ST* values are higher than 0.8 for these outputs but the *CV* values of these outputs are lower than 0.1. If only the results provided by sensitivity analysis are used, *Hinit* would have been considered as an important parameter to be determined, but considering also the coefficient of variation allows stating that this parameter is not as important as previously

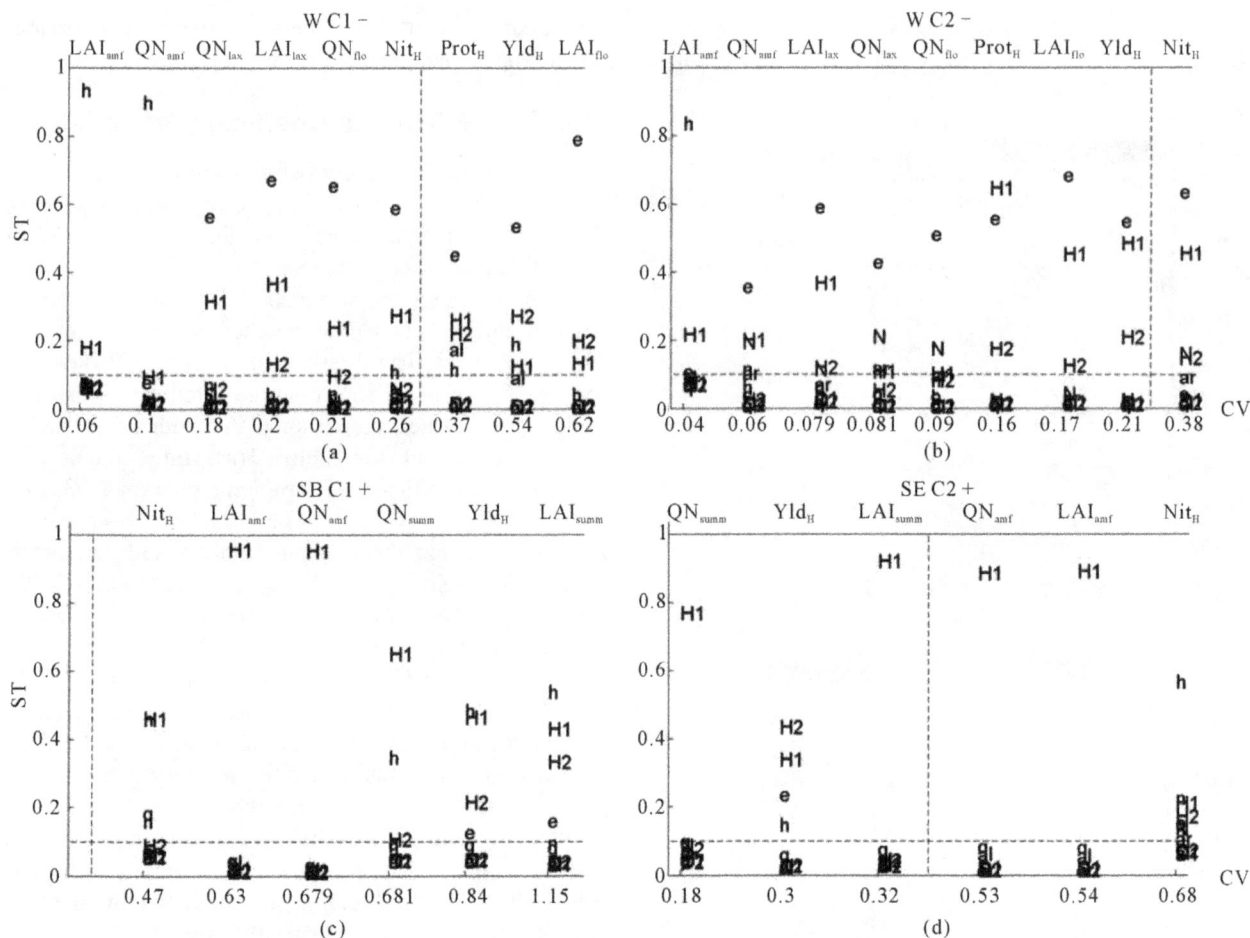

Figure 5. Coefficient of variation CV of each output variables and the corresponding total effects ST of the 13 soil parameters. The horizontal dashed line is situated at ST = 10% and the vertical dashed line is situated at CV = 1/3. The outputs are simulated for (a) wheat crop, dry weather and shallow soil, (b) wheat crop, wet weather and shallow soil, (c) sugar beet crop, dry weather and deep soil and (d) sugar beet crop, wet weather and deep soil. Label C1 correspond to the dry weather and C2 to the wet one, W correspond to the wheat crop and SB to sugar beet, − correspond to a shallow soil and + to a deep one. Parameter labels are presented in **Table 1**.

thought. The parameter *epc*(2), which contributes for a large part to the variance of all the output variables during all the wheat crop season (see Section 3.1), proves to be the most important parameter to be determined accurately for simulating the wheat crop when the type of soil depth is shallow.

The **Figure 5(c)** shows the results when the sugar beet crop is simulated with a dry weather and a deep soil. It reveals that all the output variables have a coefficient of variation higher than 1/3 meaning that the uncertainties on the soil parameters generate a large uncertainty on the considered variables. Among those parameters, five need to be measured accurately: *epc*(2), *HCC*(1), *HCC*(2), *Hinit* and *q*0. The main difference between the results presented in **Figures 5(a)** and **(c)** is that all the output variables are strongly affected by the measurement of the soil parameters when the sugar beet is simulated. When sugar beet is simulated with a deep soil and a wet weather (**Figure 5(d)**), only the output variables *LAI* and

QN at the stage *AMF* and *Nit* have a CV higher than 1/3 (resp. *CV* = 0.54, 0.53 and 0.68). For *LAI* and *QN* at the stage *AMF*, the only parameter having a *ST* higher than 10% is *HCC*(1). For *Nit*, a lot of parameters exceeds this threshold: *argi*, *Norg*, *q*0, *epc*(2), *HCC*(1), *HCC*(2), *Hinit* and *NO3init*. It is thus necessary to determine accurately a lot of parameters for simulating correctly the output *Nit*, while it is necessary to determine only one parameter for simulating correctly *LAI* and *QN* at *AMF*.

When sugar beet is simulated with a deep soil and a wet weather (**Figure 5(d)**), only the output variables *LAI* and *QN* at the stage *AMF* and *Nit* have a CV higher than 1/3 (resp. *CV* = 0.54, 0.53 and 0.68). For *LAI* and *QN* at the stage *AMF*, the only parameter having a *ST* higher than 10% is *HCC*(1). For *Nit*, a lot of parameters exceeds this threshold: *argi*, *Norg*, *q*0, *epc*(2), *HCC*(1), *HCC*(2), *Hinit* and *NO3init*. It is thus necessary to determine accurately a lot of parameters for simulating correctly the output *Nit*, while it is necessary to determine only one parameter for simulating correctly *LAI* and *QN* at *AMF*. The main difference between these results and those presented in **Figure 5(c)** is that, excepted for the output *Nit*, at most one parameter has to be accurately known for simulating correctly the sugar beet crop in deep soil and wet conditions. The parameter *HCC*(1), which contributes for a large part to the variance of all the output variables during all the sugar beet crop season (see Section 3.1), proves to be the most important parameter to be measured accurately for simulating the sugar beet crop when the soil is deep.

4. CONCLUSIONS

Global sensitivity analysis is an interesting tool for ranking parameters with respect to their contribution to the variance of the output variables of a model. However, the only use of sensitivity indices proves to be unsatisfactory for deciding which parameters should be accurately measured in a given configuration. Only the combination of uncertainty and sensitivity analysis is relevant to reach this goal. We propose in this study a simple and easy to use method that combines these two analysis in order to select the parameters that needs particular accuracy for simulating a set of variables of interest with an acceptable precision. This method, which can be easily applied to any crop model and group of parameters, has three steps: 1) compute the global sensitivity indices for each uncertain parameter; 2) compute the coefficient of variation of the outputs of interest from the set of simulations performed at step 1); and 3) select the parameters to be accurately measured for simulating correctly these outputs by setting thresholds on sensitivity indices and coefficients of variation. Of course the results of this method are strongly linked to the uncertainties hypothesized for the parameters and special attention must be paid to this aspect. Coefficients of variation and sensitivity indices thresholds should be adapted to each case depending on the level of measurements constraints and of the accuracy wishes for model output simulations.

We apply this method to the crop model STICS for selecting soil parameters that need to be measured at a field scale. Practically this needs the knowledge of the conditions under which the crop grows (weather, type of soil depth, agricultural techniques...) and it has been shown here that the results depend on these conditions. Moreover, the field scale variability of soil parameters, assumed in this application, gives a larger importance to parameters related to water availability than those related to mineralization. Concerning non-permanent soil parameters such as initial conditions, the application of the method needs thus to be based on future scenarios. This application shows that the number of STICS soil parameters to be measured accurately for simulating cor-

rectly the output variables considered here for wheat and sugar beet crops (given the parameters uncertainties used and in the configurations studied) can be significantly reduced by the use of this method. This is of particular interest given the time and financial cost of soil measurements.

REFERENCES

[1] Batchelor, W.D., Basso, B. and Paz, J.O. (2002) Examples of strategies to analyze spatial and temporal yield variability using crop models. *European Journal of Agronomy*, **18**, 141-158.

[2] Gabrielle, B., Roche, R., Angas, P., Cantero-Martinez, C., Cosentino, L., Mantineo, M., Langensiepen, M., Henault, C., Laville, P., Nicoullaud, B. and Gosse, G. (2002) A priori parameterisation of the CERES soil-crop models and tests against several European data sets. *Agronomie*, **22**, 119-132.

[3] Houlès, V., Mary, B., Guérif, M., Makowski, D. and Juste, E. (2004) Evaluation of the crop model STICS to recommend nitrogen fertilization rates according to agro-environmental criteria. *Agronomie*, **24**, 1-9.

[4] Launay, M. and Guérif, M. (2003) Ability for a model to predict crop production variability at the regional scale: An evaluation for sugar beet. *Agronomie*, **23**, 135-146.

[5] Tremblay, M. and Wallach, D. (2004) Comparison of parameter estimation methods for crop models. *Agronomie*, **24**, 351-365.

[6] Varella, H., Guérif, M., Buis, S. and Beaudoin, N. (2010) Soil properties estimation by inversion of a crop model and observations on crop improves the prediction of agro-environmental variables. *European Journal of Agronomy*, **33**, 139-147.

[7] Flenet, F., Villon, P. and Ruget, F.O. (2003) Methodology of adaptation of the STICS model to a new crop: Spring linseed (Linum usitatissimum, L.). *STICS Workshop*, Camargue, FRANCE, 367-381.

[8] Hadria, R., Khabba, S., Lahrouni, A., Duchemin, B., Chehbouni, A., Carriou, J. and Ouzine, L. (2007) Calibration and validation of the STICS crop model for managing wheat irrigation in the semi-arid Marrakech/Al Haouzi plain. *Arabian Journal for Science and Engineering*, **32**, 87-101.

[9] Singh, A.K., Tripathy, R. and Chopra, U.K. (2008) Evaluation of CERES-Wheat and CropSyst models for water-nitrogen interactions in wheat crop. *Agricultural Water Management*, **95**, 776-786.

[10] Guérif, M. and Duke, C. (1998) Calibration of the SUCROS emergence and early growth module for sugar beet using optical remote sensing data assimilation. *European Journal of Agronomy*, **9**, 127-136.

[11] Ferreyra, R.A., Jones, J.W. and Graham, W.D. (2006)

Parameterizing spatial crop models with inverse modeling: Sources of error and unexpected results. *Transactions of the Asabe*, **49**, 1547-1561.

[12] Irmak, A., Jones, J.W., Batchelor, W.D. and Paz, J.O. (2001) Estimating spatially variable soil properties for application of crop models in precision farming. *Transactions of the ASAE*, **44**, 1343-1353.

[13] Nemes, A., Timlin, D.J., Pachepsky, Y.A. and Rawls, W.J. (2009) Evaluation of the Rawls *et al.* (1982) Pedotransfer Functions for their Applicability at the US National Scale. *Soil Science Society of America Journal*, **73**, 1638-1645.

[14] King, D., Daroussin, J. and Tavernier, R. (1994) Development of a soil geographic database from the soil map of the European communities. *Catena*, **21**, 37-56.

[15] Bourennane, H., King, D., Couturier, A., Nicoullaud, B., Mary, B. and Richard, G. (2007) Uncertainty assessment of soil water content spatial patterns using geostatistical simulations: An empirical comparison of a simulation accounting for single attribute and a simulation accounting for secondary information. *Ecological Modelling*, **205**, 323-335.

[16] Samouelian, A., Cousin, I., Tabbagh, A., Bruand, A. and Richard, G. (2005) Electrical resistivity survey in soil science: A review. *Soil & Tillage Research*, **83**, 173-193.

[17] Braga, R.P. and Jones, J.W. (2004) Using optimization to estimate soil inputs of crop models for use in site-specific management. *Transactions of the ASAE*, **47**, 1821-1831.

[18] Varella, H., Guérif, M. and Buis, S. (2010) Global sensitivity analysis measures the quality of parameter estimation: The case of soil parameters and a crop model. *Environmental Modelling & Software*, **25**, 310-319.

[19] Bouman, B.A.M. (1994) A framework to deal with uncertainty in soil and management parameters in crop yield simulation: A case-study for rice. *Agricultural Systems*, **46**, 1-17.

[20] St'astna, M. and Zalud, Z. (1999) Sensitivity analysis of soil hydrologic parameters for two crop growth simulation models. *Soil & Tillage Research*, **50**, 305-318.

[21] Saltelli, A., Chan, K. and Scott, E.M. (2000) Sensitivity analysis. John Wiley and Sons, New York.

[22] Aggarwal, P.K. (1995) Uncertainties in crop, soil and weather inputs used in growth-models: Implications for simulated outputs and their applications. *Agricultural Systems*, **48**, 361-384.

[23] Blasone, R.S., Madsen, H. and Rosbjerg, D. (2008) Uncertainty assessment of integrated distributed hydrological models using GLUE with Markov chain Monte Carlo sampling. *Journal of Hydrology*, **353**, 18-32.

[24] Lawless, C., Semenov, M.A. and Jamieson, P.D. (2008) Quantifying the effect of uncertainty in soil moisture characteristics on plant growth using a crop simulation model. *Field Crops Research*, **106**, 138-147.

[25] Tolson, B.A. and Shoemaker, C.A. (2008) Efficient prediction uncertainty approximation in the calibration of environmental simulation models. *Water Resources Research*, **44**, W04411.

[26] Van der Keur, P., Hansen, J.R., Hansen, S. and Refsgaard, J.C. (2008) Uncertainty in simulation of nitrate leaching at field and catchment scale within the odense river basin. *Vadose Zone Journal*, **7**, 10-21.

[27] Campolongo, F. and Saltelli, A. (1997) Sensitivity analysis of an environmental model an application of different analysis methods. *Reliability Engineering & System Safety*, **57**, 49-69.

[28] Gomez-Delgado, M. and Tarantola, S. (2006) GLOBAL sensitivity analysis, GIS and multi-criteria evaluation for a sustainable planning of a hazardous waste disposal site in Spain. *International Journal of Geographical Information Science*, **20**, 449-466.

[29] Lamboni, M., Makowski, D., Lehuger, S., Gabrielle, B. and Monod, H. (2009) Multivariate global sensitivity analysis for dynamic crop models. *Field Crops Research*, **113**, 312-320.

[30] Makowski, D., Naud, C., Jeuffroy, M.H., Barbottin, A. and Monod, H. (2006) Global sensitivity analysis for calculating the contribution of genetic parameters to the variance of crop model prediction. *Reliability Engineering & System Safety*, **91**, 1142-1147.

[31] Pathak, T.B., Fraisse, C.W., Jones, J.W., Messina, C.D. and Hoogenboom, G. (2007) Use of global sensitivity analysis for CROPGRO cotton model development. *Transactions of the ASABE*, **50**, 2295-2302.

[32] Saltelli, A., Tarantola, S. and Campolongo, F. (2000) Sensitivity analysis as an ingredient of modeling. *Statistical Science*, **15**, 377-395.

[33] Brisson, N., Launay, M., Mary, B. and Beaudoin, N. (2008) Conceptual basis, formalisations and parameterization of the STICS crop model. Quae, Versailles.

[34] Brisson, N., Ruget, F., Gate, P., Lorgeou, J., Nicoullaud, B., Tayot, X., Plenet, D., Jeuffroy, M.H., Bouthier, A., Ripoche, D., Mary, B. and Juste, E. (2002) STICS: A generic model for simulating crops and their water and nitrogen balances. II. Model validation for wheat and maize. *Agronomie*, **22**, 69-92.

[35] Launay, M., Graux, A.-I., Brisson, N. and Guérif, M. (2009) Carbohydrate remobilization from storage root to leaves after a stress release in sugar beet (*Beta vulgaris* L.): Experimental and modelling approaches. *The Journal of Agricultural Science*, **147**, 669-682.

[36] Cariboni, J., Gatelli, D., Liska, R. and Saltelli, A. (2004) The role of sensitivity analysis in ecological modelling. *4th Conference of the International Society for Ecological Informatics*, Busan, 24-28 October 2004, 167-182.

[37] Chan, K., Tarantola, S., Saltelli, A. and Sobol, I.M. (2001) Variance-based methods. In: Saltelli, A., Chan, K. and

Scott, E.M., Eds., *Sensitivity Analysis*, Wiley, New York, 167-197.

[38] Ratto, M., Young, P.C., Romanowicz, R., Pappenberger, F., Saltelli, A. and Pagano, A. (2007) Uncertainty, sensitivity analysis and the role of data based mechanistic modeling in hydrology. *Hydrology and Earth System Sciences*, **11**, 1249-1266.

[39] Saltelli, A. and Bolado, R. (1998) An alternative way to compute Fourier amplitude sensitivity test (FAST). *Computational Statistics & Data Analysis*, **26**, 445-460.

[40] Saltelli, A., Tarantola, S. and Chan, K.P.S. (1999) A quantitative model-independent method for global sensitivity analysis of model output. *Technometrics*, **41**, 39-56.

[41] Guérif, M., Beaudoin, N., Durr, C., Machet, J.M., Mary, B., Michot, D., Moulin, D., Nicoullaud, B. and Richard, G. (2001) Designing a field experiment for assessing soil and crop spatial variability and defining site specific management strategies. *Proceedings 3rd European Conference on Precision Agriculture*, Montpellier, 677-682.

15

Storm phosphorus concentrations and fluxes in artificially drained landscapes of the US Midwest

Philippe Vidon[1*], Hilary Hubbard[2], Pilar Cuadra[2], Matthew Hennessy[2]

[1]Department of Forest and Natural Resources Management, The State University of New York College of Environmental Science and Forestry, SUNY-ESF, Syracuse, USA; *Corresponding Author
[2]Department of Earth Sciences, Indiana University-Purdue University, IUPUI, Indianapolis, USA

ABSTRACT

This study investigates phosphorus (P) concentrations and fluxes in tile drains, overland flow, and streamflow at a high temporal resolution during 7 spring storms in an agricultural watershed in Indiana, USA. Research goals include a better understanding of 1) how bulk precipitation and antecedent moisture conditions affect P concentrations and fluxes at the watershed scale; 2) how P concentrations and fluxes measured in tile drains translate to the whole watershed scale; 3) whether P losses to the stream are significantly affected by overland flow. Results indicate that bulk precipitation and antecedent moisture conditions are not good predictors of SRP or TP losses (either concentration or flux) to the stream. However, along with previously published storm data in this watershed, results indicate a threshold-based behavior whereby SRP and TP fluxes significantly increase with precipitation when bulk precipitation exceeds 4 cm. Although total SRP and TP fluxes are very much driven by flow, SRP and TP fluxes are somewhat limited by the amount of P available for leaching for most storms. On average, SRP fluxes in tile drains are 13% greater than in the stream, and stream SRP fluxes account for 45% of TP fluxes at the watershed scale. Our results indicate that when P is the primary concern, best management practices aimed at reducing P losses via tile drains are likely to have the most effect on P exports at the watershed scale.

Keywords: Total Phosphorus; Dissolved Reactive Phosphorus; Scale; Precipitation; Sub-Surface Drainage; Export Rate

1. INTRODUCTION

Continuous increases in human population and associated activities (use of fertilizers, waste water treatment plants) in the United States in the last 50 years have led to a significant increase in phosphorous (P) release to surface water in many regions [1]. This addition of phosphorus to freshwater systems has been tied to excessive algae growth and taste and odor problems in many freshwater environments, and to overall ecosystem degradation in estuaries and coastal areas [1-3]. Within this context, the US Midwest has been identified as a major contributor to P losses in the Mississippi River Basin (MRB) [1]. There has therefore been much interest in the past few decades in better understanding the processes regulating P losses to streams in artificially drained landscapes of the US Midwest where subsurface drainage is common, and where large inputs of P to streams have been reported [1,4].

In Illinois, Royer *et al.* (2006) showed that most P exports to streams occur during high flow periods (>80% of annual P export occurred during extreme discharge events) and that P losses to streams often exhibit a strong seasonal pattern with most P export occurring between mid-January and June [4]. Consistently with that finding, Vidon *et al.* (2008) showed that stream total phosphorus (TP) concentrations in excess of 0.125 mg/L occurred more often (58% - 79% of the time) during high flow periods (*i.e.* winter and spring) than in the summer (33% - 64%) for a series of agricultural streams in Indiana [5]. Understanding the processes regulating P losses to streams in late winter-early spring during high flow events is therefore of primary importance in order to reduce P losses to streams without negatively affecting crop yield in artificially drained landscapes of the US Midwest. In that regard, a better estimate of total phosphorus (TP) and soluble reactive phosphorus (SRP) fluxes to stream is extremely important as fluxes, more so than concentration, ultimately control P loadings to the Mississippi

River and the Gulf of Mexico.

However, in spite of th e recognized importance of high flow periods in regulating P losses at the watershed scale, most studies reporting P fluxes do so on a seasonal or annual basis [4,6,7]. Studies that report P losses on a storm event basis, often focus on tile drains [8] or take place in other regions such as Denmark [9,10] or Ontario, Canada [11,12]. For instance, Vidon & Cuadra (2011) report SRP and TP fluxes for two tile drains in a tile drained watershed in central Indiana, and showed that these fluxes were extremely variable from storm to storm, with SRP and TP fluxes ranging from 0.1 - 18.3 g/ha/storm for SRP, and 1 - 86.4 g/ha/storm for TP [8]. In Denmark, Kronvang *et al.* (1997) report particulate phosphorus (TP-SRP) losses in tiledrains for 17 storms over a two-year period varying approximately between 1 and 23 g P/ha/storm [9]. Many studies also re port P losses at the watershed scale and show that stream bank erosion and/or overland flow often significantly contribute to P losses, but these studies rarely in tegrate plot scale m easurements (tile drains) with watershed scale measurements (stream) during storms [13-16]. This lack of integration between P exports in tile drains and associated P exports in streams strongly limits our ability to scale the results obtained at the plot scale (tile drain) to the entire watershed (stream). It also limits our ability to determine to what extent P losses in the stream mirror P losses in tile drain s, and to establish the relative importance of tile drain flow and overland flow in reg ulating P losses a th e watershed scale. Such inform ation is critically needed to inform eitherthe use of best management practices (BMP) known to affect the relative importance of o verland flow vs. infiltration (and therefore tile drain flow) (e.g. cover crop, tile drain spacing, tillage practices, etc.), or the implementation of BMPs designed, at least in part, to reduce direct overland flow contributions to th e stream (e.g. stream/riparian zone restoration, riparian zone management).

Finally, focusing efforts during storm events is of primary importance because most P losses to streams occur during high flow events [4], and m ost climate change models predict not only an increase in temperature in the coming years, but also a n increase in the intensity and frequency of storm events in many areas around t he globe, including the US Midwest [17-19]. Understanding the relationship between precipitation characteristics (e.g. bulk precipitation), antecedent moisture conditions, and P concentrations and fluxes in the US Midwest is th ere-fore of prim ary importance to determine how future changes in precipitation patterns might affect P losses to streams in the MRB. From a regulatory stand-point, better estimates of P fluxes on a daily or sub-daily basis are also needed to help States develop better total maximum daily load estimates for nutrient—impaired streams and

rivers [4].

Key research questions addressed in this study include: 1) how do bulk precipitation and antecedent moisture conditions affect P concentrations and fluxes at th e watershed scale? 2) how do P concentrations and f luxes measured in tile d rains translate at th e whole watersh ed scale? and 3) are P losses to the stream significantly affected by overland flow (when it occurs), and what is the relative importance of SRP in TP losses at the watershed scale? The implications of ou r results for watershed management are also discussed. In order to achieve these objectives, we measured bulk precipitation, antecedent moisture conditions, and SRP concentrations and f luxes for seven storms in two tile drains, the stream, overland flow (if any), and precipitation in Leary Weber Ditch, an artificially drained watershed representative of ag ro-ecosystems of the US Midwest. We also measured TP concentrationsin the stream for four selected st orms to establish the relative importance of SRP in TP for a variety of hydrological conditions.

2. MATERIAL AND METHODS

2.1. Site Description

Leary Weber Ditch (LWD) (7.2 km^2) is located in the larger Sugar Creek watershed, approximately 20 km east of Indianapolis, Indiana (**Figure 1**). Climate at the site is classified as temperate continental and humid. The average annual temperature for central Indiana is 11.7°C with an average January temperature of –3.0°C and an average July temperature of 23.7°C. The long-term average annual precipitation (1971-2000) is 100 cm [20]. Soils in the watershed are dominated by well-buffered poorly drained loams or silt lo ams, and typically belong to the Crosby-Brookston association. Crosby-Brookston soils are generally deep, very poorly drained to somewhat poorly drained with a silty cl ay loam texture in the first 30 cm of the soil profile. Soils in LWD are suited for row crop agriculture such as c orn and soybean but require artificial drainage to lower the water tab le, removing ponded water, adding nutrients and ensuring good soil tilth. Conventional tillage and a co rn/soybean rotation has been implemented consistently for t he last 20 y ears in LWD. Each year, approximately 50% of the watershed is planted with corn, and the remaining portion is planted with soybean. Soybean is generally planted early May, and glyphosate applied mid-May. Phosphorus application on soybean generally averages 112 kg·ha^{-1}·yr^{-1}. For corn, fertilizer as anhydrous ammonia is generally applied at a rate of 180 kg·N·ha^{-1}·yr^{-1} and herbicides atrazine and acetochlor are generally applied mid-May. Potash (K$_2$O) is applied post-harvest on soybean fields at a rate of approximately 220 kg·ha^{-1}. LWD (87% row crop, 6% pasture, 7% non-agricultural land use) is representative of many watersheds in the US Midwest where poorly drained soils dominate, and where artificial drainage is co m-monly used to lower the water table [21].

2.2. Field and Laboratory Measurements

A total of 7 storms were monitored between February and June in 2009 and 2010. Bulk precipitation for the storms studied was measured using a network of 7 rain gages distributed throughout the watershed. The two tile-drains monitored for this study (TD1 and TD2) are located in t he headwaters of the watershe d (**Figure 1**). Each tile-drain is 20.3 cm ID and located approximately 120 cm below the ground s urface. TD1 extends 660 m from the stream and drains a n area approximately 8.1 ha in size. TD 2 extends 710 m from the stream and drains an area approximately 6.1 ha in size. Each tile drain was equipped with a Doppler velocity meter (ISCO 2150) for continuous discharge measurements, and a In-Situ LTC probe (level-temperature-conductivity). Whenever possi-ble (i.e. when the stream water level was below the tile drain), discharge was also measured by hand using the bucket method to validate discharge measurements ob-tained with the Doppler velocity meters. No significant differences between manual and a utomated discharge measurments were found. The occurrence of overland flow was measured using a H-flume inserted into the ground, equipped with a In-Situ LT (level-temperatue) logger (In-Situ Inc.). Stream stage at the outlet of the watershed was measured using an In-Situ LTC probe (In-Situ Inc.). Discharge was measured biweekly using a handheld Doppler velocity meter (Sontek) so a rating curve could be established. A total of 8 riparian zone wells were also installed between the field edge and the stream to capt ure antecedent water table depth at the field edge before each storm, as well as riparian ground-water quality.

Water samples for soluble reactive phosphorus (SRP) analysis were collected in tile drains 1 and 2 (TD1 ans TD2), in overland flow (if any), and in th e stream using auto samplers (ISCO 6712). In the stream, water samples for total phosphorus analysis (TP) were collected for storms 1, 2, 4 and 6 only, to minimize project cost. In tile drains, the sample collection line from each ISCO sam-pler was located at least 1 m into the tile-drains, and Doppler velocity measurements confirmed that no flow reversals occurred in the til e-drains during the storms studied, therefore indicating that tile sa mples were no t contaminated by stream water when the tiles were sub-merged during storms. Samplers used to collect water samples in the stream and the two tile d rains were trig-gered manually before the beginning of each storm, and generally set t o collect water sa mples every 20 minutes during the rising limb of the hydrograph or the first 24 hours of the storm. Each 1L sample was a composite of 3

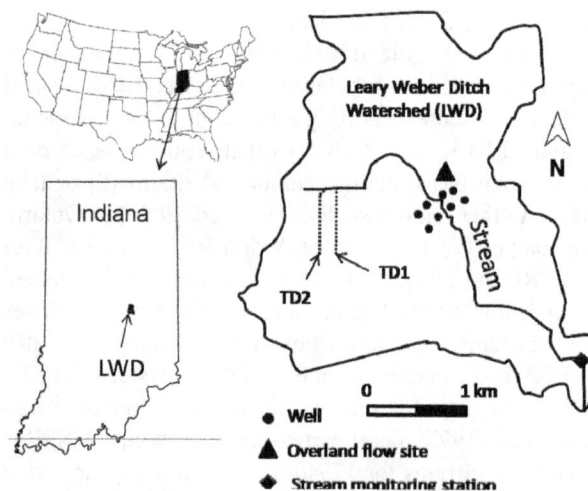

Figure 1. Experimental site location. TD1 and TD2 correspond to the two tile drains monitored for this study in 2009 and 2010.

samples taken 20 minutes apart (1 bottle per hour = 24 hours). Sampling interval was ex tended to 2 hours (3 samples taken 40 minutes apart per bottle) on the falling limb of the hydrograph. Although all water samples col-lected on the rising limb of the hydrograph and around peak flow were analyzed, not all sa mples were necessa-rily analysed on the falling limb of each hydrograph to limit cost. Additional water samples were also collected in riparian groundwater we lls (immediately before each storm) and in rain ga ges (immediately after each storm) to measure riparian and precipitation water chemitry for each of the storms studied.

Water samples were never left m ore than 24 hours in the field and w ere immediately filtered using GF/F What-man 0.7 μm filter upon return to th e laboratory (except TP samples). Triplicate analysis of 10% of all sam ples and analysis of check standards every 10 samples were performed to assess measurement error, and check for the accuracy and precision of measurement techniques. The standard error on reported solute values was typically less than 10% for all solutes. Both SRP and TP (after a persulfate digestion) concentrations were det ermined col-orimetrically using standard m ethods [22] on a K onelab 20 Photometric Analyzer (EST Analytical).

2.3. Hydrological Dataanalysis and Flux Calculations

For this study, the start of each event was de fined when a perceptible rise in discharge in the stream was observed. The end of the event was defined when flow in the stream returned to pre-event flow values or when a new event started, which ever occurred first. Seven and four-teen day antecedent discharges (7 dQ and 14 dQ, respec-tively) in the stream were calculated as the m ean dis-charge during the 7 and 14 days preceeding each event.

Solute fluxes in gram of P per storm were calculated for each storm by first multiplying the concentration of the sample for each sampling interval (mg/L) by the average discharge for that interval (L/s) and a unit conversion factor. Fluxes reported here in g/ha/storm were obtained by dividing the solute flux for each storm (g/storm) by the contributing area to each tile-drain (m²) or the stream (m²) and a unit conversion factor. Solute export yields (g/ha/hr) before each storm were calculated as the flux in the hour preceding the beginning of the storm. Solute export yields (g/ha/hr) during storms are calculated as the average hourly solute fluxes over the duration of the storm. Significant differences between groups were established using student t-tests. Significance was established at $p < 0.05$.

3. RESULTS

The 7 storms studied ranged from 1.02 cm to 4.45 cm in bulk precipitation (**Figure 2**), with maximum average daily stream flows varying between 290 - 480 L/s for storms when overland flow occurred (storms 3, 5, and 6), and between 87 - 164 L/s for storms without overland flow (storms 1, 2, 4 and 7). Mean tile flow during storms varied from 0.5 L/s (storm 1, TD1) to approximately 13 L/s for storms 3 and 6 in TD2, with peak tile flow generally occurring with or immediately before the peak in stream flow (data not shown). Water table depth at the field edge in the hours preceding the beginning of each storm ranged from 127 - 167 cm below ground surface (BGS) for storms 1, 2, 4, and 7, and from 97 - 125 cm

BGS for storms 3, 5 and 6. Antecedent flow conditions (*i.e.* 7 and 14 day antecedent flow) were highest for storms with higher maximum average daily flow and overland flow (*i.e.* storms 3, 5 and 6) than for other storms (**Figure 2**). Bulk precipitation amounts were however not consistently higher for storms 3, 5, and 6 (1.02 cm < bulk precipitation < 4.45 cm) than for storms 1, 2, 4 and 7 (2.03 cm < bulk precipitation < 2.67 cm).

Soluble reactive phosphorus (SRP) concentrations immediately before and during storms 1 - 7 are presented in **Table 1** for the stream, tile drains 1 and 2 (TD1 and TD2), precipitation and riparian zone groundwater. Overland flow (OLF) only occurred for storms 3, 5, and 6, so SRP concentrations in OLF are only reported for these storms. Before the storms, SRP concentrations were the most variable in riparian zone wells with SRP varying by one order of magnitude between storms 6 and 7 (0.01 mg/L), and storm 1 (0.15 mg/L). SRP concentration was high in the stream at baseflow before storm 1 (0.30 mg/L) but remained in the 0 - 0.02 mg/L range for all other storms. When TD1 and TD2 were flowing before the storms, SRP concentrations varied between 0.01 - 0.02 mg/L. SRP concentrations in precipitation varied by one order of magnitude from storm to storm (0.01 - 0.49 mg/L).

During storms, mean SRP concentrations in TD1 and TD2 varied between 0.01 mg/L storm 2) and 0.04 mg/L (storm 3). In the stream, mean SRP concentrations (0.01 - 0.04 mg/L) were generally 2 to 4 times higher than at baseflow. The only exception is for storm 1, which started

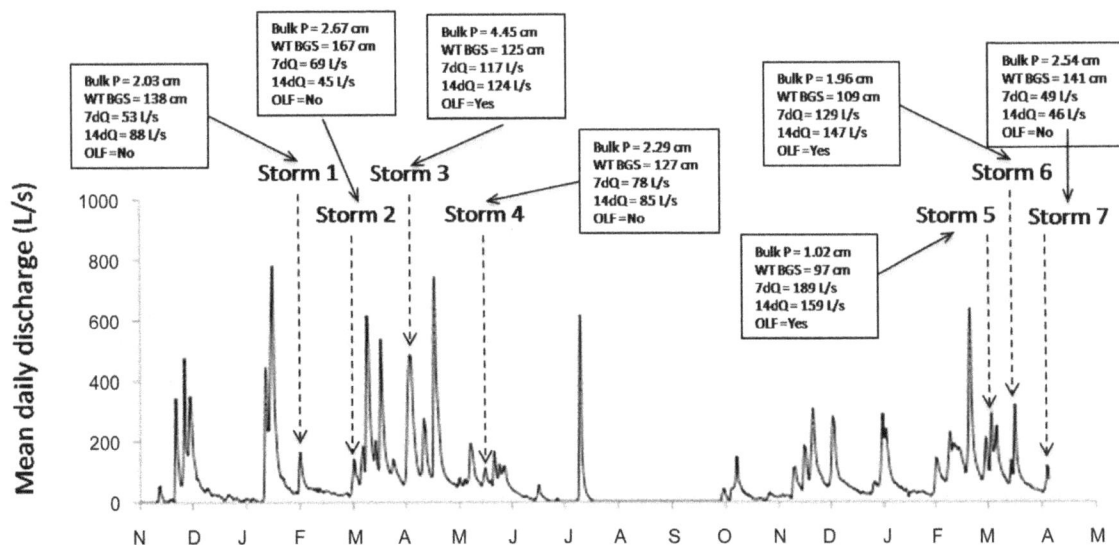

Figure 2. Mean daily discharge (L/s) in the stream at the outlet of the study watershed (Leary Weber Ditch) between November 2008 and May 2010. Storm 1 (February 26, 2009), storm 2 (April 1, 2009), storm 3 (April 29, 2009), storm 4 (June 11, 2009), storm 5 (March 29, 2010), storm 6 (April 8, 2010), and storm 7 (April 26, 2010) are the storms during this period for which water samples were collected in the watershed. Bulk precipitation amounts (Bulk P), antecedent water table depth below ground surface (WT BGS), 7-day antecedent discharge (7 dQ), 14-day antecedent discharge (14 dQ), and the occurrence of overland flow (OLF) are also indicated for each storm.

Table 1. Mean soluble reactive phosphorus (SRP) concentrations before storm 1 (February 26, 2009), storm 2 (April 1, 2009), storm 3 (April 29, 2009), storm 4 (June 11, 2009), storm 5 (March 29, 2010), storm 6 (April 8, 2010), and storm 7 (April 26, 2010) in the stream, tile drain 1 (TD1), tile drain 2 (TD2), and riparian groundwater. Mean SRP concentrations during storms 1 - 7 in precipitation, TD1, TD2, the stream, and overland flow (OLF) are also indicated. Values in parenthesis indicate one standard deviation (n/a = not available).

Mean SRP (mg/L)	Pre-storm				During storm				
	Stream	TD1	TD2	RZ	Precip.	TD1	TD2	Stream	OLF
Storm 1	0.30 (0.19)	n/a	0.01 (0.00)	0.15 (0.13)	0.01 (n/a)	n/a	0.02 (0.01)	0.04 (0.04)	-
Storm 2	0.01 (0.00)	0.01 (0.00)	n/a	0.02 (0.01)	0.13 (n/a)	0.01 (0.00)	n/a	0.02 (0.00)	-
Storm 3	0.01 (0.00)	0.02 (0.01)	0.01 (0.00)	0.02 (0.01)	0.49 (n/a)	0.04 (0.02)	0.04 (0.02)	0.04 (0.03)	0.17 (0.09)
Storm 4	<BDL	0.01 (0.00)	0.01 (0.00)	0.03 (0.02)	0.01 (n/a)	0.01 (0.00)	0.02 (0.01)	0.01 (0.00)	-
Storm 5	0.01 (0.00)	0.01 (0.00)	0.02 (0.00)	0.21 (0.60)	0.02 (n/a)	0.02 (0.00)	0.02 (0.01)	0.02 (0.01)	0.58 (1.70)
Storm 6	<BDL	0.02 (0.01)	No flow	0.01 (0.00)	0.01 (n/a)	0.03 (0.01)	0.04 (0.01)	0.02 (0.03)	0.26 (0.02)
Storm 7	0.02 (0.02)	No flow	No flow	0.01 (0.00)	0.06 (n/a)	0.03 (0.02)	0.03 (0.01)	0.03 (0.02)	-

on the tail end of a 770 L/s mean daily discharge storm (**Figure 2**), and where high SRP concentrations before the storm were reported in the stream (0.30 mg/L, **Table 1**). SRP concentrations in overland flow were consistently one order of magnitude higher (0.17 - 0.58 mg/L) than in the stream for storms 3, 5 and 6 (*i.e.* the 3 storms for which overland flow occurred). High temporal resolution SRP concentration patterns in the stream, TD1, and TD2 (**Figure 3**) revealed that SRP concentration patterns in TD1 and TD2 (when both were available) were generally similar to each other, but often different from SRP concentration patterns observed in the stream. For storms 1 and 2, stream SRP concentration showed a quick increase in concentration as discharge peaked. Conversely, a decrease in SRP concentration was observed in TD2 as discharge peaked for storm 1 (no data for TD1 for this storm). For storm 2, no clear changes in SRP concentration were observed in TD1 (no data for TD2 for this storm) as stream discharge peaked. For storm 3, both SRP in the stream and tile drains showed a sharp increase in concentration during the first peak in discharge. A smaller peak in SRP concentration was observed in the stream during the second peak in discharge. Grab samples in TD1 and TD2 during the second peak in discharge (no ISCO sampler data for this peak owing to equipment malfunction) are consistent with the occurrence of a second peak in SRP concentration in tile drains at this time. For storm 4, a steady increase in SRP concentrations in the stream and both tile drains was observed over the duration of the storm. For storm 5, although overall higher SRP concentrations were observed on the rising limb of the stream hydrograph in the stream (and to some extent in TD1 and TD2) than immediately before the beginning of this storm, individual SRP measurements remained highly variable from one sample to the next for this storm. For both storms 6 and 7, stream SRP showed a clear increase in concentration as a function of flow. However, in tile drains, SRP concentrations

only slowly increased with flow (and remained high on the falling limb) for storm 6, and did not show a consistent increase in concentration as a function of flow for storm 7.

TP concentrations measured in the stream for storms 1, 2, 4 and 6 were 0.14 mg/L, 0.06 mg/L, 0.05 mg/L, and 0.04 mg/L, respectively (>100 samples) (**Figure 4**). For storms 1, 2, and 4, TP concentrations were significantly ($P < 0.05$) higher than SRP concentrations. Although TP concentrations were also higher than SRP concentrations in the stream for storm 6, they were not statistically significantly different ($P > 0.05$). TP concentrations generally showed a clear increase in concentration as a function of flow for all storms and peaked at the same time as SRP. The only exception is for storm 4 where SRP peaked after the peak in discharge, and where TP concentrations were extremely variable from one sample to the next.

SRP (all storms) and TP (stream only for storms 1, 2, 4 and 6) fluxes (g/ha/storm) and yields (g/ha/hr) are presented in **Table 2**. Baseflow SRP and TP yields are also reported for the stream. The average stream SRP flux for storms 1 - 7 was 2.01 g/ha/storm, with a maximum flux of 8.76 g/ha/storm for storm 3, and a minimum value of 0.39 g/ha/storm for storm 4. In tile drains, average SRP fluxes were 2.37 g/ha/storm and 2.57 g/ha/storm in TD1 and TD2, respectively. Maximum SRP fluxes (8.20 g/ha/storm in TD1; 6.94 g/ha/storm in TD2) were also observed for storm 3 in tile drains. Average TP fluxes for storms 1, 2, 4, and 6 were 1.98 g/ha/storm, which is approximately twice as high as the average SRP flux for these four storms (0.89 g/ha/storm). Stream SRP yield was on average 8 times larger during storms, than immediately before the storm. The highest increase in SRP yield in the stream was observed for storm 3, where SRP yield increased form <0.001 g/ha/hr immediately before the storm began to 0.128 g/ha/hr during the storm. Although the average SRP yields in TD1 and TD2 were

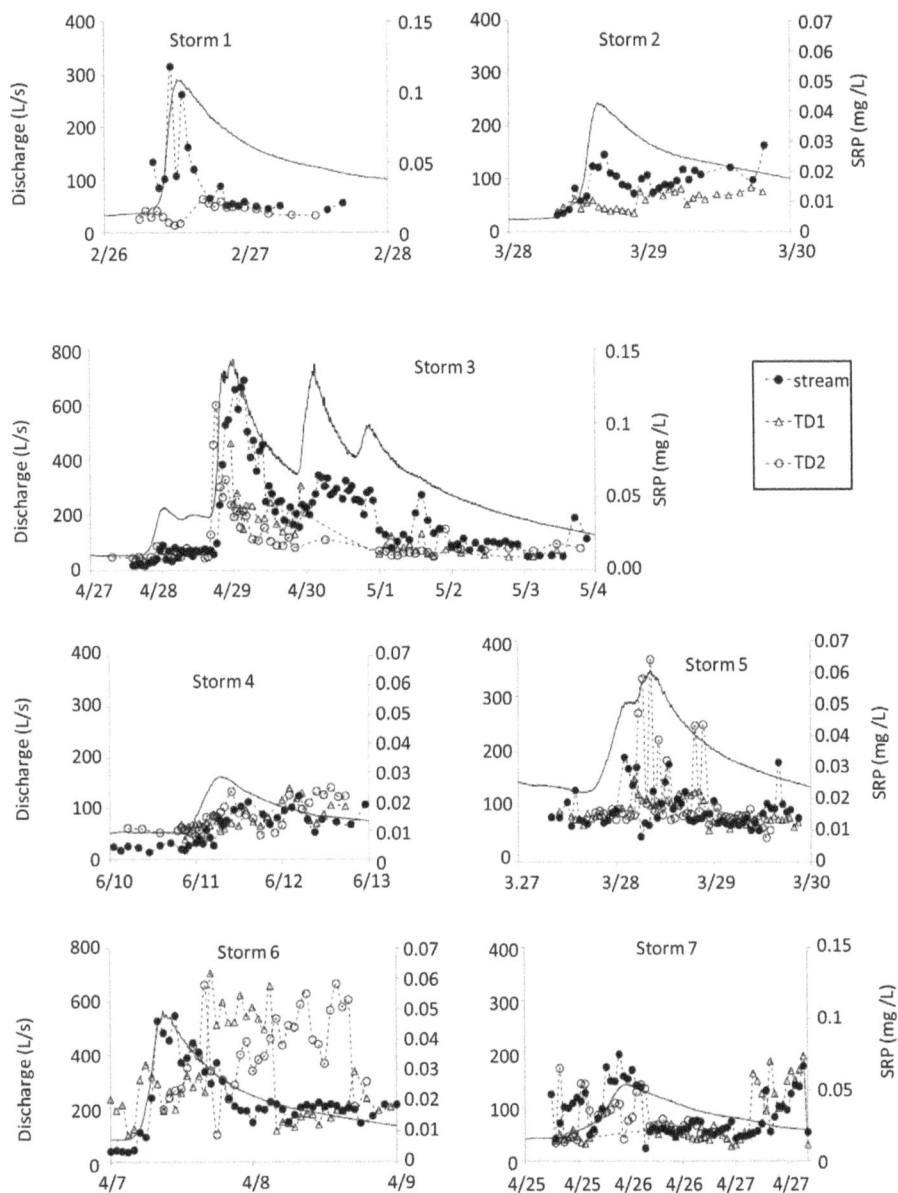

Figure 3. Stream discharge (L/s) (solid line), and soluble reactive phosphorus (SRP) concentrations in the stream (watershed outlet) and in tile drain 1 (TD1) and tile dr ain 2 (TD2) for storm1 (February 26, 2009), storm 2 (April 1, 20 09), storm 3 (April 29, 2009), storm 4 (June 11, 2009), storm 5 (March 29, 2010), storm 6 (April 8, 2010), and storm 7 (April 26, 2010).

slightly lower than in the stream for storms 1 and 2, the average SRP yield in tile drains (0.049 g/ha/hr) for storms 1 - 6 (no flux data in tile drains for storms 7) was, on average, 23% higher than in the st ream (0.040 g/ha/hr). If fluxes (g/ha/storm) are compared simply on a storm basis (as opposed to an hourly basis), SRP fluxes in tile drain s for storms 1 - 6 (2.47 g/ha/storm) are on average, 13% greater than in the stream (2.19 g/ha/storm). For storms 1, 2, 4, and 6 for which TP data are available, average TP yield for all 4 storms was 7.5 times higher during the storm (0.054 g/ha/hr) than immediately before (0.007 g/ha/hr).

Double mass curves (cumulated SRP or TP flux vs. cumulated discharge) indicate changes in the export rate of SRP or TP over the duration of a storm as a function of flow (**Figure 5**). For storm 3, the export rate of SRP over time (indicated by the slope of the double mass curve) varies over the course of t he event, with a sharp increase in export rate at the beginning of the event (first 8 mm of di scharge), followed by a st eady export rate (linear curve) for the remaining of the event. A similar pattern was observed for storms 1 and 7 , and to a l esser degree for storm 6. For storms 2, 4 and 5, the rate of SRP export was constant throughout the duration of these

Table 2. Soluble reactive phosphorus (SRP) flu xes (g N/ha/storm) and yields (g/ha/hr) for storm 1 (Febru ary 26, 2009), storm 2 (April 1, 2009), storm 3 (April 29, 2009), storm 4 (June 11, 2009), storm 5 (March 29, 2010), stor m 6 (April 8, 20 10), and storm 7 (April 26, 2010) in tile drain 1 (TD1), tile drain 2 (TD2), and the stream (watershed outlet). Total phosphorus (TP) fluxes (g N/ha/storm) and yields (g/ha/hr) for storms 1, 2, 4, and 6 in the stream. (Note: Fluxes in TD1 and TD2 for storm 7 were not cal cu-lated because discharge data in these tile drains were not available for this storm).

	SRP Flux in g/ha/storm				TP Flux in g/ha/storm	
	Stream (base flow)	Stream (storm flow)	TD1	TD2	Stream (base flow)	Stream (storm flow)
Storm 1		1.25	n/a	0.58		3.25
Storm 2		0.50	0.32	n/a		1.72
Storm 3		8.76	8.20	6.94		n/a
Storm 4		0.39	0.39	0.47		1.14
Storm 5		0.84	0.93	2.13		n/a
Storm 6		1.41	2.03	2.71		1.82
Storm 7		0.90	n/a	n/a		n/a

	SRP Yield in g/ha/hr				TP Yield in g/ha/hr	
	Stream (base flow)	Stream (storm flow)	TD1	TD2	Stream (base flow)	Stream (storm flow)
Storm 1	0.009	0.037	n/a	0.018	0.010	0.096
Storm 2	0.001	0.014	0.008	n/a	0.006	0.051
Storm 3	<0.001	0.128	0.148	0.123	n/a	n/a
Storm 4	0.001	0.007	0.008	0.010	0.007	0.022
Storm 5	0.009	0.019	0.020	0.052	n/a	n/a
Storm 6	0.002	0.037	0.034	0.068	0.006	0.048
Storm 7	0.008	0.015	n/a	n/a	n/a	n/a

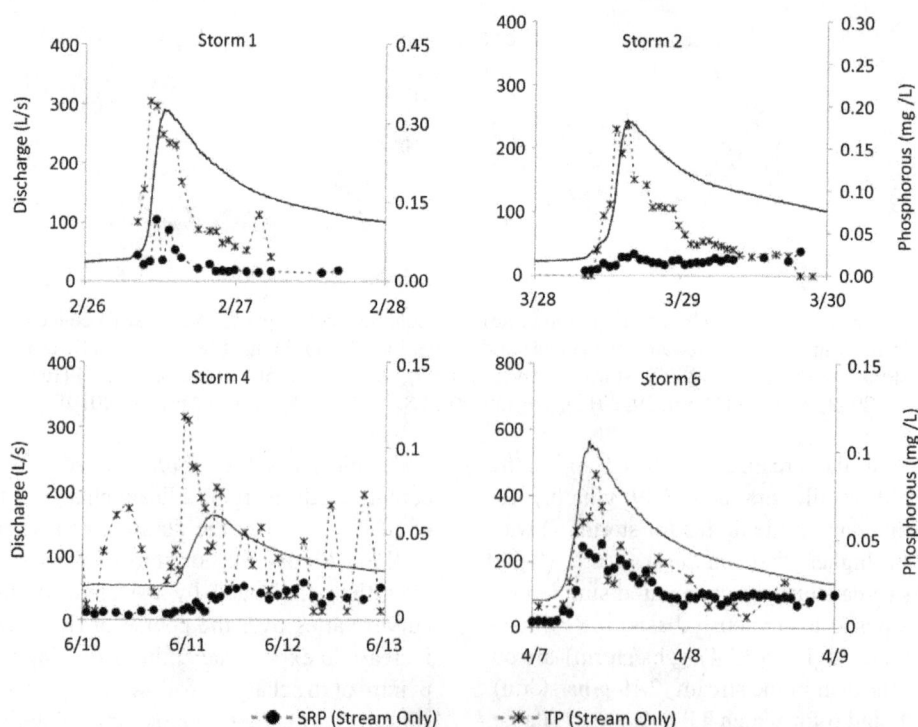

Figure 4. Stream discharge (L/s) (solid line), and soluble reactive phosphorus (SRP) and to tal phosphorus (TP) concentrations in the stream (watershed outlet) for storm 1 (February 26, 2009), storm 2 (April 1, 2009), storm 4 (June 11, 2009), and storm 6 (April 8, 2010).

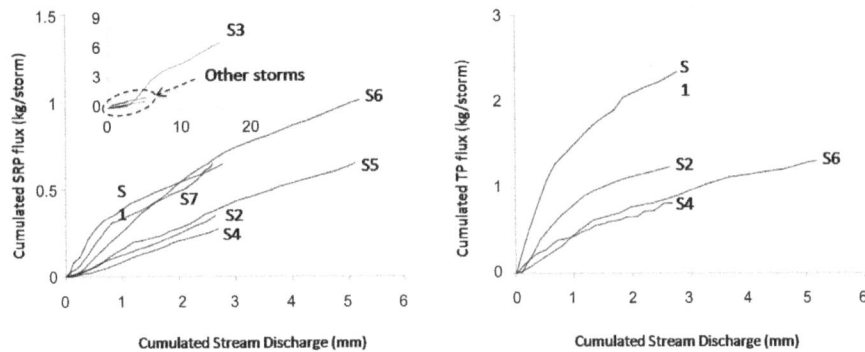

Figure 5. Double mass curves showing cumulated stream discharge (mm) versus cumulated soluble reactive phosphorus flux (kg/storm) in the stream (watershed outlet) for storms 1 - 7 (left), and cumulated total phosphorus flux (kg/storm) in the stream for storms 1, 2, 4 and 6 (right).

storms. For TP, the export rates were highly non linear for storms 1 and 2, with a progressive decrease of the rate of TP export over the duration of these storms. Although less pronounced than for storms 1 and 2, the export rates of TP over the duration of storms 4 and 6 also progressively declined over the duration of these two storms.

4. DISCUSSION

In spite of gaps in the data set because of equipment malfunction (see results section for details), this data set is the first to present concentration and flux measurements for both SRP and TP in tile drains (in duplicate), overland flow, and stream flow across scale (plot scale to whole watershed scale) for a series of spring storms in the US Midwest. This offers a unique opportunity to empirically address key questions about the dynamics of P in artificially drained landscapes of the US Midwest at a time (late winter-spring) when most P losses to the Mississippi River Basin occur [4].

4.1. How Do Bulk Precipitation and Antecedent Moisture Conditions Affect P Concentrations and Fluxes at the Watershed Scale?

Results indicate that high antecedent moisture conditions (storms 3, 5 and 6) are associated with the occurrence of overland flow and high mean daily discharge in the stream. The three highest mean SRP concentrations were however observed for storms 1, 3 and 7. When all the storms are combined, SRP concentrations are not significantly correlated (P > 0.05) to bulk precipitation, or any of the measures of antecedent moisture conditions used in this study (*i.e.* water depth before the storm, 7 and 14 day antecedent flow conditions). High SRP concentrations (storms 1, 3, 7) in the stream are also not consistently associated with storms with overland flow (storms 3, 5 and 6) or significantly correlated (P > 0.05)

to SRP concentrations in the stream at baseflow immediately before the storm. Similarly, mean storm TP concentrations measured for storms 1, 2, 4 and 6 are not significantly correlated (P > 0.05) to hydrological conditions. This suggests that bulk precipitation and antecedent moisture conditions are likely not strong controls on SRP (or TP) concentrations in the stream in spring. Similarly, SRP concentration in tile drains are not significantly correlated to hydrological conditions.

In term of fluxes, the highest SRP flux is associated, in both tile drains and the stream, with the storm with the highest antecedent moisture conditions and highest bulk precipitation amount (storm 3) (**Figure 2**). Statistically speaking, SRP fluxes in the stream are significantly correlated to bulk precipitation (r = 0.83, P < 0.05), but when storm 3 is removed, the correlation disappears (r = 0.30, P > 0.05). A similar pattern is observed in tile drains. This suggests the existence of a threshold below which SRP fluxes are not influenced by bulk precipitation. Non-linear behavior in tile drains in terms of P export has been previously reported in this watershed [8]. In that study, the authors stress the non linear behavior of tile drains, especially in term of SRP exports, with SRP fluxes between 1 to 3 order of magnitude higher for storms associated with more than 6 cm of bulk precipitation than for those with less than 3 cm of bulk precipitation. **Figure 6** combines the SRP and TP fluxes (g/ha/storm) in the stream, TD1, and TD2 reported here for 2009 and 2010 (7 storms, **Table 2**), with those reported in Vidon & Cuadra (2011) in TD1 and TD2 for a series of storms in 2008 (4 storms) [8]. Together, these results suggest the existence of a threshold around 4 cm of bulk precipitation below which SRP and TP fluxes are not correlated to bulk precipitation. Above this threshold, a significant increase in SRP and TP flux is observed as bulk precipitation increases. This stresses the non-linear behavior of the watershed in terms of P export, and the existence of a threshold above and below which hydrological response

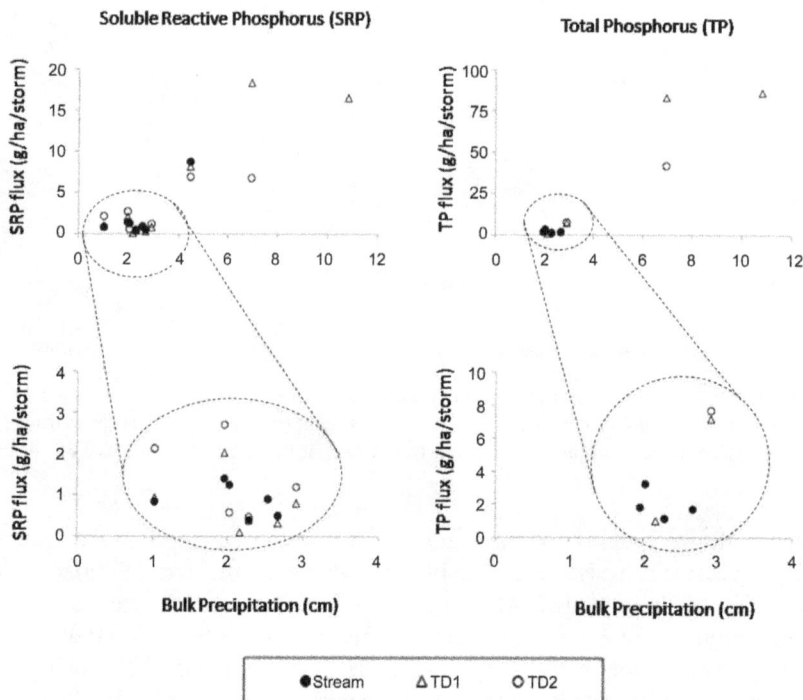

Figure 6. Bulk precipitation amounts (cm) vs. so luble reactive phosphorus (SRP) fluxes (left) and total phosphorus fluxes (right) in kilo gram per hectare in tile drain 1 (TD1), tile drain 2 (TD2), and the stream for storms 1 - 7 in 2009 and 2010 (this study), and for a series of four 2008 storms in TD1 and TD2 as reported in Vidon and Cuadra (2011).

drastically changes. One should therefore be cautious in generalizing relationships between P fluxesand bulk precipitation or antecedent moisture conditions as those may change depending on individual storm characteristics.

The analysis of the double mass curves for SRP and TP (**Figure 5**) revealed that for storms 2, 4, a nd 5, the pool of SRP available for leaching was not limiting to SRP export. For other storms (*i.e.* storms 1, 3, 6 and 7) and for all storms for TP, data indicate a progressive exhaustion of the pool of SRP (and TP) available for leaching as these storms progressed. Indeed, a linear relationship between the cumulated discharge (mm) and the cumulated solute flux (kg) indicates that the rate of leaching remains constant over the duration of the storm. When a l ogarithmic shaped curve best fits the double mass curve (e.g. storm 1 or 2 for TP), this suggests that the leaching rate of the solute decreases towards the end of the storm. This indicates a progressive exhaustion of the solute pool available for leaching over the duration of the storm. Our results therefore suggest that for 4 out of 7 storms for SRP, and for 4 out of 4 storms for TP, SRP and TP pools are limiting to SRP and TP exports.

4.2. How Do P Concentrations and Fluxes Measured in Tile Drains Translate at the Whole Watershed Scale?

Results indicate no consi stent differences between

mean SRP concentrations between TD1 and TD2, but also, and more importantly, no significant differences in mean SRP concentrations between the tile d rains (TD1 and TD2) and the stream (**Table 1**). However, the timing of SRP concentration changes as a fu nction of flow are not necessarily similar between tile drains and the stream (**Figure 3**). This suggests that although SRP concentration patterns as a fun ction of flow m ay vary, tile d rain data could potentially be used in heavily tile-drained watersheds of the US Midwest to estim ate mean stream SRP concentrations when needed.

When SRP fluxes in the stream (2.19 g/ha/storm) are compared with those reported in TD1 (2.37 g/ha/storm) and TD2 (2.57 g/ha/storm) for storms 1 - 6, fluxes in tile drains are approximately 13% higher than in the stream. SRP yields (**Table 2**) are 23% higher in tile drains than in the stream. This difference between fluxes (on a st orm basis) and yields (on a hourly basis) stems from the flashier behavior of tile drains relativ e to the stream. However, regardless of whether yields or fluxes are used for comparison, P exports in tile drains remain higher (either on a per storm or hourly basis) than in the stream. Because 100% of the contributing area to each tile drain is agricultural (corn/soybean), it is likely that differences in land use contribute to the larger export rate of SRP in tile drains than in the stream at the whole watershed scale (87% row crop, 6% pasture, 7% other (road + residential)). Further, the mean flow path length is by definition shorter at the field scale (tile drain) than at the whole watershed scale. This difference in drainage intensity as a function of scale also likely contributes to the differences observed between SRP fluxes in tile drains and the stream.

Considering that most studies reporting SRP and/or TP fluxes at the watershed scale generally focus on either tile drains [8,9] or stream [4] but rarely both, it may be tempting for watershed managers trying to identify general patterns of P exports based on published work to use P flux information obtained in tile drains to estimate P fluxes at the watershed scale or vice-versa. Our data however suggest that higher fluxes (per hectare) will occur in tile drains (at the plot scale) than in the stream (at the watershed scale). Although we do not suggest that SRP fluxes in tile drains should be assumed to be consistently 13% higher than in the stream, we believe that in the absence of simultaneous flux estimates in tile drains and stream flow, one should apply a correction factor to available flux data to compensate for the overall difference in SRP fluxes observed between the stream and tile drains. We however believe that more studies are needed to better constrain the variables affecting the relationship between P fluxes in tile drains and P fluxes at the whole watershed scale. For instance, differences in land use between watersheds are likely to affect this ratio of 13% between SRP fluxes in tile drains and in the stream. Difference in tile drain diameter and spacing from one watershed to the next will also affect the drainage efficiency at the plot scale, and therefore the relationship between SRP fluxes in tile drains and in the stream at the whole watershed scale. Finally, this ratio was established based on 7 storms, in late winter-early spring, when the soil was bare or crops barely starting to emerge from the ground. It is likely that changes in vegetation development stage will affect this relationship over the course of an entire year. Consequently, although we recommend that a correction factor be applied when using tile drain flux data to establish whole watershed P losses (or vice-versa), we do not suggest that this correction factor be always assumed to be 13%.

4.3. To What Extent Does Overland Flow Affect P Losses to the Stream, and What Is the Relative Importance of SRP in TP?—Implication of Results for Watershed Management

Although SRP concentrations in overland flow are generally one order of magnitude higher than in tile drains for storms when overland flow occurs, this does not translate into higher stream SRP concentrations for these storms (**Table 1**). This suggests that at least in the watershed studied, SRP contributions to stream from overland flow are negligible from a mass balance standpoint, suggesting that efforts aiming at reducing SRP losses to streams in artificially drained landscapes should primarily focus on reducing SRP losses in tile drains. Previous work in the watershed indicates that overland flow (when it occurs) contributes, on average, 18% of stream flow for a series of storms ranging from 2.6 cm to 23.8 cm in bulk precipitation [23]. This work however also indicates that for storms associated with <5 cm in precipitation (as in our study), overland flow generally only contributes to <5% of stream flow [23]. Although 7 storms are monitored for SRP, our study only provides limited TP data (only 4 storms and for the stream only) but stream TP concentrations are not higher for storm 6, during which overland flow occurred, than for storms 1, 2, and 4, for which no overland flow occurred. This is also consistent with overland flow not being a significant contributor to P losses (in terms of mass) to the stream, at least in spring for the range of storms studied (1.02 cm < bulk precipitation < 4.45 cm).

In terms of the relative concentration of SRP in TP, our results suggest that for the >100 samples collected in the stream for which both SRP and TP are available, SRP typically accounts for 39% of TP. When fluxes are compared, SRP fluxes represent 45% of TP fluxes. This is consistent with results reported by McDowell & Wilcock (2004) in New Zealand, where stream SRP concentrations represented approximately 27% of TP concentrations in summer/fall and 35% of TP concentrations in winter/spring [15]. For three Illinois streams, Royer *et al.* (2006) reported that over a 12-month period, SRP fluxes represented 44% of TP fluxes [4]. In a series of storms in 2008, Vidon & Cuadra (2011) indicate that in TD1 and TD2, SRP fluxes typically accounted for 10% - 22% of TP fluxes [8].

From a watershed management standpoint, these results indicate that efforts should be made to reduce both SRP and particulate phosphorus (PP) (PP = TP – SRP) losses to tile drains and the stream. Indeed, PP represents at least 60% of TP over multiple storms (in terms of concentration), and flux data indicate that PP losses are larger than SRP losses. However, SRP is more bioavailable than PP [24], and is therefore more likely to have a direct impact on ecosystem primary productivity (e.g. algae blooms); so from an ecosystem perspective, SRP is likely more important than PP. Our results also stress that although SRP concentrations are generally much higher in overland flow than in tile drains (**Table 1**); the occurrence of overland flow does not have any significant impact on P losses (concentration or flux) at the watershed scale for moderate size storms (<5 cm bulk precipitation). Consequently, although erosion control measures may have positive effects on water quality with respect to sediment losses [25] and pesticides losses [26]; when P is the primary concern, our results suggest that, at least for tile drain dominated systems in spring, best management practices with the potential to reduce P losses to stream via tile drains (wetlands intercepting tile drains, controlled drainage, tillage practices...) are likely to have more effects on P exports at the watershed scale than best management practices designed, at least in part, to reduce direct overland flow contributions to the stream (e.g. stream/riparian zone restoration, riparian zone management).

5. CONCLUSION

Overall, results suggest that bulk precipitation and antecedent moisture conditions are not good predictors of SRP or TP losses (either concentration or flux) to the stream, and that overland flow is not a significant contributor to P losses at the watershed scale for the storms studied (<5 cm bulk precipitation). Taken together with previous work in the watershed, our results suggest a threshold-based behavior for this watershed in terms of SRP and TP exports, whereby SRP and TP fluxes in either tile drains or the stream significantly increase with precipitation when bulk precipitation exceeds 4 cm, but show no significant positive correlation with bulk precipitation below that threshold. If climate change predictions hold true and that the frequency of large precipitation events increases in the coming years in the US Midwest, we will likely see a significant increase in P losses to streams (assuming land use practices remain the same). However, for most storms, our data indicate that total SRP and TP losses are somewhat limited by the amount of P available for leaching (**Figure 5**). This suggests that although P losses in artificially drained landscapes of US Midwest are primarily driven by hydrology (high P flux during high flow conditions), further source reduction could potentially have a significant effect on overall P losses. Our results also indicate that SRP fluxes per unit drainage area are on average 13% higher in tile drains than at the watershed scale on a storm basis (and 23% higher on a hourly basis). Differences in land use and drainage intensity are logical explanation for these differences, but scaling ratios are not commonly reported in the literature. As discussed previously, these ratios (either storm or hourly ratios) should certainly be used with caution as they are likely to change with seasons, land use, and tile drain spacing/diameter, but when no empirical data linking plot scale flux estimates to whole watershed flux estimate are available, we propose that these ratios could be used as scaling ratios between tile drain flux observations and whole watershed flux estimates. It is important to note that over the course of an entire year, changes in crop development stage and antecedent moisture conditions play a critical role in regulating the hydrological response to precipitation of tile drained watersheds in the US Midwest [27] or other regions such as Oregon [28,29], or Ontario [11,12]. Our results are therefore primarily applicable to the late winter/spring period in the US Midwest, when most P losses to streams on an annual basis occur [4].

6. ACKNOWLEDGEMENTS

The project described in this publication was supported by grant/ cooperative agreement number # 08HQGR0052 to P. Vidon from the United States Geological Survey (USGS). Its contents are solely the responsibility of the authors and do not necessarily represent the official views of the USGS. Additional funding was also provided by an Indiana University-Purdue University, Indianapolis RSGF grant to P. Vidon and a Mirsky Fellowship to P.E. Cuadra. The authors would like to thank Lani Pascual, Vince Hernly and Bob E. Hall for help in the field and the laboratory, and Jeff Frey and Nancy Baker from the USGS Indianapolis Office for their help in the development phase of the project.

REFERENCES

[1] Alexander, R., Smith, R.S., Schwarz, G.E., Boyer, E.W., Nolan, J.V. and Brakebill, J.W. (2008) Differences in phosphorus and nitrogen delivery to the Gulf of Mexico from the Mississippi River Basin. *Environmental Science and Technology*, **42**, 822-830.

[2] Carpenter, S.R., Caraco, N.F., Correll, D.L., Howarth, R.W., Sharpley, A.N. and Smith, V. (1998) Nonpoint pollution of surface waters with phosphorus and nitrogen. *Ecological Applications*, **8**, 559-568.

[3] David, M.B. and Gentry, L.E. (2000) Anthropogenic inputs of nitrogen and phosphorus and riverine export for Illinois, USA. *Journal of Environmental Quality*, **29**, 494-508.

[4] Royer, T.V., David, M.B. and Gentry, L.E. (2006) Timing of riverine export of nitrate and phosphorus from agricultural watersheds in Illinois: Implications for reducing nutrient loading to the Mississippi River. *Environmental Science and Technology*, **40**, 4126-4131.

[5] Vidon, P., Tedesco, L.P., Pascual, D.L., Campbell, M.A., Casey, L.R., Wilson, J. and Gray, M. (2008) Seasonal changes in stream water quality along an agricultural/urban land-use gradient. *Proceedings of the Indiana Academy of Sciences*, **117**, 107-123.

[6] Cooke, S.E. and Prepas, E.E. (1998) Stream phosphorus and nitrogen export from agricultural and forested watersheds on the Boreal Plain. *Canadian Journal of Fisheries and Aquatic Sciences*, **55**, 2292-2299.

[7] Coulter, C.B., Kolka, R.K. and Thompson, J.A. (2004) Water quality in agricultural, urban and mixed land use watersheds. *Journal of the American Water Resources Association*, **40**, 1593-1601.

[8] Vidon, P. and Cuadra, P.E. (2011) Phosphorus dynamics in tile-drain flow during storms in the US Mid west. *Agricultural Water Management*, **98**, 532-540.

[9] Kronvang, B., Laubel, A. and Grant, R. (1997) Suspended sediment and particulate phosphorus transport and delivery pathways in an arable catchment, Gelbaek stream, Denmark. *Hydrological Processes*, **11**, 627-642.

[10] Laubel, A., Jacobsen, O.H., Kronvang, B., Grant, R. and Andersen, H.E. (1999) Subsurface drainage loss of particles and phosphorus from field plot experiments and a tile-drained catchment. *Journal of Environmental Quality*, **28**, 576-584.

[11] Macrae, M.L., English, M.C., Schiff, S.L. and Stone, M. (2007) Intra-annual variability in the contribution of tile drains to basin discharge and phosphorus export in a first-order agricultural catchment. *Agricultural Water Management*, **92**, 71-182.

[12] Macrae, M.L., English, M.C., Schiff, S.L. and Stone, M. (2010) Influence of antecedent hydrological conditions on patterns of hydrochemical export from a first-order agricultural watershed in Southern Ontario, Canada. *Journal of Hydrology*, **389**, 101-110.

[13] Sharpley, A.N., Gburek, W.J., Folmer, G. and Pionke, H.B. (1999) Sources of phosphorus exported from an agricultural watershed in Pennsylvania. *Agricultural Water Management*, **41**, 77-89.

[14] Sekely, A.C., Mulla, D.J. and Bauer, D.W. (2002) Streambank slumping and its contribution to the phosphorus and suspended sediment loads to the Blue Earth River, Minnesota. *Journal of Soil and Water Conservation*, **57**, 243-250.

[15] McDowell, R.W. and Wilcock, R.J. (2004) Particulate phosphorus transport within streamflow of an agricultural catchment. *Journal of Environmental Quality*, **33**, 2111-2121.

[16] Gentry, L.E., David, M.B., Royer, T.V., Mitchell, C.A. and Starks, K.M. (2007) Phosphorus transport pathways to streams in tile-drained agricultural watersheds. *Journal of Environmental Quality*, **36**, 408-415.

[17] Karl, T.R. and Knight, R.W. (1998) Secular trends of precipitation amount, frequency, and intensity in the United States. *Bulletin of the American Meteorological Society*, **79**, 231-241.

[18] Davis Todd, C.E., Harbor, J.M. and Tyner, B. (2006) Increasing magnitudes and frequencies of extreme precipitation events used for hydraulic analysis in the Midwest. *Journal of Soil and Water Conservation*, **61**, 179-184.

[19] Milly, P.C.D., Dunne, K.A. and Vecchia, A.V. (2005) Global pattern of trends in streamflow and water availability in a changing climate. *Nature*, **438**, 347-350.

[20] NOAA (2005) Climatological data, Indianapolis. National Oceanic and Atmospheric Administration, National Climatic Data Center.

[21] Lathrop, T.R. (2006) Environmental setting of the sugar creek and leary weber ditch basins, Indiana, 2002-04. U.S. Geological Survey Scientific Investigations Report, Reston.

[22] Clesceri, L.S., Greenberg, A.E. and Eaton, A.D. (1998) Standard methods for the examination of water and waste water. 20th Edition, American Public Health Association, Washington DC.

[23] Baker, N.T., Stone, W.W., Wilson, J.T. and Meyer, M.T. (2006) Occurrence and transport of agricultural chemicals in leary Weber ditch basin, Hancock County, IN, 2003-04. US Geological Survey Scientific Investigations Report, Reston.

[24] Ekholm, P. (1991) Bioavailability of phosphorus. In: Svendsen, L.M. and Kronvang, B., Eds, *Phosphorus in the Nordic Countries. In Methods Bioavailability, and Measures, Nord, Vol.* 47, Nordic Council of Ministers, Copenhagen, 109-120.

[25] McKergrow, L.A., Weaver, D.M., Prosser, I.P., Grayson, R.B. and Reed, A.E. (2003) Before and after riparian management: Sediment and nutrient exports from a small agricultural catchment, Western Australia. *Journal of Hydrology*, **270**, 253-272.

[26] Shipitalo, M.J. and Owens, L.B. (2006) Tillage system, application rate, and extreme event effects on herbicide losses in surface runoff. *Journal of Environmental Quality*, **35**, 2186-2194.

[27] Vidon, P., Hubbard, L.E. and Soyeux, E. (2009) Seasonal solute dynamics across land uses during storms in glaciated landscape of the US Midwest. *Journal of Hydrology*, **376**, 34-47.

[28] Poor, C.J. and McDonnell, J.J. (2007) The effect of land use on stream nitrate dynamics. *Journal of Hydrology*, **332**, 54-68.

[29] Wigington, P.J., Moser, T.J. and Lindeman, D.R. (2005) Stream network expansion: A riparian water quality factor. *Hydrological Processes*, **19**, 1715-1721.

16

Outlook of future climate in northwestern Ethiopia

Dereje Ayalew[1,2*#], Kindie Tesfaye[1], Girma Mamo[3], Birru Yitaferu[4], Wondimu Bayu[5]

[1]College of Agricultural and Environmental Sciences, Haramaya University, Haramaya, Ethiopia;
[*]Corresponding Author
[2]College of Agricultural and Environmental Sciences, Bahir Dar University, Bahir Dar, Ethiopia
[3]Ethiopian Institute of Agric-Research, Coordinator, National Agro meteorology Research, Nazreth, Ethiopia
[4]Amhara Region Agricultural Research Institute, Bahir Dar, Ethiopia
[5]ICARDA-Ethiopia National Project Officer International Center for Agricultural Research in the Dry Areas, Bahir Dar, Ethiopia

ABSTRACT

Climate change is described as the most universal and irreversible environmental problem facing the planet Earth. Whilst climate change is already manifesting in Ethiopia through changes in temperature and rainfall, its magnitude is poorly studied at regional levels. The objective of this study was to assess and quantify the magnitude of future changes of climate parameters using Statistical Downscaling Mode (SDSM) version 4.2 in Amhara Regional State which is located between 8°45′N and 13°45′N latitude and 35°46′E and 40°25′E longitude. Daily climate data (1979-2008) of rainfall, maximum and minimum temperatures were collected from 10 observed meteorological stations (predictand). The stations were grouped and compared using clustering method and Markov chain model, whereas the degree of climate change in the study area was estimated using the coupled HadCM3 general circulation model (GCM) with A2a emission scenarios (Predictors). Both maximum and minimum temperatures showed an increasing trend; and the increase in mean maximum and minimum temperature ranges from 1.55°C - 6.07°C and from 0.11°C - 2.81°C, respectively in the 2080s compared to the base period considered (1979-2008). The amount of annual rainfall and number of rainy days also decreased in the study Regions in the 2080s. The negative changes in rainfall and temperature obtained from the HadCM3 model in the current study are alarming and suggest the need for further study with several GCM models to confirm the current results and develop adaptation options.

Keywords: Amhara Regional State; Climate

[#]Current address: International Maize and Wheat Improvement Center (CIMMYT), Addis Ababa, Ethiopia.

Change; Ethiopia; HadCM3; Statistical Downscaling

1. INTRODUCTION

Climate has profound effects on the biophysical resources of the planet. It is the major factor controlling the patterns of vegetation structure, productivity, and plant and animal species composition [1]. Many plants can successfully reproduce and grow only within a specific range of temperature and respond to specific amounts and seasonal patterns of rainfall, and fail to survive if climate changes. Thus, there is now a substantial concern over the global problem of climate change and its current and future impacts. The impact of changing climate, such as rising global average temperatures and increases in frequency and severity of extreme events, droughts and floods, are already affecting human well-being, biodiversity and ecosystems, economies and societies worldwide [2]. Seasons are shifting, temperatures are climbing and sea levels are rising around the globe. Meanwhile, our planet must still supply us and all living things with air, water, food and safe places to live [3].

Africa contains one-fifth of all known species of plants, mammals, and birds, as well as one-sixth of amphibians and reptiles. However, because of climate change, these ecosystems are threatened [4,5]. Similarly, the varied ecology, edpahic and climatic conditions in Ethiopia which accounts for the wide diversity of biological resources both in terms of flora and fauna is declining because of climate variability [5,6]. The negative impacts associated with climate change are also compounded by many factors, including widespread poverty, human diseases, and high population density, which is estimated to double the demand for food, water, and livestock forage within the next 30 years [7]. For example; climate variability and extreme weather events, such as high temperatures and erratic rainfall, are critical factors in initiating malaria epidemics and increase the spread of infectious diseases especially in the highlands of Ethiopia [3,8,9]. In addition, [10,11] noted that rainfall variability and associated droughts have historically been major causes of food shortages and famines in Ethiopia.

Being able to predict seasonal climate can help to mi-nimize the possibility of "climate surprises" in order to reduce impacts to society and ecosystems. With these predictions, decision makers can be provided with reliable scientific information about possible extreme climate events. Many sectors of society use these predictions, including agriculture, fishing, forestry, energy, insurance, public health, water resources, recreation, transportation, health, and construction. For example, with an increased risk of drought, we know that some possible impacts include water shortages, early onset of the wildfire season, and lower or no crop yields. This in turn could lead to increased disaster assistance payments, higher food prices, and disrupted transportation on internal waterways. With advance knowledge of drought, water resource managers can adjust the timing of water releases from reservoirs, and farmers can alter the type of crops they plant [12].

While most of us feel the effects of climate variation, many businesses, services and activities depend on climate prediction to prepare adequately, to manage risk, protect the environment, and to save lives [13]. A prediction process, begins with observing and accurately measuring the most recent and current environmental conditions. The tools of the predictor can therefore be based on statistics from historical climate data which have been analyzed to show the relationships between the Earth's surface conditions and climate [12].

Numerical models (General Circulation Models or GCMs), are currently most credible tools available for simulating the response of the global climate system [14]. However, because of their coarse spatial resolution, GCMs to date are unable to provide reliable climatic information at regional and local scales [15]. As suggested in different guidelines and documentation developed by the IPCC [16,17], the climate change information required for many impact studies is at a much finer spatial scale than that provided by GCMs. Because of this limitation, other alternative methods called down-scaling techniques using statistical downscaling model (SDSM) have been developed in recent years to obtain fine resolution climate change information at regional/local scales. The SDSM calculates statistical relationships, based on multiple linear regression techniques, between large-scale (the predictors) and local (the predictand) climate [15]. These relationships are developed using observed climate data and, assuming that these relationships remain valid in the future, and hence can be used to obtain downscaled local information for some future time period by driving the relationships with GCM-derived predictors [16,18,19].

Hadley Centre Coupled Model (HadCM3) outputs established on the A2 scenario is the most commonly used GCM to develop climate change scenarios in a given area and regional impact assessment studies [20, 21]. Its good simulation of current climate without using flux adjustments was a major advance at the time it was developed and it still ranks highly compared to other models [22]. Hence, HadCM3 was one of the major models used in the IPCC Third and Fourth Assessments. Further, (e.g. [23]) used HadCM3 models in order to develop climate change scenarios of rainfall, minimum and maximum temperatures in arid regions of Chile in order to analyze the future potential impact of climate change on stream flow. In addition, HadCM3 GCM model out-puts established on the A2 scenario was used to down- scale climate change scenarios at regional levels [24]. Further more, [25] also used the outputs of HadCM3 A2a and B2a scenarios for climate change impact analysis using crop modeling.

Accurate and timely climate information and predictions can help many sectors of society circumvent the impacts posed by climate variations. This reduces the risk of economic setbacks and ecological damage. Therefore, the aim of this paper was to downscale and quantify the magnitude of future changes of maximum and minimum temperatures and rainfall at a local level in the Amhara Regional State using the HadCM3A2a GCM models.

2. MATERIALS AND METHODS

2.1. Description of the Area

Amhara Region is located between 8°45′N and 13°45′N latitude and 35°46′E and 40°25′E longitude in North West Ethiopia. The total area of the region is estimated at 156,960 km^2, which is divided into 11 administrative zones and 105 districts [26] (see **Figure 1**).

2.2. Climate of the Study Area

The climate of the Amhara Region is affected significantly by variation in altitude, its latitudinal position, prevailing winds, air pressure and circulation and its proximity to the sea. Traditionally, the climate of the Region is divided in to Kola (hot zone) which represents and cover 31% of the Region with altitude below 1500 meters above sea level (a.s.l.) and Woyina Dega (warm zone) which covers 44% of the Region encaompasses areas between 1500 - 2500 m a.s.l and Dega (cold zone) which covers 25% of the Region representing areas between 2500 - 4620 m a.s.l [9]. The annual mean temperature of the Region is between 15°C and 21°C. But in valleys and marginal areas the temperature exceeds 27°C [27].

3. DATA SETS AND METHODOLOGY

3.1. Observed Data

Thirty years observed rainfall data (1979-2008) was

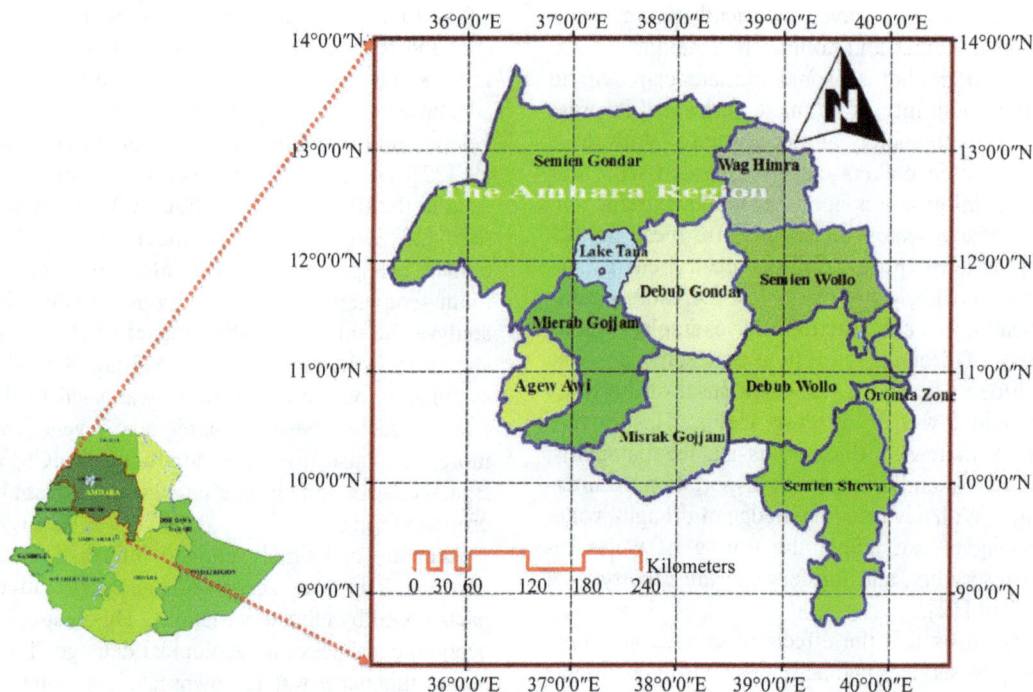

Figure 1. Geographical location of Amhara Regional State with reference to national map of Ethiopia.

collected from National Meteorological Agency (NMA) of Ethiopia for 10 meteorological stations in the Amhara Region. The meteorological stations were clustered by hierarchical clustering method with quantitative data set and Squared Euclidean Distance with wards methods using StatistiXL software.

3.2. General Circulation Model Data (GCM)

The GCM data were collected from the IPCC Data Distribution Center (DDC) which was established in 1998 by the World Metrological Organization and the United Nations Environmental Program, following a recommendation by the task group on Data and Scenario Support for Impact and Climate Assessment [28]. However, because the daily fields are only available directly from the respective modeling centers, the climatic data used for SDSM had been collected from the Canadian Institute for climate studies website for model output of HadCM3 [29]. The predictor variables were supplied on a grid basis of size 2.5° latitude × 3.75° longitude so that the data were downloaded from the nearest grid box to the study area.

As shown in **Figure 2**, Hence the nearest grid box for the HadCM3 model, which represents the study area, is the one at 10.95°N latitude and 37.85°E longitude (Y = 31 Latitude 10°N and X = 12 Longitude 41.25°E).

3.3. Baseline Period Data

One of the criteria used in evaluation of the performance of any useful downscaling method is whether the

historic (observed) condition can be replicated or not [16]. It is therefore imperative that the methods used for transforming the results of climate models to meteorological stations will generate precipitation rainfall and temperature series that should have the same statistical properties as observed meteorological data for use in climate modeling [30]. The years from 1/1/1979 to 31/12/2008 were used as a baseline period for the study [5]. Thus, the HadCM3 data were downscaled for the baseline period A2 emission scenarios and the statistical properties (mean and variance) of the downscaled data were compared with observed data of the same period.

4. METHODS OF CLUSTERING ANALYSIS AND HOMOGENITY TEST

4.1. Clustering Analysis

Clustering (or numerical classification) uses a data array, or matrix, to classify (or cluster) the cases (rows) of data into groups according to values of their attributes (columns). Geographical location (latitude, longitude and elevation) and climatologically mean monthly rainfall (1979-2008) of all stations were used to calculate their standardized standard deviation. Hence using Hierarchical Clustering method the stations were grouped; there was no prior knowledge of grouping (or any prior grouping was ignored). The data were separated data into groups based on individual cases using clustering analysis [31]. Therefore, the data was summarized and stations were grouped using tree graph (dendrogram) (**Figure 3**).

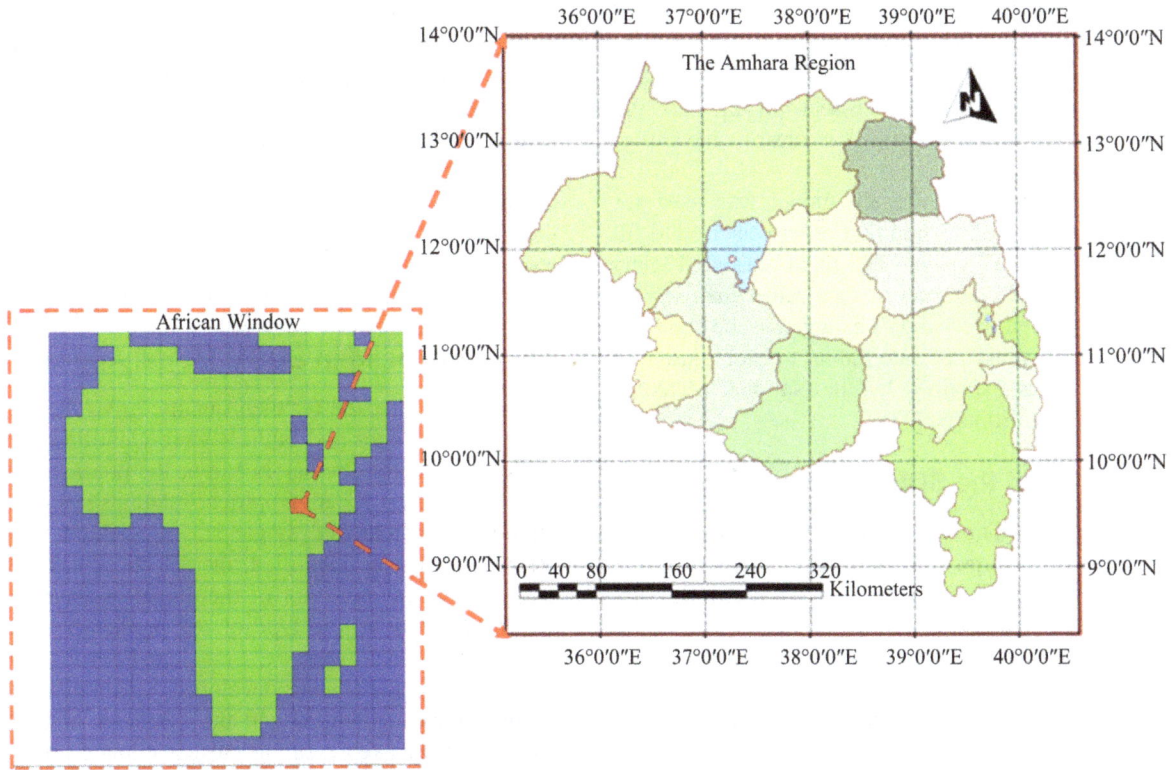

Figure 2. The African continent window with 2.5° latitude × 3.75° longitude grid size from which the grid box for the study area was selected (shown in red/or back if printed in black and white).

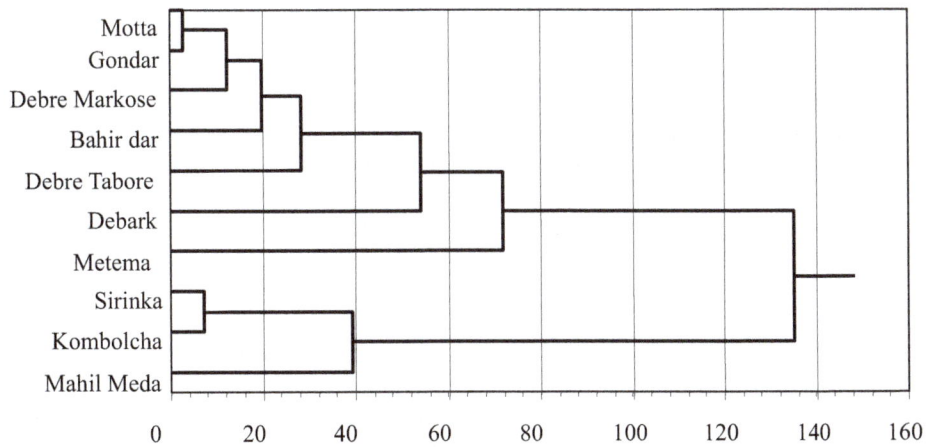

Figure 3. Dendrogram of meteorological stations in the Amhara Regional State.

4.2. Selection of Representative Station

Since the downscaling method is statistical using a site specific tool called SDSM, there was a need to select one representative station which was statistically correlated with the rest of the stations in each cluster. According to [31], the software StatistiXL was used for clustering partial correlation of each station belonging to each group and the software gave all statistical tests needed. The station with better correlation coefficient in each group was selected for the study.

4.3. Homogeneity Test

Markov chain model is one of the facilities in Instat + V3.36 for fitting the probabilities of rain and to the rainfall amounts. A model was first fitted to the chance of rain and then to the amounts. First order Markov chain model [32], uses the probability of rain or dry day in the given observational data. A threshold value 0.1 mm/day was used as dry day and all days with rainfall value greater than this value were considered as rainy day. Following this it was possible to examine whether all the stations in each cluster have similar rainfall seasons or not. Moreover, first order Markov chain model clearly gave the probability of getting rain on a daily basis and the chance of getting rain in each season [32]. This was used to identify the dry and rainy seasons of each cluster in the study area.

The zero order Markov chain model was used for testing the homogeneity while first order was used to fill missing daily rainfall values. The main reason to select the first order to fill the missing data rather than second order was that first order didn't exaggerate the resulting values and gave more accurate model to each cluster of the study area as explained by the National Meteorological Agency of Ethiopia [33].

4.4. Settings of the SDSM

According to [16], the concept of regional climate being conditioned by the large-scale state may be written as:

$$R = F (L)$$

where,

R represents the predicted (regional or local climate variables),

L represents the predictor (a set of large-scale climate variables),

F is a deterministic/Stochastic function conditioned by L and has to be found empirically from observed or modeled data sets.

The climate scenario for future period were developed from statistical downscaling using GCM HadCM3 predictor variable for the emission scenarios for 100 years based on the mean of 20 ensembles; and the analyses was done based on three 30-years periods centered on the 2020s (2011-2040), 2050s (2041-2070) and 2080s (2071-2099).

4.5. Settings Used for SDSM

For the observed and the National Centre for Environmental Prediction (NCEP) data the year length was set to be the default (366 days), which allows 29 days in February in leap years. However, as HadCM3 have modeled years that do only consist of 360 days, the default value was changed to 360 days. The base period used for the model was from 1/1/1979 to 31/12/ 2008.

T he event threshold value is important to treat trace values during the calibration period. For the temperature parameter, this value was set to be 0 while for daily rainfall calibration purpose this parameter was fixed to be 0.1 mm/day so that trace rain days below this thresh-old value were considered as a dry day. Missing data were replaced by –999.

For the daily temperature values, no transformation was used as it is normally distributed and the model was non-conditional. However, for the daily rainfall, the fourth root transformation was used as its data were skewed and

as its model was conditional. The range of variation of the downscaled daily weather parameters was controlled by fixing the variance inflation. The default value, which is 12, was used for the daily temperature values; where as for daily precipitation this value was set as 18, in order to magnify the variation. The bias correction, which compensates for any tendency to over or underestimate the mean of conditional processes by the downscaling model, was set to be 0.8 for daily rainfall, and the default 1.0 for daily temperature (indicating no bias correction).

4.6. Steps Used in SDSM Model Approach

Data were checked for their quality; gross data errors, missing data codes and outliers were identified prior to model calibration. Predictors and/or the predictand were also transformed. Then, empirical relationships between gridded predictors and single site predictands (station precipitation) were identified; the selection was done at most care as the behavior of the climate scenario completely depends on the type of the predictors selected. After this, models were calibrated and results for the different time slices were developed. Calibrated models (using independent data) and the synthesis of artificial time series data representing current climate conditions were verified. Then both derived SDSM scenarios and observed climate data were interrogated with the analyses data screen of SDSM and monthly statistics produced were plotted using the compare results screen. Then, rapid assessment of downscaled versus observed, and/or current versus future climate scenarios were compared.

Finally, the outputs of the Hadley Centre Coupled Model (HadCM3) with A2a scenario, which is the most commonly used GCM for downscaling future climate of an area (e.g. [23]), were used to develop climate change scenarios in the region.

According to [16], the regression weights produced during the calibration process were applied to the time series outputs of the GCM model. This is based on the assumption that the predictor-predictand relationships under the current condition remain valid under future climate conditions too. Twenty ensembles of synthetic daily time series data were produced for A2a SRES scenarios for a period of 139 years (1961 to 2099). The final product of the SDSM downscaling method was then found by averaging the twenty independent stochastic GCM ensembles of A2a and B2a emissions scenarios. Statistical significance of the mean values of maximum and minimum temperatures found at different time scales were tested using a paired t-test mean comparison at 0.05 alpha level.

5. RESULTS AND DISCUSSIONS

5.1. Clustering Analysis

A dendrogram (tree graph) is a common means of graphically summarizing a clustering pattern using Statisti XL. The dendrogram usually starts with all individuals as separate clusters ("tips") and shows the combination of fusions back to a single "root". The order of individuals shown in the dendrogram was the order in which the groups enter the clustering [31]. In this study, the maximum distance or similarity to classify as one group is 60 (**Figure 3**).

All available stations, which were 10 before clustering, were clustered and classified in to three groups and each group was designated as Cluster/Region A (Kombolcha, Mahil-Meda and Sirinka), Cluster/Region B (Metema) and Cluster/Region C (Motta, Gondar, Debre Markose, Debre Tabore and Debark) randomly.

The designation of each cluster was done just to rep-

resent climatologically homogeneous areas within the study area (**Figure 3**). The geographical locations of the stations grouped and the representative clusters used for the study are given in **Figure 4**. Based on the clusters, climatologically the Amhara region has a mean maximum temperature of 26.0°C, 35.8°C and 23.6°C for Cluster A, B and C, respectively, but inter-annual variability of rainfall was observed in all clusters (**Figures 5** and **6**).

Based on the maximum and minimum temperature data, the mean monthly temperature over the Amhara Region tend to increase until the beginning of June in all regions and decreased during the main rainy season ((June to September) (**Figure 6**)). Contrarily, T_{min} had greater value during the main rainy season, and the mean minimum temperature over Amhara Region was 11.6°C,

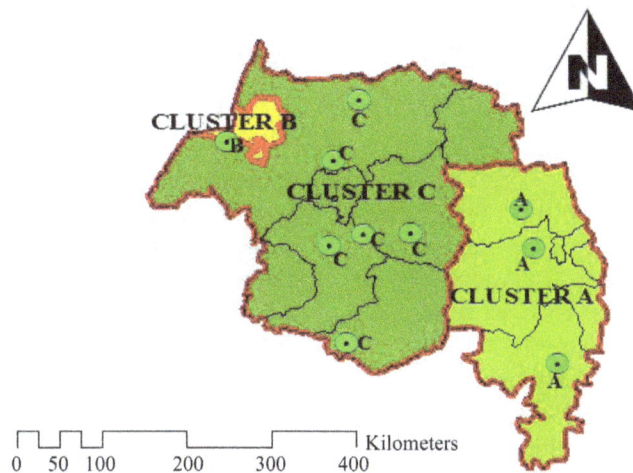

Figure 4. Three climate regions in the Amhara Regional State clustered based on 30 years climate data (1979-2008).

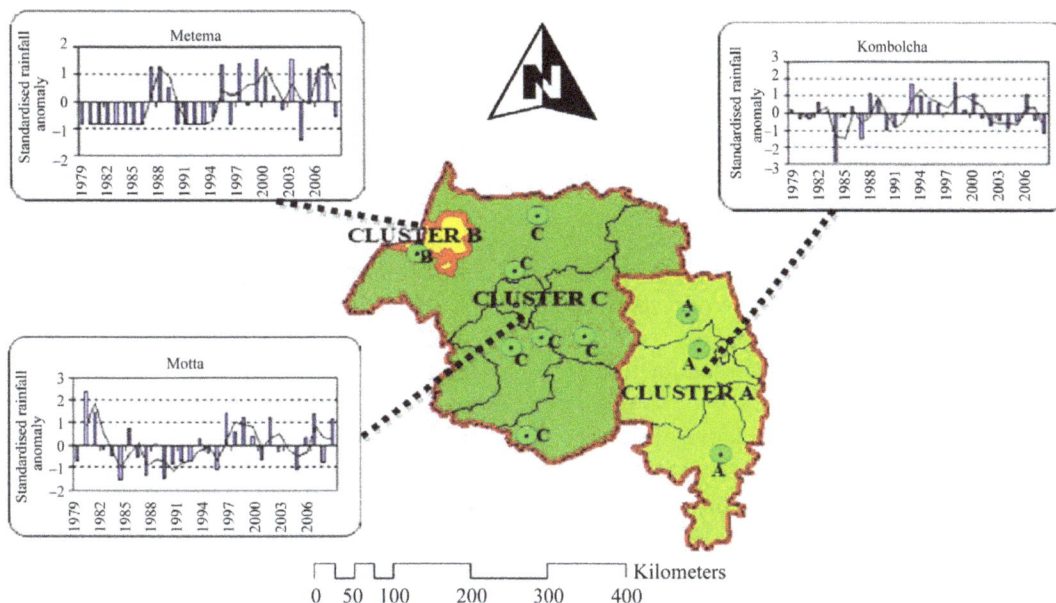

Figure 5. Rainfall variability in three clustered regions in the Amhara Regional State for the period 1979-2008.

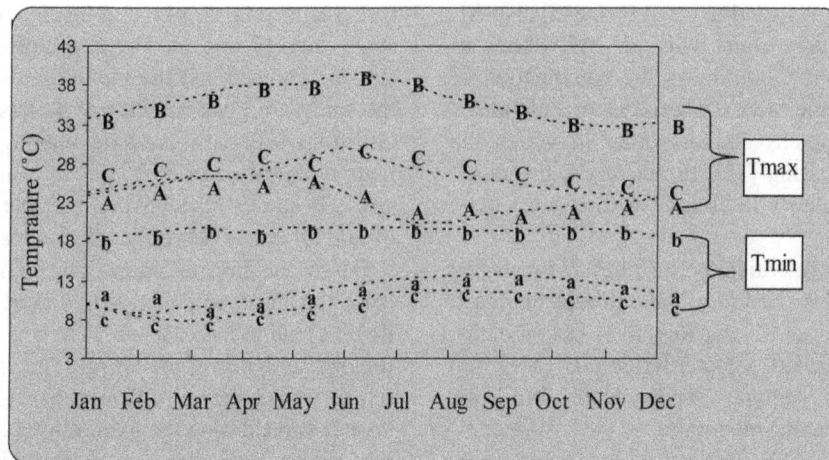

Figure 6. Monthly average temperature of three clustered regions (A, B and C) in the Amhara Regional State for the period 1979-2008. Letters: A, B, C and a, b, c represent respective clusters for maximum and minimum temperatures.

19.3°C and 10.0°C for Clusters A, B and C, respectively.

5.2. Homogeneity of Stations

Homogeneity test analysis showed that Cluster C and Cluster B had only one main rainy season. The rainfall in Cluster C is long with early onset and late end date of the rain, while in Cluster B rainfall is characterized by late onset and early cessation of rain (**Figure 7**).

Cluster A has two rainy seasons which are locally known as "belg" (small rainy season) and "kiremit" (main rainy season). The two rainy seasons have no distinct dry period on their translation; hence, the small rainy season is soon after followed by the main rainy season. In addition, the length of main rainy season is similar to Cluster C.

5.3. Selection Representative Stations

Partial correlations of each station belonging to each group were analyzed. In each group, stations with better correlation coefficient were selected (highlighted with grey/black if it is printed in black and white in **Table 1**). Kombolcha and Motta meteorological stations had good correlation in their respective Cluster (**Table 1**). While, Metema meteorological stations didn't correlate with other observatory stations, hence it stood alone as Cluster B. Henceforth, only these stations were used for further analysis.

These stations were taken as representative of the climate stations with respect to average monthly rainfall (priority was given for rainfall than temperature). Therefore, temperature data was taken from the selected representative station out of respective Cluster of a group. Consequently, the climate scenarios were only developed for these selected stations.

5.4. Predictor Variables Selected

The type and explanations of the predictors, which showed better correlation with the daily maximum temperature, daily minimum temperature, and daily rainfall predictands at $p < 0.05$ significance level are shown in **Table 2**. Partial correlations indicate that the corresponding predictor had the strongest association with the predictand.

5.5. Calibration and Validation

The simulated maximum and minimum temperatures in Clusters A and C had better agreement with the observed results than rainfall/precipitation (**Table 3**). Though simulated values of rainfall showed a lesser agreement as compared to the maximum temperature and minimum temperatures in all clusters, the result was quite acceptable due to the fact that precipitation is a conditional process [14]. Conditional processes like precipitation are dependent on other intermediate processes like the occurrence of humidity, cloud cover, and/or wet-days. Unconditional processes like temperature; however, are not regulated by other intermediate processes.

In addition, as indicated in the SDSM manual [14], local temperatures are largely determined by regional forcing whereas precipitation series display more "noise" arising from local factors. Hence, larger differences can be observed in precipitation ensemble members than that of temperature.

On the other hand, the minimum and maximum temperatures over Cluster B simulation showed a very poor agreement with the observed one (**Table 3**). Even though it is difficult to exactly point out the reason behind, this could be due to low/inferior data quality used for the comparison.

Validation was done based on 15 year simulation from 1993 to 2008. The validation statistics for maximum and minimum temperatures and rainfall in all clusters are shown in **Table 4**.

The correlation factors that were found during the calibration step are more or less maintained during the validation period too, even better agreement was found here.

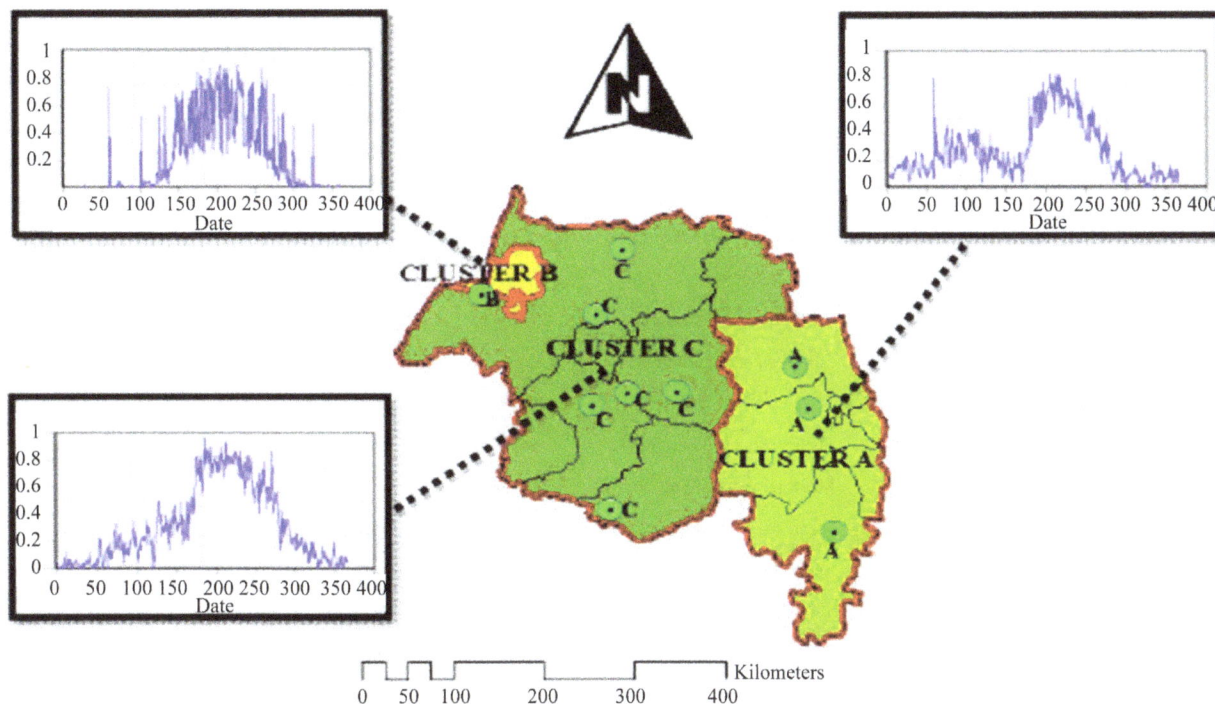

Figure 7. Homogeneity test using Markov chain model (zero order) for three cluster regions in the Amhara Regional State.

Table 1. Correlation matrix of stations average monthly rainfall in three cluster regions of the Amhara Regional State.

Cluster A	1	2	3
Kombolcha (1)	1.000	0.983	0.977
Mahil Meda (2)	0.983	1.000	0.954
Sirinka (3)	0.977	0.954	1.000

Cluster B	1	2	3
Metema	1.000	0.983	0.977

Cluster C	1	2	3	4	5	6
Bahir Dar (1)	1.000	0.901	0.976	0.965	0.982	0.997
Debark (2)	0.901	1.000	0.945	0.821	0.838	0.915
Debre Makos (3)	0.976	0.945	1.000	0.954	0.959	0.983
Debre Tabore (4)	0.965	0.821	0.954	1.000	0.989	0.961
Gondar (5)	0.982	0.838	0.959	0.989	1.000	0.979
Motta (6)	0.997	0.915	0.983	0.961	0.979	1.000

Table 2. List of predictor variables that gave better correlation results at p < 0.05.

	Predictand	Predictorors (NCEP Reanalysis)	Notation	Partial r[1]
Cluster A	Precipitation/rainfall	Surface vorticity	ncepp_zaf.dat	0.11
		Relative humidity at 850 hPa height	ncepr850af.dat	0.18
		Mean temperature at 2 m	Nceptempaf.dat	−0.11
	Maximum Temperature	Surface airflow strength	Ncepp_faf.dat	0.19
		Surface divergence	Ncepp_zhaf.dat	−0.12
		500 hPa zonal velocity	Ncepp5_uaf.dat	−0.3
		850 hPa zonal velocity	Ncepp8_uaf.dat	0.26
		Surface specific humidity	Ncepshumaf.dat	−0.17
		Mean temperature at 2 m	Nceptempaf.dat	0.43
	Minimum Temperature	Surface airflow strength	Ncepp_faf.dat	0.15
		500 hPa zonal velocity	Ncepp5_uaf.dat	0.20
		850 hPa zonal velocity	Ncepp8_uaf.dat	−0.15
		Surface specific humidity	Ncepshumaf.dat	0.31
		Mean temperature at 2 m	Nceptempaf.dat	0.47
Cluster B	Precipitation/rainfall	Mean sea level pressure	Ncepmslpaf.dat	0.38
		Surface vorticity	ncepp_zaf.dat	0.12
		850 hPa geopotential height	Ncepp850af.dat	0.35
		500 hPa geopotential height	Ncepp500af.dat	0.15
	Maximum Temperature	850 hPa meridional velocity	Ncepp8_vaf.dat	0.22
		850 hPa geopotential height	Ncepp850af.dat	−0.24
		Mean temperature at 2 m	Nceptempaf.dat	0.14
		Surface meridional velocity	Ncepp_vaf.dat	−0.10
		Surface vorticity	ncepp_zaf.dat	−0.10
	Minimum Temperature	500 hPa wind direction	Ncepp5thaf.dat	0.11
		850 hPa vorticity	Ncep_zaf.dat	0.14
		850 hPa wind direction	Ncepthaf.dat	0.16
		500 hPa vorticity	Ncepp5_zaf.dat	−0.15
Cluster C	Precipitation/rainfall	850 hPa divergence	Ncepp8zhaf.dat	0.26
		Near surface relative humidity	Nceprhumaf.dat	0.19
		Surface specific humidity	Ncepshumaf.dat	−0.11
		Mean sea level pressure	Ncepmslpaf.dat	0.15
	Maximum Temperature	850 hPa airflow strength	Ncepp8_faf.dat	−0.12
		Relative humidity at 500 hPa	Ncepr500af.dat	−0.26
		Mean temperature at 2 m	Nceptempaf.dat	0.49
		Mean sea level pressure	Ncepmslpaf.dat	−0.24
		Surface wind direction	Ncepp_thaf.dat	−0.18
	Minimum Temperature	850 hPa zonal velocity	Ncepp8_uaf.dat	0.19
		850 hPa vorticity	Ncepp8_zaf.dat	−0.10
		850 hPa geopotential height	Ncepp850af.dat	0.11

[1]The partial correlation coefficient (r) shows the explanatory power that is specific to each predictor. All are significant at $p \leq 0.05$. hpa: is a unit of pressure, 1 hPa = 1 mbar = 100 Pa = 0.1 kPa.

Table 3. Calibration statistics of daily rainfall and maximum and minimum temperatures for three clustered regions in the Amhara Regional State.

Cluster	Predictand	R^2		Standard Error	
		Unconditional	Conditional	Unconditional	Conditional
A	Precipitation/rainfall	0.183	0.046	0.432	0.460
	Maximum temperature	0.539		1.903	
	Minimum temperature	0.651		1.718	
B	Precipitation/rainfall	0.146	0.170	0.392	0.461
	Maximum temperature	0.382		2.261	
	Minimum temperature	0.192		3.746	
C	Precipitation/rainfall	0.205	0.191	0.455	0.447
	Maximum temperature	0.623		1.290	
	Minimum temperature	0.615		1.465	

Table 4. Validation statistics of daily rainfall and maximum and minimum temperatures for three clustered regions.

Cluster	Predictand	R^2		Standard Error	
		Unconditional	Conditional	Unconditional	Conditional
A	Precipitation/rainfall	0.141	0.143	0.65	0.457
	Maximum temperature	0.632		1.531	
	Minimum temperature	0.626		2.387	
B	Precipitation/rainfall	0.157	0.152	0.056	0.544
	Maximum temperature	0.594		1.397	
	Minimum temperature	0.648		2.435	
C	Precipitation/rainfall	0.188	0.171	0.354	0.356
	Maximum temperature	0.520		1.724	
	Minimum temperature	0.518		1.417	

5.6. Projected Climate Changes

According to [34], extensive area specific research conducted in various parts of the world, had concluded that, even a slight warming in average surface temperatures increased the likelihood of extreme weather (hail, lightning, tornadoes, heat waves and/or damaging winds). The findings in this paper also showed similar conditions. Relative to the base climate period of 1979-2008, the statistical downscaled values of SDSM projections for the period 2080s showed that T_{max} could rise by up to 2.3°C, 4.9°C and 1.8°C, while rainfall could decrease by 27.2%, 42.3% and 12.2% over Clusters A, B and C, respectively (**Table 5**); all SDSM downscaled value were significant (p < 0.05). Although this finding provided important information for preparing developmental strategies and policy interventions in the region, the results are based only on oneGCMs output and opns the door for testing the result with more GCM results.

5.7. Analysis of Variability and Extremes of Basic Downscaled Parameters (Tmax, Tmin, and Pcpn)

5.7.1. Daily Maximum Temperature (Tmax)

Temperature is normally distributed [5], and hence the distributions of T_{max} and T_{min} were examined for changes in mean and variance, and how changes in these two statistics affected the occurrence of records of hot or cold weather. The relevant statistics for T_{max} from observed and downscaled data showed there was an increase in mean value of maximum temperature and its variability in Clusters A, B and C. **Table 6** shows the increase in mean value of Tmax up to 2.06°C, 6.07°C and 1.55°C with its increase in variability (standard deviation) of about 18.83%, 13.15% and 8.51% in Clusters A, B and C, respectively in the 2080s. This result is in better conformity with [5] that also projected an increase in mean global temperature of up to 5.8°C by the end of the 21st

Table 5. Downscaled maximum and minimum temperatures and rainfall in three clustered regions of Amhara Regional State.

Tri-decades	Tmax (Absolute change)			Tmin (Absolute change)			Pcpn (Percentage change)		
	2020's	2050's	2080's	2020's	2050's	2080's	2020's	2050's	2080's
Cluster A	0.5	1.3	2.3	0.8	1.8	3.6	−8.28	−19.3	**−27.2**
Cluster B	2.6	3.1	4.9	4.6	3.3	0.3	−12.9	−28.2	**−42.3**
Cluster C	0.4	1.0	1.8	0.2	0.3	0.5	−4.2	−3.3	**−12.2**

2020's = 2010-2039, 2050's = 2040-2069, and 2080's = 2070-2099.

Table 6. Summered statistics mean of maximum temperature in three clustered regions of Amhara Regional State.

Cluster	Year	Statistics			
		Mean	Std	Maximum	Minimum
A	1979-2008	26.44a	2.23	33.20	17.70
	2020's	26.71b	2.58	35.54	14.81
	2050's	27.57c	2.55	35.67	16.38
	2080's	28.50d	2.65	37.83	19.97
	abcd; significant at 0.001 level				
B	1979-2008	36.91a	2.66	47.85	27.81
	2020's	38.71b	2.74	49.46	28.23
	2050's	40.08c	2.85	51.22	31.97
	2080's	42.98d	3.01	55.06	35.95
	abcd; significant at 0.001 level				
C	1979-2008	23.78a	1.41	28.01	17.35
	2020's	24.00b	1.46	28.38	15.87
	2050's	24.60c	1.48	28.99	15.53
	2080's	25.33d	1.53	29.51	18.10
	abcd; significant at 0.001 level				

Means followed by different letters within a cluster region are significantly different from each other at 5% p level.

century. The possible reason for rise in temperature according to [35] may be associated with the dramatic warming of oceans over the Atlantic and /or Indian oceans which further warm the sea surface temperature in the two oceans and result in low pressure formation in the area and decrease in moist air advection towards Ethiopia.

This might be the case for the increase in number of days with $T_{max} > 28°C$, $T_{max} > 30°C$ and $T_{max} > 32°C$ in all clusters by 2080's (**Figure 8**). Further, [5] reported that climate change may increase the frequency of ENSO warm phases in Africa. The present study also showed the more likely occurrence of heat wave in all clusters for the projected years than before. The projected climate in Cluster A showed that heat wave was more likely to occur starting from 2020's, and it would be a common weather event for Cluster B. However, Cluster C will stay safe and experiences wave free days even by 2080's (**Figure 8**).

5.7.2. Daily Minimum Temperature

The present study revealed that the average daily minimum temperature in Amhara region generally was increasing. For example, a rise in mean daily minimum temperature of about 2.80°C, 0.11°C and 0.64°C are noted in Clusters A, B and C by 2080s, respectively (**Table 7**). This finding is in agreement with the historical climate of the country. According to [36], annual minimum temperature has been increased by about 0.37°C every 10 years over the past 55 years. Although, the result in Clusters B and C showed less change in mean minimum temperature than Cluster A from the base period, it indicates a shift towards warmer conditions in the future. In general, the projected climate in the three clusters indicated a decreas in the number of cold days (**Figure 9**)

Days with minimum temperature less than 7°C are projected to become less common, in all clusters in the future climate scenario (**Figure 9**). This is in line with [5] and [37] who reported increasing trend in minimum

Table 7. Summered statistics mean of minimum temperature in three cluster regions in Amhara Regional State.

Cluster	Year	Statistics for Tmin over Amhara region			
		Mean	Std	Maximum	Minimum
A	1979-2008	12.32a	2.13	16.71	4.11
A	2020'S	12.84b	2.24	17.39	2.37
A	2050'S	13.92c	2.22	18.57	3.47
A	2080'S	15.13d	2.26	20.42	7.32
	abcd; significant at 0.001 level				
B	1979-2008	19.00	2.70	30.3	11.03
B	2020'S	19.00	1.42	23.84	12.68
B	2050'S	18.97	1.43	24.54	12.90
B	2080'S	19.11	1.45	24.59	12.07
	NS				
C	1979-2008	9.90a	2.36	14.27	3.64
C	2020'S	9.97b	2.41	14.54	4.21
C	2050'S	10.27c	2.40	14.49	4.25
C	2080'S	10.54d	2.38	14.96	4.42
	abcd; significant at 0.001 level				

Means followed by different letters within a cluster region are significantly different from each other at 5% p level.

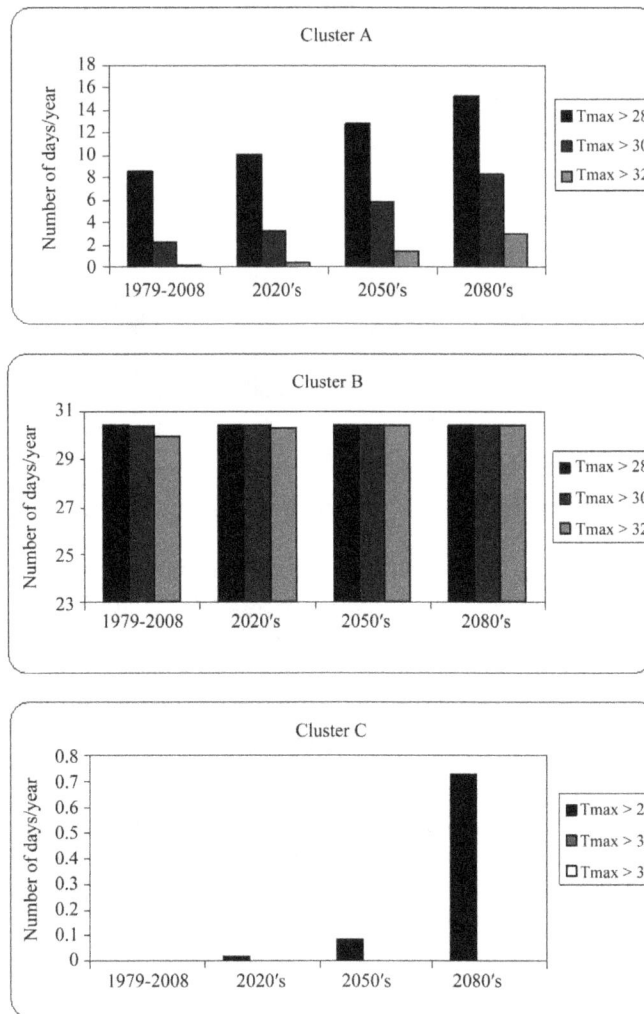

Figure 8. Projected number of hot days per year in three cluster regions in Amhara Regional State.

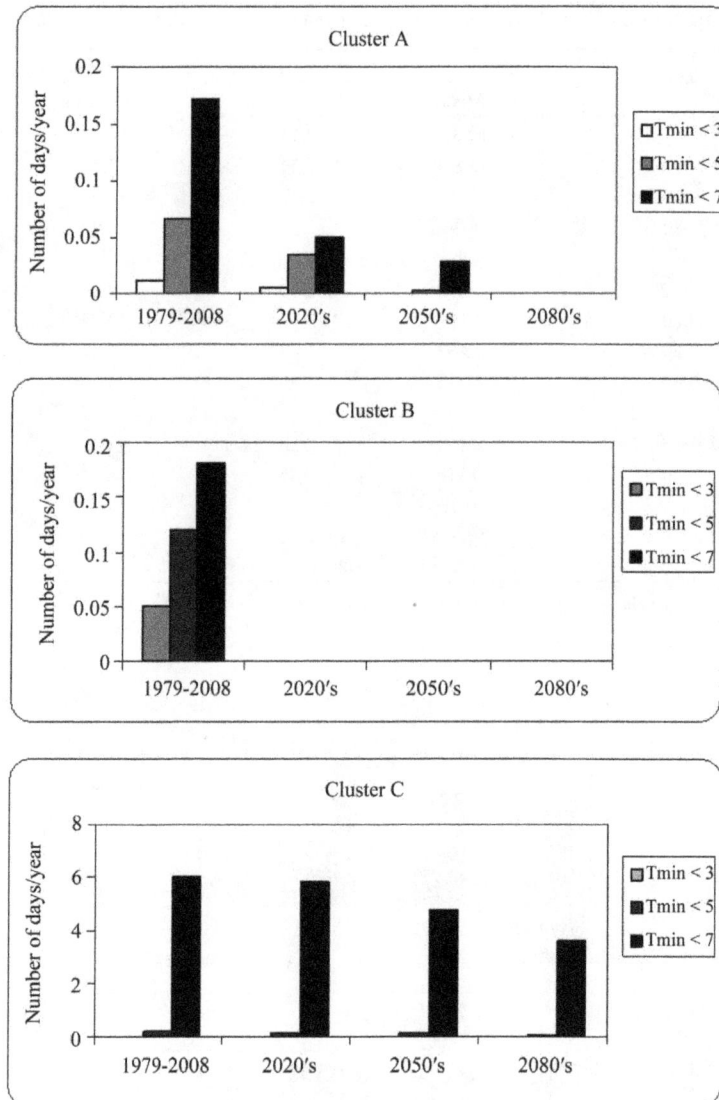

Figure 9. Projected number of cold days per year in three cluster regions in Amhara Regional State.

temperature over Ethiopia with decreasing the number of days with minimum temperature in the future.

5.7.3. Rainfall Variability and Extremesin Future Daily Pcpn

Rainfall is not usually well approximated by normal distributions [5]. The projected scenario in the current study in all clusters in the study region showed that there was a decrease in the number of positive anomalies and an increased of negative anomalies compared with the base period (1979-2008). In the base period climate, there were 13, 15 and 17 positive anomalies and 15, 15 and 13 negative anomalies in Clusters A, B and C, respectivelly As compared to 12, 10 and 13 "positive" anomalies and 14, 18 and 17 negative anomalies respectively in the 2080s (**Figure 10**).

This decrease in positive anomaly and increase in ne-

gative anomaly lead to an extended periods of below average rainfall in the study areas. Rainfall variability in general and decrease in positive anomalies in particular is in accordance with the decrease in projected annual rainfall by 27.2%, 42.3% and 12.2% in 2080s in cluster A, B and C, respectively (**Table 5**). [36] also indicated that Ethiopia's average annual rainfall has shown a very high level of variability. Further, a study to analyze farmers' perceptions of climate change in the Nile Basin of Ethiopia by [38] indicated that temperature has increased and the level of rainfall has declined. Furthermore, [5, 39,40], add that the climate change induced warming of the Indian Ocean is likely to lead to persistent droughts in east Africa in the coming years; hence the monsoon winds that bring seasonal rain to sub-Saharan African could be 10% - 20% drier than the 1950-2000 averages.

Similarly, the finding in this study showed the more

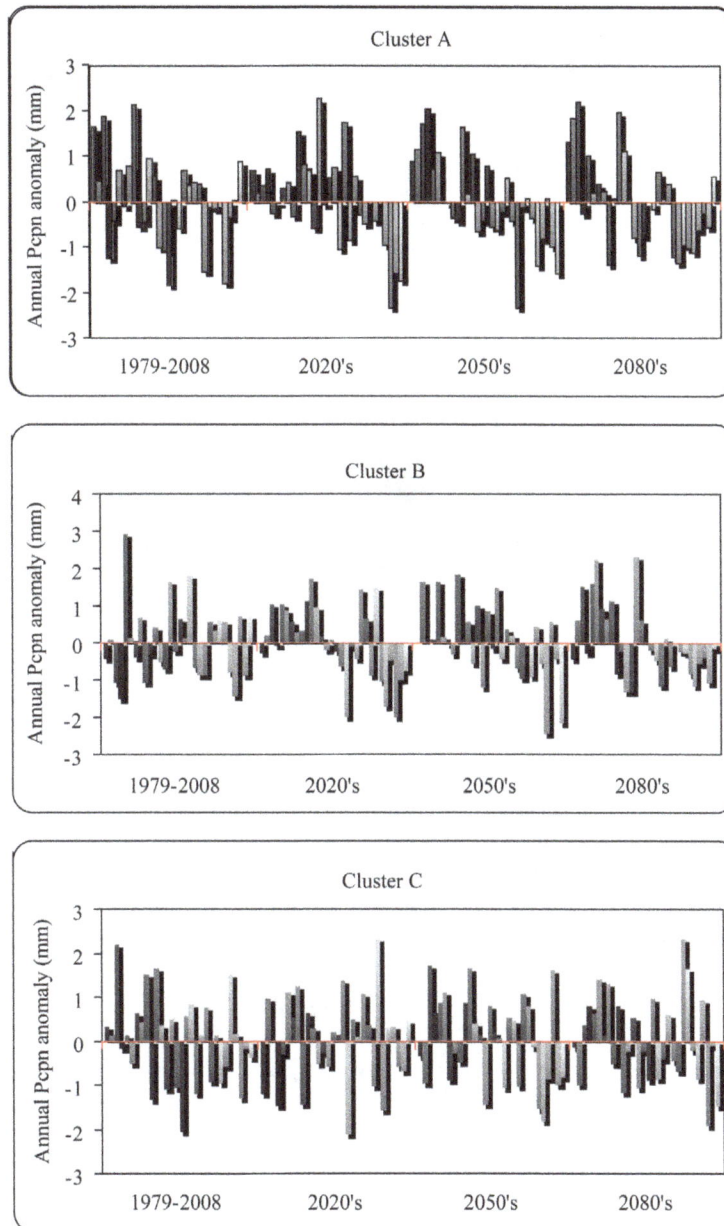

Figure 10. Annual rainfall anomaly for the period (1979-2008), 2020s, 2050s and 2080s in three cluster regions in the Amhara regional state.

likely occurrence of persistent drought, (anomalies of ≥−0.84. It means that the number of rainy days per year in future climate will decreased dramatically (**Figure 10**); which could shrink household farm production by up to 90 percent of a normal year output [41].

As shown in **Table 5**, there was high decrease in percentage change of rainfall in all clustered regions. This could be due to a dramatic decrease in the number of rainy days in the 2080s as compared to the base period (**Figures 11** and **12**). This finding is in agreement with the finding of [42] who noted an increase in the frequency of drought occurrence in the past few dacades in

Ethiopia.

Although the annual rainfall amount decreases (**Figure 12**), the number of rainy days in the projected time periods at Cluster B did not decrease in equal magnitude as Clusters A and C (**Figure 11**).

6. CONCLUSIONS AND RECOMMENDATIONS

6.1. Conclusions

Downscaled temperature results for the Amhara region using both A2a emission scenario showed increasing

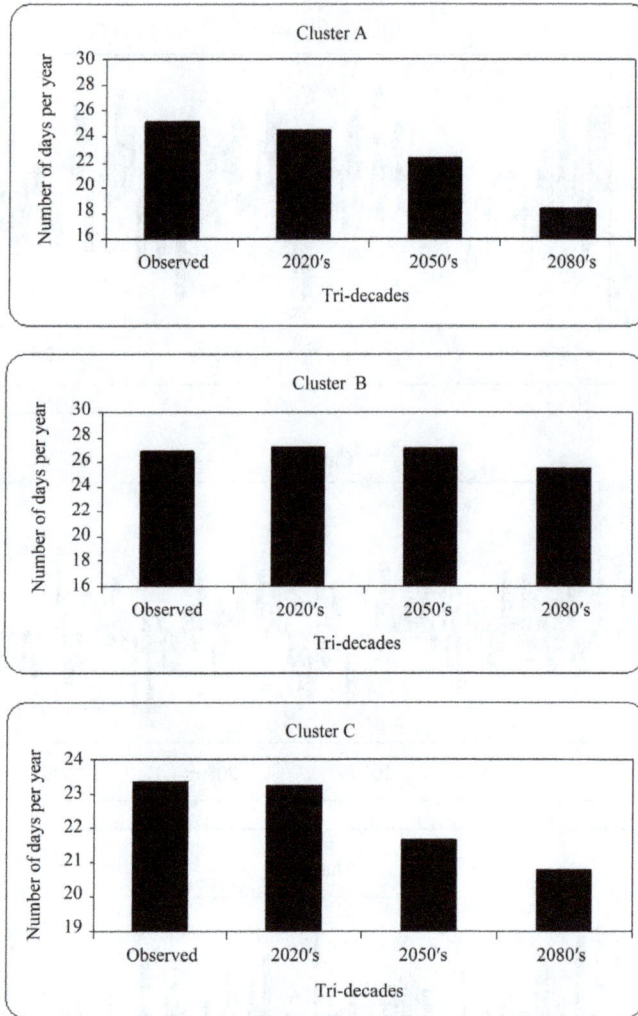

Figure 11. Projected number of rainy days per year in three cluster regions in the Amhara Regional State.

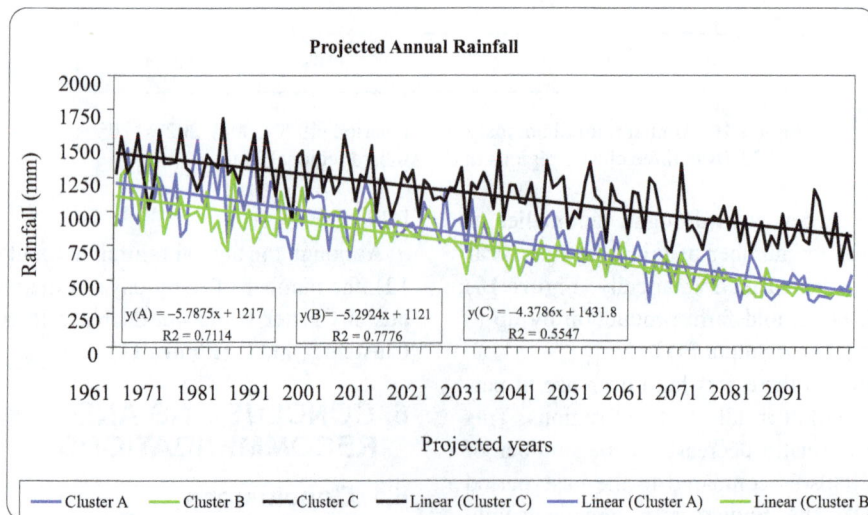

Figure 12. Future trends of rainfall in Clusters A, B and C in three cluster regions in the Amhara Regional State.

trend for the three tri-decadal periods centered on the 2020s, 2050s, and the 2080s. Maximum temperature could rise up by 2.3°C, 4.6°C and 1.80°C, while minimum temperature increased by 3.6°C, 0.3°C and 0.50°C in Clusters A, B and C, respectively in 2080s. Unlike temperature, rainfall results showed decreasing trend in all projected years. A percentage decrease in rainfall of about 27.2%, 42.3% and 12.2% was noted in Clusters A, B and C, respectively in the 2080s.

Generally, the effects of global warming due to greenhouse gas emissions and sea surface warming is reflected in the study area with an increase in projected temperatures and a decrease in rainfall which resulted in increased number of projected hot days, fewer number of projected cold days and decreased number of projected rainydays. These events may lead to increased occurrence of drought and water shortage over the Amhara Regional State.

6.2. Recommendations

Generally, based on the results of this study the following recommendations are forwarded: The projected changes in temperature and rainfall in the Amhara region urges decision makers to incorporate climate change scenarios in devising sustainable regional coping strategies, including: water harvesting technologies, supplementary irrigation, improved seeds which can tolerate moisture and temperature stresses, afforestations and soil conservation techniques, preventing free grazing and related strategies and measures.

The result of any model depends on the quality of the input data; therefore, to make the evaluation of climate change studies more complete, it is recommended to conduct similar studies with large number of meteorological stations and GCM models.

7. ACKNOWLEDGEMENTS

We express our sincere thanks to the Ministry of Education and Haramaya University for their financial support to carry out this work. We also express our sincere thanks to Bahir Dar University for allowing the first author to pursue his PhD study. Equally important was the support given by the Swiss NCCR North-South program and linking this particular research with the global climate implications. The National Meteorological Agency is also acknowledged for providing most of the daily rainfall and temperature data used for the study.

REFERENCES

[1] Gitay, H., Suarez, A., Watson, R.T. and Dokken, D.J. (2002) Climate change and biodiversity. IPCC (Intergovernmental Panel on Climate Change) technical paper V, 86.

[2] UNEP (United Nations Environment Programme) and CMS (Convention on the Conservation of Migratory Species of Wild Animals) (2006) Migratory species and climate change: Impacts of a changing environment on wild animals UNEP/CMS Secretariat, Bonn, Germany.

[3] The nature conservancy website.

[4] World Wide Fund for Nature (2006) Climate change impacts on East Africa: A review of the scientific literature, Gland. WWF, Morges.

[5] IPCC Technical Summary (2001) Climate change 2001: The dcientific basis. In: Houghton, J.T., Ding, Y., Griggs, D.J., Noguer, M., van der Linden, P.J., Dai, X., Maskell, K. and Johnson, C.A., Eds., Technical Summary of the Working Group I Report, Cambridge University Press, Cambridge, 94.

[6] Biodiversity Strategy and Action plan (2005) Biodiversity Strategy and Action plan. BSAP. Final document.

[7] Davidson, O., Halsnaes, K., Huq, S., Kok, M., Metz, B., Sokona, Y. and Verhagen, J. (2003) The development and climate nexus: The case of sub-Saharan Africa. Climate Policy, 3, S97-S113.

[8] Zhou, G., Minakawa, N., Githeko, A.K. and Yan G. (2004) Association between climate variability and malaria epidemics in the East African highlands. Proceedings of the National Academy of Sciences of the United States of America, 101, 2375-2380.

[9] Craig, M.H., Kleinschmidt, I., Nawn, J.B., Le Sueur D. and Sharp, B.L. (2004) Exploring 30 years of malaria case data in KwaZulu-Natal, South Africa: Part I. The impact of climatic factors. Tropical Medicine and International Health, 9, 1247-1257.

[10] Wood, A. (1977) A preliminary chronology of Ethiopian droughts. In: Dalby, D., Church, R.J.H. and Bezzaz, F., Eds., Drought in Africa, Vol. 2, International African Institute, London, 68-73.

[11] Pankhurst, R. and Johnson, D.H. (1988) The great drought and famine of 1888-92 in northeast Africa. In: Johnson, D.H. and Anderson, D.M., Eds., The Ecology of Survival: Case Studies from Northeast African History, Lester Crook Academic Publishing, London, 47-72.

[12] NOAA (The National Oceanic and Atmospheric Administration) research website.

[13] Australian Bureau of Meteorology.

[14] Wilby, R.L. and Dawson, C.W. (2007) SDSM—A decision support tool for the assessment of regional climate change impacts. Environmental Modeling Software, 17, 145-157.

[15] Wilby, R.L., Dawson, C.W. and Barrow, E.M. (2002) Statistical downscaling model SDSM, version 4.1. Department of Geography, Lancaster University, Lancaster.

[16] Wilby, R.L. and Dawson, C.W. (2004) Using SDSM version 3.1—A decision support tool for the assessment of regional climate change impacts. User Manual, Leicester.

[17] Dibike, Y.B. and Coulibaly, P. (2005) Downscaling precipitation and temperature with temporal neural networks.

Department of Civil Engineering, and School of Geography and Geology, McMaster University, Hamilton.

[18] Kilsby, C.G., Jones, P.D., *et al.* (2007) A daily weather generator for use in climate change studies. *Environmental Modelling and Software*, **22**, 1705-1719.

[19] Pryor, S.C., Schoof, J.T. and Barthelmie, R.J. (2005) Empirical downscaling of wind speed probability distributions. *Journal of Geophysical Research*, **110**, Article ID: D19109.

[20] Gordon, C., Cooper, C., Senior, C.A., Banks, H.T., Gregory, J.M., Johns, T.C., Mitchell, J.F.B. and Wood, A. (2000) The simulation of SST, sea ice extents and ocean heat transports in a version of the Hadley Centre coupled model without flux adjustments. *Climate Dynamics*, **16**, 147-168.

[21] Pope, V.D., Gallani, M.L., Rowntree, P.R. and Stratton, R.A. (2000) The impact of new physical parameterizations in the Hadley Centre climate model—HadAM3. *Climate Dynamics*, **16**, 123-146.

[22] Reichler, T. and Kim, J. (2008) How well do coupled models simulate today's climate? *Bulletin of the American Meteorological Society*, **89**, 303-311.

[23] Souvignet, M. and Heinrich J. (2008) Future temperatures and precipitations in the arid northern-central Chile: A multi-model downscaling approach. *6th Alexander von Humboldt International Conference*, 24 June 2010.

[24] Zeray, L., Roehrig, J. and Alamirew, D. (2006) Climate change impact on Lake Ziway watershed water availability, Ethiopia. *Conference on International Agricultural Research for Development*, Tropentag, 11-13 October 2006.

[25] Tachieobeng, E., Gyasi, E., Abekoe, S.A.M. and Ziervogel, G. (2010) Farmers' adaptation measures in scenarios of climate change for maize production in semi-arid zones of Ghana. *2nd International Conference on Climate, Sustainability and Development in Semi-Arid Regions*, Fortaleza, 16-20 August 2010.

[26] Central Statistics Authority (2005) The federal democratic Republic of Ethiopia statistical abstract. CSA, Addis Ababa.

[27] Federal Democratic Republic of Ethiopia (1997) The conservation strategy of Ethiopia: The resources base, its utilization and planning for sustainability, vol. I. Environmental protection authority in collaboration with the ministry of economic development and cooperation. FDRE, Addis Ababa.

[28] Intergovernmental Panel on Climate Change (2011) The

IPCC Data Distribution Center.

[29] Canadian Institute for Climate Studies (CICS).

[30] Houghton, J.T. (2001) Climate change the scientific basis: Contribution of working group I to the third assessment report of the intergovernmental panel on climate change. Cambridge University Press, Cambridge.

[31] Robert A. and Wither P. (2007) StatistiXL, version 1.8, a powerful statistics and statistical analysis add-in for Microsoft Excel. Microsoft, Washington DC.

[32] Stern, R., Knock, J., Rijks, D. and Dale, I. (2002) Instat+ (interactive statistics package). Statistics services center. University of Reading, Reading.

[33] National Meteorological Services Agency (1985) Climate and agro climatic resources of Ethiopia. Addis Ababa.

[34] Francis, D. and Hengeveld, H. (1998) Extreme weather and climate change.

[35] NCAR (National Centre for Atmospheric Research) (2005) A continent split by climate change: New study projects drought in southern Africa. Boulder Co., Nairobi.

[36] National Meteorological Services (2007) Climate change national adaptation program of action (NAPA) of Ethiopia. NMS, Addis Ababa.

[37] National Meteorological Services Agency (2001) Initial national communication of Ethiopia to the United Nations framework convention on climate change (UNFCCC). NMSA, Addis Ababa.

[38] Temesgen, D. (2007) Measuring the economic impact of climate change on Ethiopian agriculture: Ricardian approach. World Bank policy research paper No. 4342. World Bank, Washington DC.

[39] Hulme, M., Doherty, R., Ngara, T., New, M. and Lister, D. (2001) Taking action against climate change in Ethiopia and South Africa.

[40] Funk, C., Senay, G., Asfaw, A., Verdin, J., Rowland, J., Michaelson, J., Eilerts, G., Korecha, D. and Choularton, R. (2005) Recent drought tendencies in Ethiopia and equatorial-subtropical eastern Africa. FEWS-NET, Washington DC.

[41] World Bank (2003) Risk and vulnerability assessment draft report. World Bank, Addis Ababa.

[42] Ketema, T. (1999) Test of homogeneity, frequency analysis of rainfall data and estimate of drought probabilities in dire Dawa, eastern Ethiopia. *Ethiopian Journal of Natural Resources*, **1**, 125-136.

17

Impact of climate change on agriculture during winter season over Pakistan

Khalid M. Malik[1*], Arif Mahmood[2], Dildar Hussain Kazmi[3], Jan Muhammad Khan[4]

[1]Director, National Agromet Center, Pakistan Meteorological Department, Islamabad, Pakistan;
[*]Corresponding Author
[2]Director General, Pakistan Meteorological Department, Islamabad, Pakistan
[3]Meteorologist, National Agromet Center, Pakistan Meteorological Department, Islamabad, Pakistan
[4]Director, Pakistan Meteorological Department, Islamabad, Pakistan

ABSTRACT

This study has been carried out to investigate the impact of climate change over Pakistan and its surrounding areas (60° - 80° E and 20° - 40° N) during winter seasons (December-February). Variability in three meteorological parameters such as: rainfall; air temperature; and moisture transport, has been investigated. Global Precipitation Climatology Center (GPCC) data for precipitation and National Centre for Environmental Prediction (NCEP) reanalysis data for computation of Moisture Flux Convergence (MFC) and temperature have been used for the period of 49 years (1961 to 2009). The study period has been divided into three phases on basis of precipitation anomaly i.e., before climate change scenario (1961-1985), transition period (1986-1999) and after climate change scenario (2000-2009).Variability in precipitation has been observed in three different ways such as, slightly increase in magnitudes, decrease in rainy days and shifting of precipitation pattern towards south of the country. Moisture transport from the surrounding has decreased with increase in precipitation which is indirectly associated with decreases in mass deposit on the glaciers. Increase in temperature is more prominent over upper and lower part as compared to the central parts of the country. Uncertainty in precipitation has also been observed. Shift of precipitation over southern parts showed positive impact over agriculture sector. As a result, Rabi crop yield has increased during last decade over southern parts of the country.

Keywords: Cloud Burst; Seasonal Temperature; Moisture Transport; Shift in Precipitation

1. INTRODUCTION

Agriculture has always been the most important sector of Pakistan's economy. In the agro climatic classification of Pakistan, more than two-third of Pakistan lies in semi-arid to arid zones [1]. About 70% of our population is living in rural areas, where most of them along with livelihood depend on agriculture production. Approximately 50% of the total national labor force is directly engaged in agriculture [2]. Therefore, majority of the p eople living in arid and semi-arid areas are totally depending on agro-pastoral activities for their survival. Currently in climate change scenari o Pakistan-like other developing world, is faced with the challenges of (being affected by variability of precipitation and risen temperatures) land degradation or desertification and other environmental problems like soil erosion, loss of so il fertility, flash floods, salinity, deforestation and associated loss of biodiversity and carbon sequestration [3]. Agriculture is the most vulnerable sector to climate change. Productivity of agriculture is being affected by a number of climatic factors and some indirect factors including rainfall pattern, temperature hike, changes in sowing and harvesting dates, water availability, evapo-transpiration and land suitability. All these elements have impact on crop yield and agricultural productivity [4].

Agriculture in Pakistan is dependent on rainfall as well as irrigation water. Water mainly meets from seasonal rainfall as well as melting of snow and ice from the glaciers. Pakistan has developed the world's largest contiguous canal irrigation system. Pakistan is covered on the north by Himalaya, Karakoram and Hindukush, which host the world's third largest snow/ice reserves. These mountains are the water tanks over the roof t hat provides water to th e reservoirs. The environment has given the operational control of t his tank in term s of temperature after the strong buildup of greenhouse gases [5].

Winter brings lot of snow over the northern mountains which melts in early summer and maintains the sustainable river flows for power generation and irrigation before the onset of the summer monsoon. In addition to solid precipitation over hilly areas, winter rain bearing systems yield substantial rainfall in sub-mountainous and low elevation plains including arid plains of Balochistan. Generally northern half gets about five times more precipitation in winter than the southern half [6]. Past analysis of precipitation data has shown a slightly decreasing trend for the northern parts of the country. On the other hand the situation for southern parts of the country is becoming better in terms of precipitation and temperature as well. But the northern belt is the major source of water to the Indus, the leading river in the country. The present increase in temperature may be major augmenting force behind the sharp decrease in this treasure of solid water. For a country which is already facing problems because of ill management in water distribution and inadequate water reservoirs, it could lead to the collapse of the local agriculture system in the time to come [6]. Higher rainfall variance seems to be the main factor behind dry-land yield fluctuations. Amount and distribution of Rainfall during crop season are important. Distribution of rainfall becomes more significant for the lands with low water holding capability and also in the seasons with adequate soil moisture available at planting [7].

Increasing temperatures may have a positive impact on agriculture in the mountain areas, for instance, through shortening of growing period for the winter season crops. Winter crops (e.g. wheat), in the high mountain areas, do not even reach to maturity in most cases and such crop is harvested premature to be used as fodder. The shortening of the growing season length due to rising temperature could be beneficial in the mountain areas as it would help the winter crops in timely maturity and as such would allow the crop to mature in the optimal period of time, with beneficial effects on crop area and yields [8].

There is high level of confidence that recent regional changes (rising tendency) in temperature have discernable impacts on precipitation, evaporation, stream flow, runoff and other elements of hydrological cycles [9]. Under increased Green House Gasses (GHG) concentrations, global climate models also exhibit enhanced intensity and shorter return periods of extreme events [10,11]. Recorded extreme events during the last decade of 20th century depict consistency in terms of intensity and frequency. The history's worst drought with extremely high air temperatures and without snow cover during winter 2001, history's worst flash floods in July 23, 2001 in Rawalpindi/Islamabad because of Cloud burst, are the few sound evidences of increased intensity of extreme events [12].

2. DATA AND METHODOLOGY

The basic water equation for the column is given as:

$$\frac{\partial W}{\partial t} + \nabla \cdot Q = E - P \qquad (1)$$

Here E represents evapotranspiration, P is precipitation, ∇ is the horizontal divergence operator and W is vertically integrated water content per unit area given by:

$$W = -\frac{1}{g}\int_{ps}^{pt} q\,\mathrm{d}p \qquad (2)$$

Whereas, Q is the vertically integrated moisture flux given by

$$Q = -\frac{1}{g}\int_{ps}^{pt} qV\,\mathrm{d}p \qquad (3)$$

Whereas q is the specific humidity in kg/kg, dp is change in pressure, p_s is surface pressure, p_t is pressure at top of an atmospheric column taken to be at 300 hPa where q becomes negligible, g is the acceleration due to gravity, 9.8 m/s^2, and V is the horizontal wind vector defined as:

$$V = ui + vj \qquad (4)$$

Here u and v are the eastward and northward wind components respectively. Bold variables indicate vector representation.

The National Center for Environmental Prediction and National Center for Atmospheric Research (NECP-NCAR) pressure level data were obtained at 6 hours intervals, 00, 06, 12 and 18 Z. Four atmospheric variables such as the specific humidity q, zonal wind u, the meridional wind v and the surface pressure Ps are used for this study. All the variables except surface pressure are located on a 144 × 73 horizontal grid with a resolution of 2.5 degrees of latitude and longitude, and at 8 levels in the vertical, 1000, 925, 850, 700, 600, 500, 400, 300 hPa. Above 300 hPa, the air is very dry and contributes little to water vapor transport. Surface pressure is also available on 144 × 73 horizontal grids.

The vertically integrated Moisture Flux Convergence (MFC) into the region was calculated by using Green's Divergence Theorem.

$$\iint \nabla \cdot Q\,\mathrm{d}s = \oint Q \cdot \bar{n}\,\mathrm{d}l \qquad (5)$$

where n is the unit outward normal vector on the domain boundaries and l is the length of the line segment along the boundary. For domain average MFC, the right side is an integral performed around the entire domain. Evaluating the right hand side of **Eq.5** around the chosen boundary, then dividing by the area of the entire domain will give average convergence rate of the domain. The boundary chosen is very coarse with a grid spacing 2.5° × 2.5°. The total land area of the domain used in this study is 3.14×10^{12} m^2. Vertically integrated MFC from surface to 300 hPa for individual grid boxes (2.5° × 2.5°) have also been computed and used for examining the spatial distribution of MFC. (**Figure 1**)

Vertical integration around the region becomes difficult due to complex terrain and uneven surface, as surface pressure does not necessarily correspond to 1000 hPa, the lowest level pressure available in the data sets. Since some points having 1000 hPa were underground while others were in the atmosphere, qV values were linearly interpolated or extrapolated as required. The Global Precipitation Climatologically Centre (GPCC) version 5 data with grid point at 0.5° resolution of latitude and longitude for precipitation and NCEP reanalysis data for geo-potential height at different levels and zonal wind at 200 hPa have been used in the study. Precipitation and temperature anomalies have been computed from long term (1901-2009) GPCC precipitation average and (1949-2009) NCEP temperature data respectively. All the analysis has been made on the winter season starting from December (of previous year) to February (of current year). This implies that precipitation of 1962 consist of average precipitation of December, 1961 to February, 1962. All the calculation monthly as well as seasonal was made on monthly basis. The calculation has been made by using the software developed by the International Research Institute, University of Columbia, which is available on line at:

3. THE STUDY PERIOD (1961-2009)

The causes of climate change in any region can be linked with variability of moisture in atmospheric and precipitation on surface. Both these factors are closely associated with variation of air temperature in the region. **Figure 2** depicted anomaly of precipitation in the domain during study period. Precipitation anomalies can easily be distributed in the three phases such as: 1) No radical change during the period 1961-1985 named as before climate change scenario, no prominent variability in precipitation is observed during this phase except for few years; 2) The period 1986-1999, during which a large variation of precipitation is observed in the domain, it is assumed that this period is considered as transition period. This period is considered as rapid climate pattern change scenario in view of surface pressure and air temperatures of Atlantic and Pacific Oceans; and 3) Drastic increase in precipitation is observed during 2000-2009 after climate change scenario. In this phase it is observed that precipitation in the region has drastically increased with increase in the possibility of uncertainty of weather conditions. The study is based on the variation of different meteorological parameters during these different phases and it has been tried to investigate causes of these variations in precipitation. The cause of this drastic increase in precipitation in the region has been framed by com-

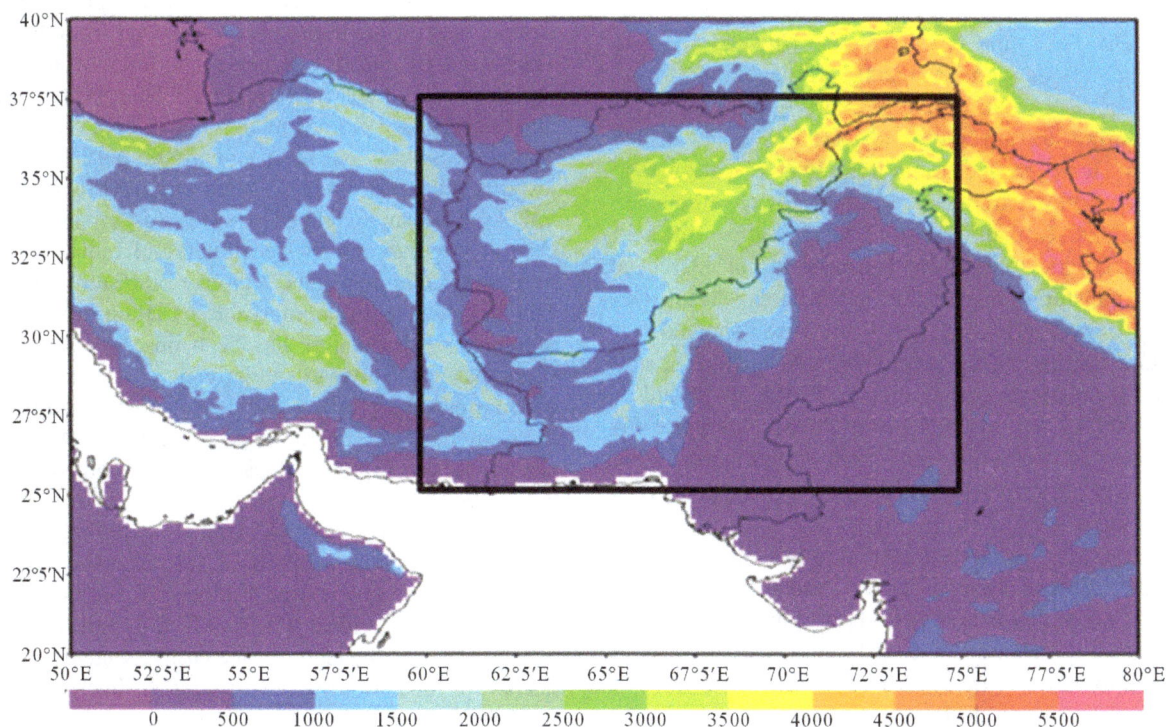

Figure 1. Boundary line in the map represents the area used for computing MFC in this study. Grids interval = 2.5° × 2.5° (longitude × latitude).

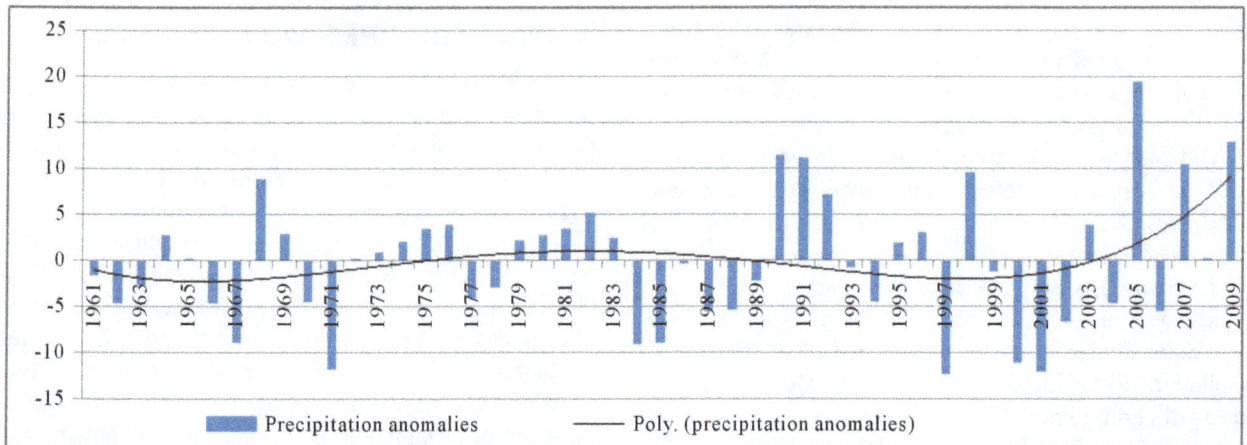

Figure 2. Anomalies of precipitation in the region during 1961-2009. Polynomial trend line represents with black steady line.

paring different meteorological parameters during these different three phases.

4. ANALYSIS OF DIFFERENT METEOROLOGICAL PARAMETERS

4.1. Moisture Flux Convergence (MFC)

Average seasonal (Dec-Feb) and five year running average MFC over entire domain has been computed and depicted in **Figure 3**. Highest moisture transported seasons were 1997-1998 and 1972-1973 with the values of 19.69 cm and 19.41 cm per month respectively. The least MFC seasons were 1999-2000 and 1996-1997 with 10.39 and 10.69 cm/month. Moisture flux convergence seasonal (Dec-Feb) time series fluctuate between 10.39 to 19.69 cm/month with an average value of 14.56 cm / month. Highest value of five years MFC has been computed during late seventy's and least was found in late ninety's. However, decreasing trend in five year average MFC is prominent from early ninety's. **Figure 4** depicted monthly and seasonal (Dec-Feb) MFC of three different periods. During the season, February is considered as highest moisture transported month whereas during December it has the least values. Comparison showed that moisture transported decreased throught out the season as well as in corresponding months during the period of 2000-2009. Five year running average MFC of entire domain showed decreasing trend from early nineties. This corresponds that moisture transport into the domain is decreasing.

In December, moisture transported has decreased during after climate change scenario, and it has remained highest during transition period. In January, moisture transported during transition and before climate change scenario is almost same whereas during after climate change scenario it decreased. In February, moisture transported during before climate change scenario is highest while during after climate change scenario it be-

came decreased. In seasonal comparison (Dec-Feb), moisture transported has been decreased after climate change scenario with 1 cm/month. In transition period (1986-1999), average moisture transported in December is highest as compared to before and after c limate change scenario because of too much variation in moisture transport during this period.

Figure 6 depicted spatial distribution of moisture flux convergence of en tire domain. Positive values indicate moisture convergence and negative values indicates moisture divergence region. The moisture divergence regions referred as source of moisture and moisture convergence region referred as sink of moisture. There are two main source of moisture for Pakistan; one is from the south *i.e.* Arabian Sea during both periods and other from the west. The source of moisture indicates the track of approaching weather systems in the region. T here are two moisture convergence regions in both before and after climate change scenarios, one is over extreme north extending towards south east up to Kashmir and other is over central Afghanistan extended towards east cro ssing from Central Pakistan and Western Punjab of India. Both periods showed almost same regions of moisture transport. This indicates that no remarkable variation in the region of moisture sink/source has been observed during the study period. However, slightly decrease in moisture transported region over Central Afghanistan and Central Pakistan has also been observed during after climate change period. **Figure 6** depicted MFC anomalies along with three different, most variability values during before climate change scenario (A), tran sition period (B), and after climate change scenario(C). It is observed that most of the values lie in (A) from 1.75 to –2.0 cm, in (B) from 3.39 to –2.3 cm and in (C) from 0 to –2 cm. It is clear evidence that during after climate change scenario, the frequency of m ore transposed moisture is in creased with reduced moisture transported in th e entire domain. The average moisture transported is 0.78 cm/month which is about 0.81 cm/month is higher than before climate

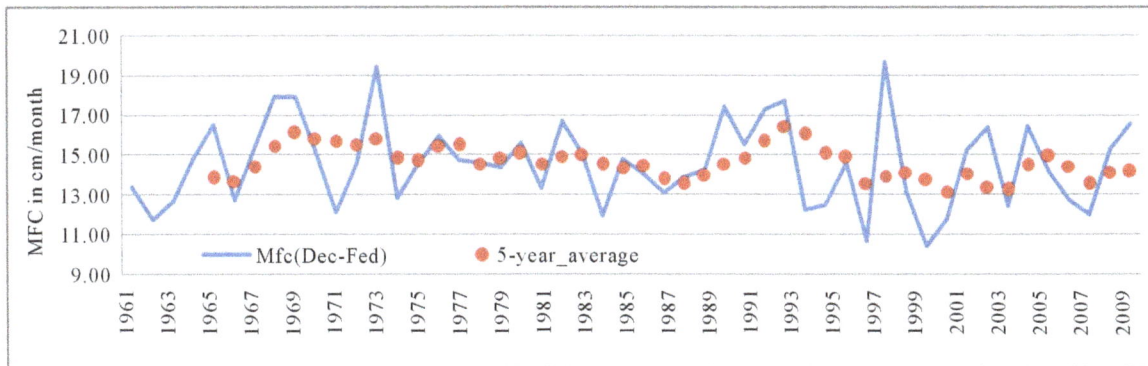

Figure 3. Seasonal (Dec-Feb) MFC (in line) and five year running average seasonal MFC (in red dots).

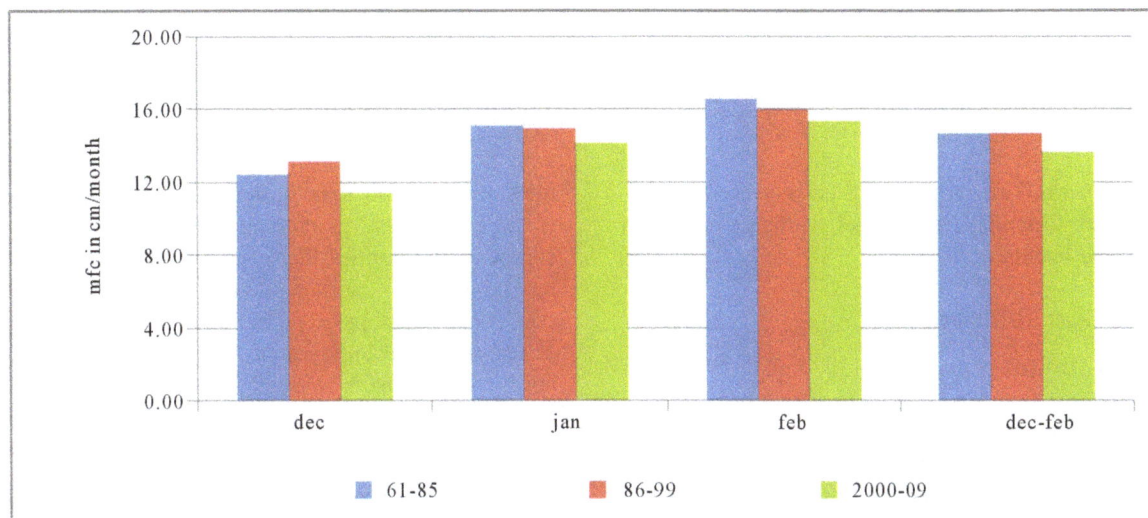

Figure 4. Comparison of Monthly and seasonal (December-February) MFC of three periods, before climate change scenario 1961-1985 (in blue), transition scenario 1986-1999 (in red) and after climate change scenario 2000-2009 (in green).

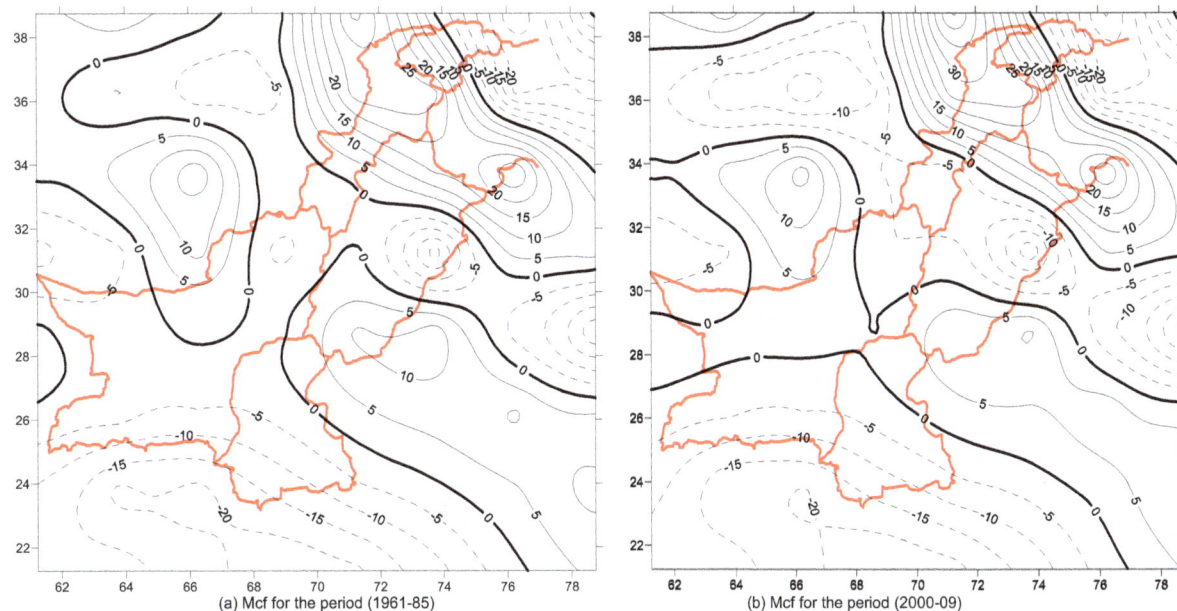

Figure 5. Spatial distribution of seasonal average MFC (December-February) for the period of (a) before climate change scenario (1961-1985) and (b) after climate change scenario (2000-2009) in cm.

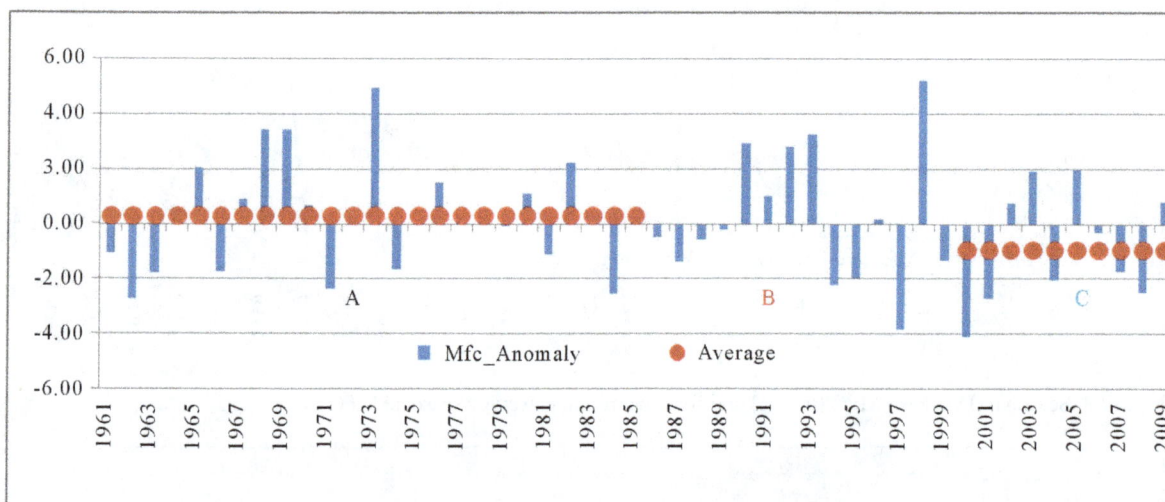

Figure 6. MFC anomalies time series (bars) and their different period averages; A) before climate change (1961-1985), B) transition period (1986-1999) and C) after climate change (2000-2009).

change scenario. The transition period showed more fluctuation in MFC anomalies is evidence that climate is going to set new reference values.

4.2. Surface Temperature

Figures 7(a)-(d) depicted monthly and seasonal temperature anomalies (departure from average 1901-2009). In December, air temperature increased all over the country except central p arts of Baluchistan extended from east to we st. Maximum decrease in tem perature was observed at Sibi and Kalat divisions, whereas, the highest increase in temperature has been observed over western and extreme northern Baluchistan, extreme north eastern Punjab and e xtreme eastern Khyber Pakhtoonkhawa. Increase in temperature over central parts of the country is not so high and is less than 1°C from long term average.

Figure 7(b) depicted January air temperature anomalies over the country. During January highest average temperature increase (more than 2°C) is ob served over northern regions along with the glaciers. Temperature has decreased over central parts of the country with maximum over Sibi, Kalat and southern Khyber Pakhtoonkhawa (KP). Overall temperature during January has been increased though out the country. **Figure 7(c)**, depicted temperature anomalies for the month of February during 2000-2009 from long term average. Temperature increasing trend has been observed over northern areas and eastern Baluchistan with east-west extension.

By comparing changes in average seasonal (Dec-Feb) air temperature (**Figures 8(a)-(b)**), it is clearly ev ident that temperature of whole domain has increased. The shift of 0°C contour line is nominal and in both scenarios it is situated almost over same areas. Furth er, it is i nteresting to note that this shift of c orresponding contour

lines of same values, on both positive and negative sides is more prominent as moving away from reference 0°C contour. This implies that less (h igher) values temperature contour lines shifted less (h igher) during the period 2000-2009 as compared to the period 1961-1985, *i.e.* shift or rise in temperature is more prominent over south and north of the domain. In Sindh, temperature has increased roughly 2°C over upper and lower parts. In Baluchistan, corresponding temperature contour has shifted towards north *i.e.* contour of 4°C which was over Quetta region during 1961-1985 periods, has shifted to north at about 100 km and crossing over the Afghanistan region during 2000-2009 period. The contour crossing over boundary of Sindh and Punjab province has s hifted at about 200 km and n ow crossing over Bahawalpur and adjoining areas of lower Punjab. In Punjab (2000-2009), increase in temperature is m ore prominent over the extreme central east and central west areas, whereas it is least ov er northern regions of the province as i t approached towards reference of 0°C contour line. The shifting of corresponding contour from before and after climate change scenario is roughly less than 50 km.

4.3. Precipitation

Figure 9 depicted monthly (a)-(c) and seasonal precipitation anomalies (d) of the region. The comparison of monthly precipitation anomalies showed mixed trend during all the months from December to January in the region. In December, precipitation has i ncreased over Baluchistan, Sindh, central and lower Punjab, whereas decreasing trend is observed over KP, Kashmir and upper Punjab. Maximum increase is more than 10 mm/month over upper Sindh and adjoining areas of Baluchistan. The Lower Punjab, Southern Baluchistan and Central Sindh showed increase in the range of 5 - 10 mm/month. In

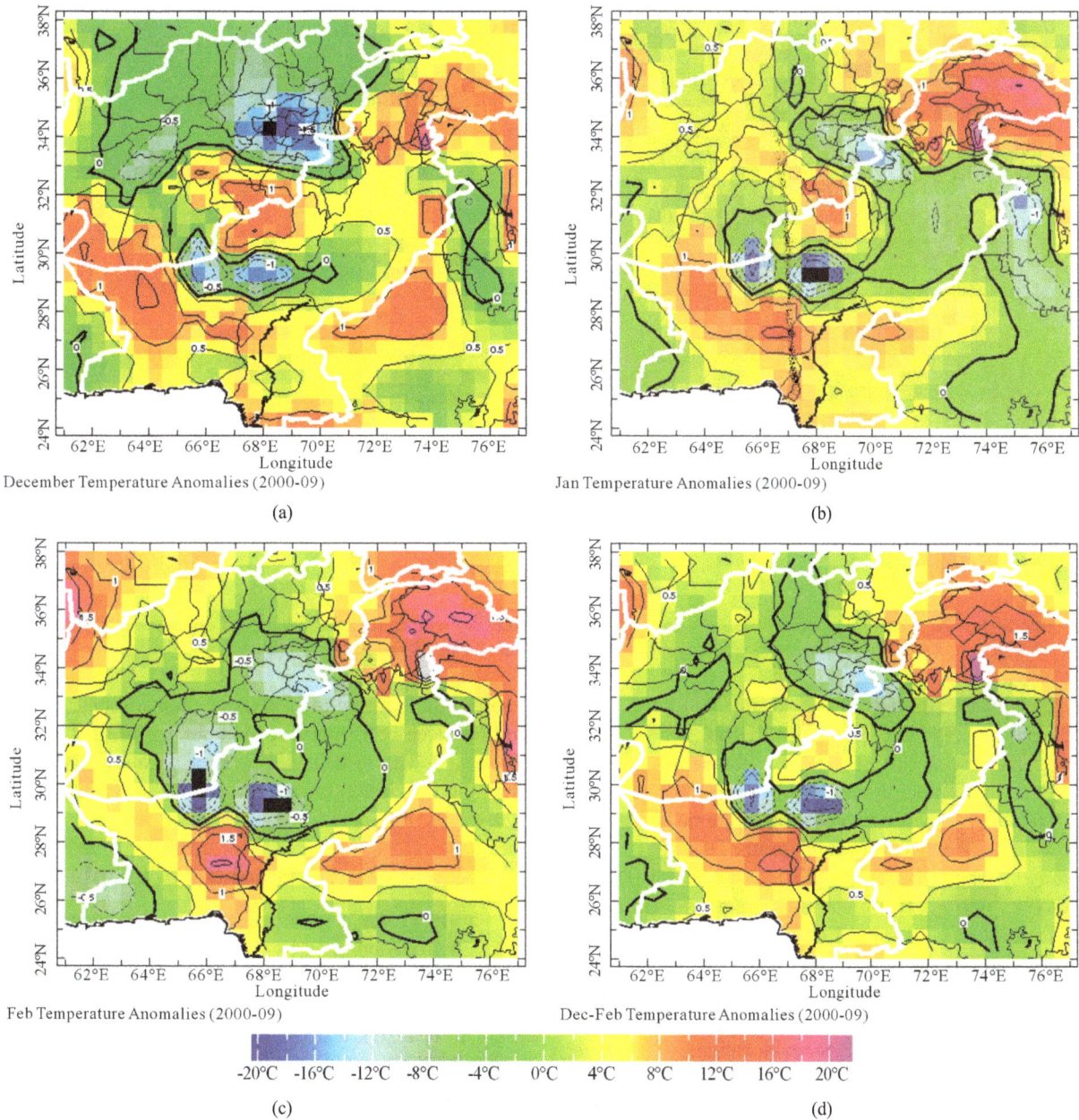

December Temperature Anomalies (2000-09)

(a)

Jan Temperature Anomalies (2000-09)

(b)

Feb Temperature Anomalies (2000-09)

Dec-Feb Temperature Anomalies (2000-09)

-20°C -16°C -12°C -8°C -4°C 0°C 4°C 8°C 12°C 16°C 20°C

(c)   (d)

Figure 7. Monthly and seasonal air temperature anomalies for the period (2000-2009) after climate change scenario, (a) For December; (b) For January; (c) For February and (d) December-February along with anomalies legend.

January decreasing trend is observed over Baluchistan and lower Sindh, while increasing trend in rest of the Pakistan. Highest increase is observed over upper KP with east west extension. Another highest positive anomaly is prominent over the Himalaya Ranges extending toward south. Some parts of Upper Punjab showed 5 - 10 mm/month increase in precipitation. Increase in precipitation over rest parts of Punjab is not remarkable. In February increasing trend is prominent all over the country except some southern parts of Baluchistan. The highest increase is observed over upper KP with two extensions, one from east to west and other is towards south.

On average no change in precipitation is observed over Sindh during February. The KP, most parts of Punjab and some north western region of Baluchistan showed more than 10 mm/month increase in precipitation.

In comparison of over all season variation shown in **Figure 9(d)**, it is observed that all over the country precipitation has increased with highest values over KP and northern regions extending towards south till southern Punjab and then further extending towards Western Baluchistan. The coastal belt showed least increase in precipitation.

Figures 10(a)-(b) depicted spatial distribution of ave-

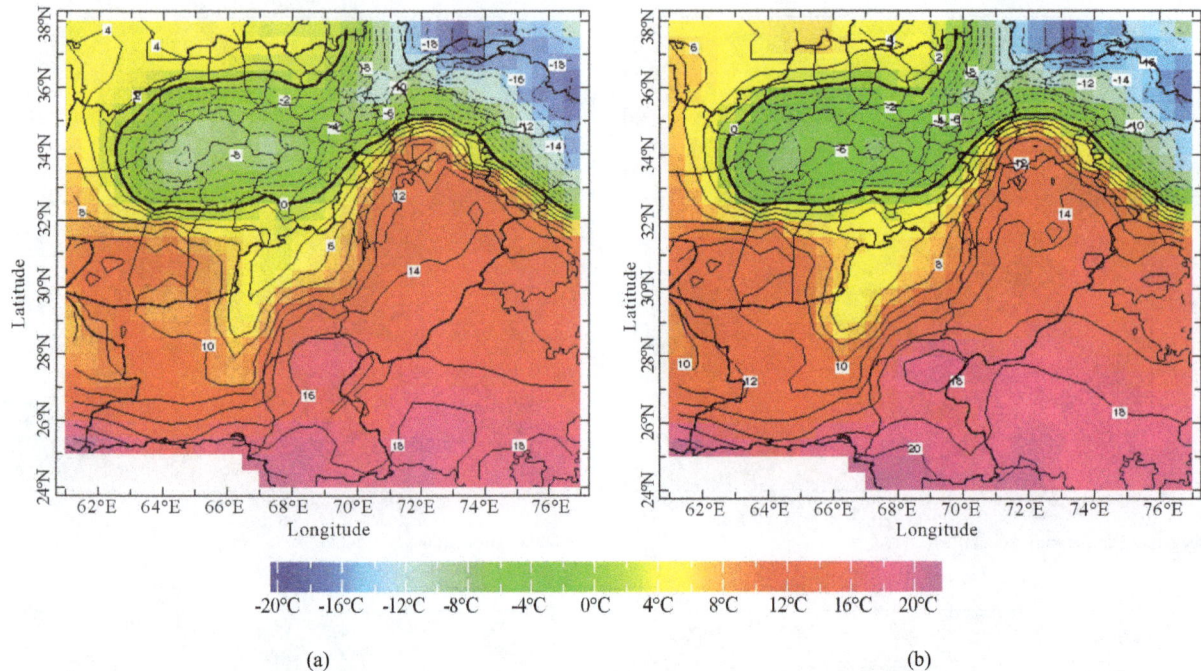

(a)   (b)

Figure 8. Spatial distribution of average seasonal (December-February) temperature for (a) before climate change scenario (1961-1985); and (b) post climate change scenario (2000-2009). C represents legend of (a) and (b).

rage precipitation during 1961-1985 and 2000-2009. Comparison of average precipitation of before and after climate change scenarios showed that the areas receiving chief amount of precipitation are the same. However, it is noticed that precipitation contours of same corresponding values have shifted towards south. This shift is more prominent from Central Punjab towards south. During 1961-1985 period, 10 mm/month contour moved northeast to southwest. It moved towards north from Central Punjab, towards south from Southern Punjab and then again towards north from Eastern Baluchistan. The same contour shifted towards south during after climate change scenarios (2000-2009) with its new location crossing over Southern Punjab and then moving across Sindh-Baluchistan boundary. Similarly, contour of 5 mm/month also shifted towards south during 2000-2009. The areas of same precipitation contour have also expanded over Afghanistan and western region of the domain. In addition, average precipitation has also increased over northern mountainous region of the country.

5. REASONS OF VARIATION IN PRECIPITATION

5.1. Geo-Potential Heights at 500 mb

In comparison of geo-potential heights at 500 mb, it is clearly evident that same corresponding contour has shifted towards south during after climate scena rios. Trough over northern Afghanistan during before climate change scenario has become more deep and shifted to-

wards south during after climate change scenario. In addition same corresponding contour has shifted south as well. It indicates that the track of western disturbances has slightly shifted towards south causes shift of precipitation towards south.

5.2. Geo-Potential Heights at 700 mb

Figure 12 depicted comparison of zonal winds at 700 mb of two periods (1961-1985) and (2000-2009). In comparison it is cl early evidence that trough become more deep during after climate change scenarios. Impact of climate change scenari os can easily be reflected during 2000-2009 period. A trough over Northern Afghanistan is prominent during after climate change sc enarios. As a result, more precipitation has been observed over Afghanistan and its extension over Pakistan will be monitored as well. This may be one of the reason of shifting of precipitation towards south of Pakistan. Trough become more deepens due to persistence of followed ridge over Northern Iran.

5.3. Zonal Winds

Comparison of winds anomalies at 200 mb during before and after climate change scenarios has been depicted in **Figure 13**. During be fore climate change sce narios, zonal winds showed decreased trend over Central Iran, Afghanistan and Upper Pakistan. However, during after climate change scenarios, zonal winds showed increasing trend over corresponding region. Strong winds at 200 mb

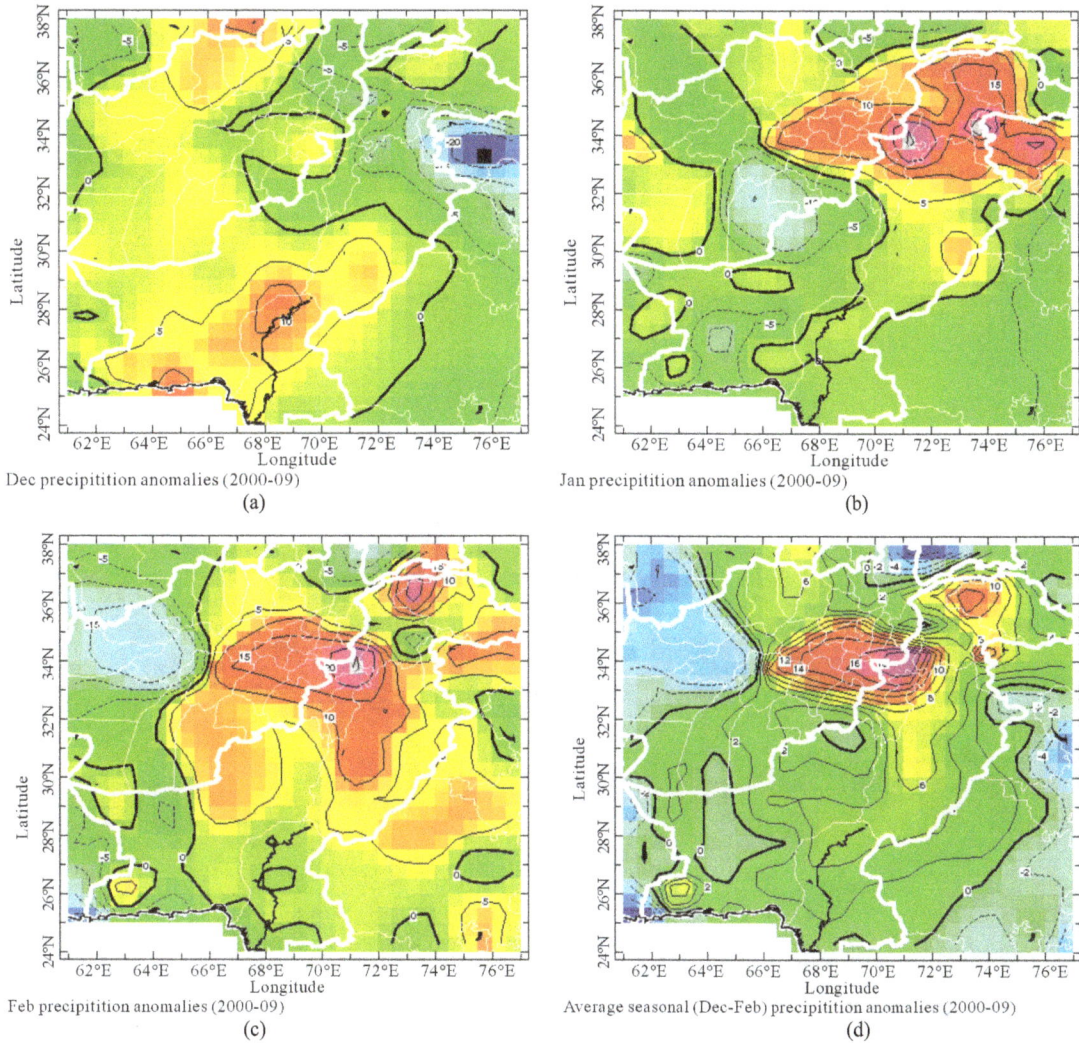

Figure 9. Monthly and seasonal departure of precipitation (mm/month) from normal, (a) for December; (b) for January, c) for February and d) for season December-February.

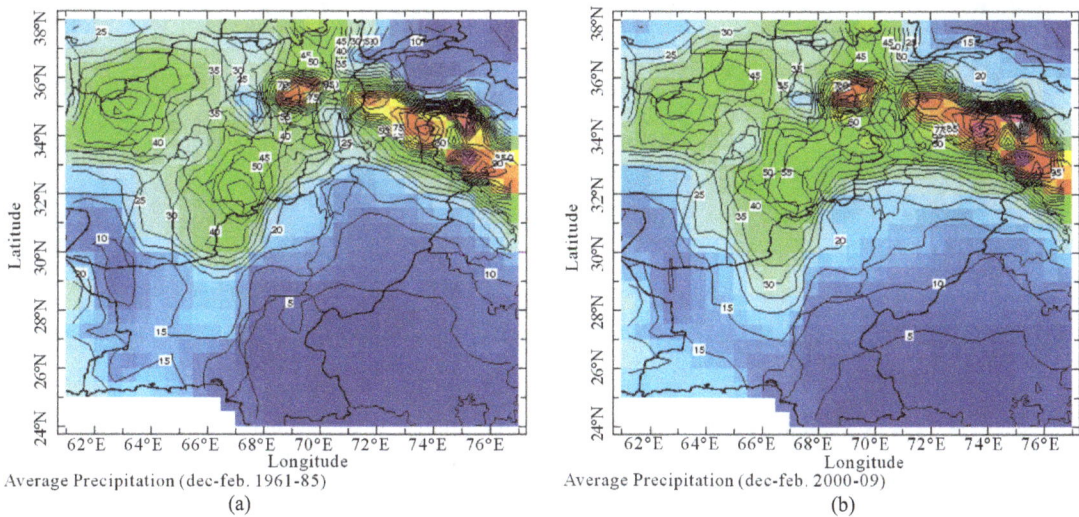

Figure 10. Average seasonal (December-February) precipitation (mm/month) during, (a) before climate change scenario (1961- 1985); and (b) after climate change scenario (2000-2009).

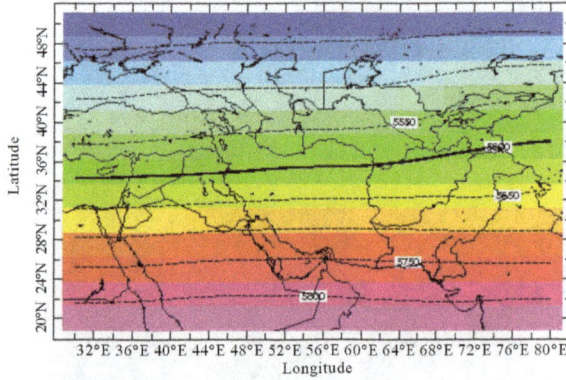

Average ght at 500 mb for the season (Dec-Feb)for the period 1961-85

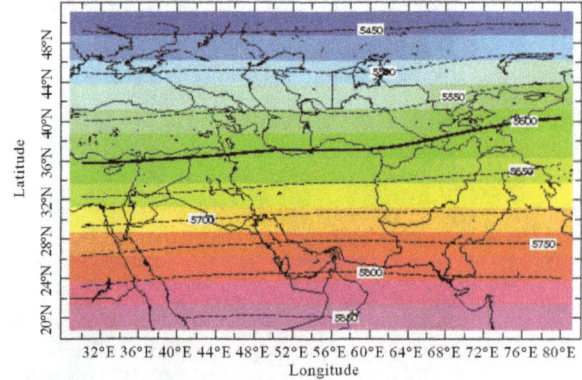

Average ght at 500 mb for the season (Dec-Feb)for the period 1961-85

Figure 11. Average seasonal (December-February) geo-potential heights at 500 mb (in meters) during, (a) before climate change scenario (1961-1985); and (b) after climate change scenario (2000-2009).

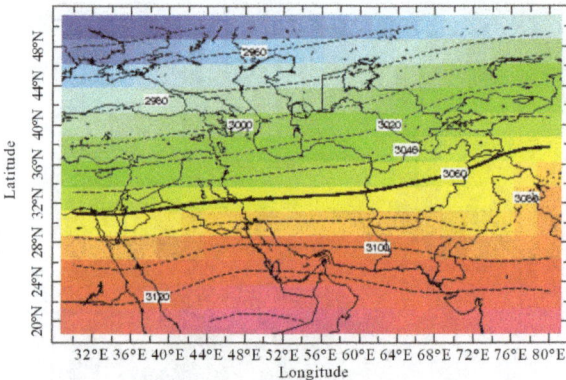

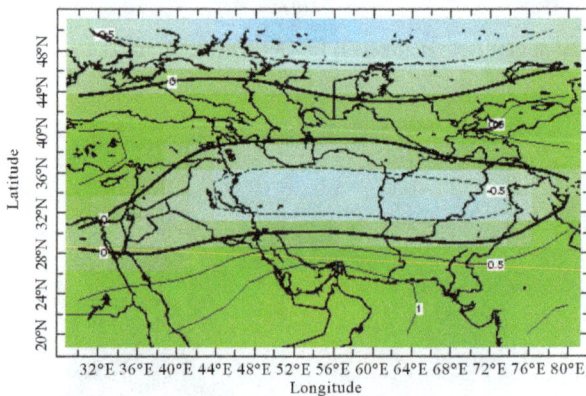

Average Ght at 700 m for the season (Dec-Feb)for the period 1961-85

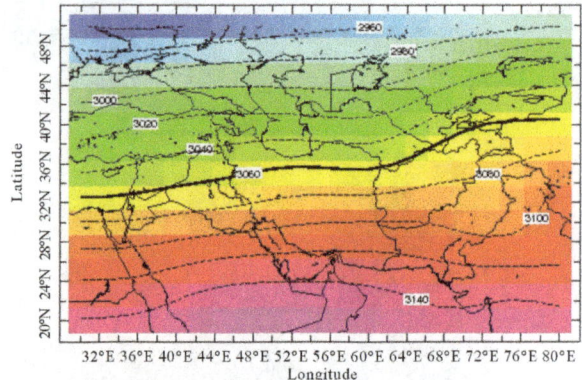

Average Ght at 700 m for the season (Dec-Feb)for the period 2000-09

Figure 12. Average seasonal (December-February) geo-potential heights at 700 mb (in meters) during, (a) before climate change scenario (1961-1985); and (b) after climate change scenario (2000-2009).

Zonal average seasonal (Dec-Feb) wind anomaly (1961-85) 200 mb

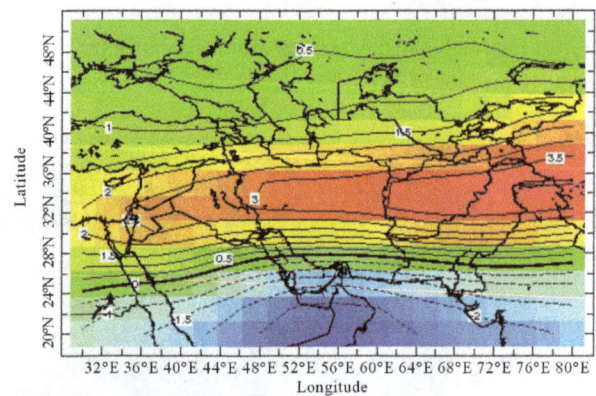

Zonal average seasonal (Dec-Feb) wind anomaly (2000-09)

Figure 13. Comparison of zonal average seasonal (December-February) winds anomalies (m/sec) of (a) before and (b) after climate change scenario.

may be associated with more precipitation and week winds are associated with less precipitation. The contour of zero change in wind speeds has already been shifted towards south during 2000-2009 period in comparison with 1961-1985 period, this indicates shift in precipitation towards south over Pakistan. **Figures 14(a)** and **(b)** depicted

comparisons of seasonal average winds for both before and after climate change scenarios.

In comparison, four factors including speed, direction, convergence area of maximum winds, and position of zonal winds are very important. However, more precipitation is associated with high winds. Comparison showed that highest zonal winds at 200 mb has shifted towards west, as a result central parts of Pakistan showed strong winds, that might be one of the cause of increase in precipitation over that region. The channel of strong winds, which squeezed during before climate change scenarios over Pakistan region has become expanded.

6. IMPACTS OF CLIMATE CHANGE ON AGRICULTURE

The province of Sindh ranks second in wheat production. Most of the lands devoted to wheat cultivation in the Lower Indus Plains are located in the irrigated districts of Nawabshah, Hyderabad, Sukker, and Tharparkar Kharipur Districts. The time series of the a nnual wheat production in the province during the last decade is shown in the **Figure 15**, which clearly indicates a rising trend in the annual production during the last decade (2000-2001 to 2009-2010) [5]. No doubt, some other agronomy factors has al so played important role for increase in yield production. Ho wever, at some extent we can correlate this inc rease in yield production with increase in precipitation into the region. The correlation between precipitation over Sindh and crop production is 0.67 with the confidence level of 95%. This indicates

that timely precipitation in the region can increase the crop production in the region.

7. CONCLUSIONS

On the basis of variation in precipitation in the entire domain, whole study period has been divided into three categories *i.e.* before (1961-1985), transition (1986-1999) and after (2000-2009) climate change periods. Association of precipitation with moisture transported into the region has been investigated. It is found that moisture transported into the domain has been dec reased with increase in precipitation during the season (December to February) after climate change. Moisture s ource region are same during both before and after climate change period whereas variation in the amount of water transported from the source region has been observed. February is considered as highest moisture transported month during the season.

Departure of after climate change period temperature from long term average (1901-2000) showed increasing trend with positive values over South and North Pakistan and decreasing trend with negative values over central parts of the country during all m onths (December to February). Increasing trend in temperature is more pro-

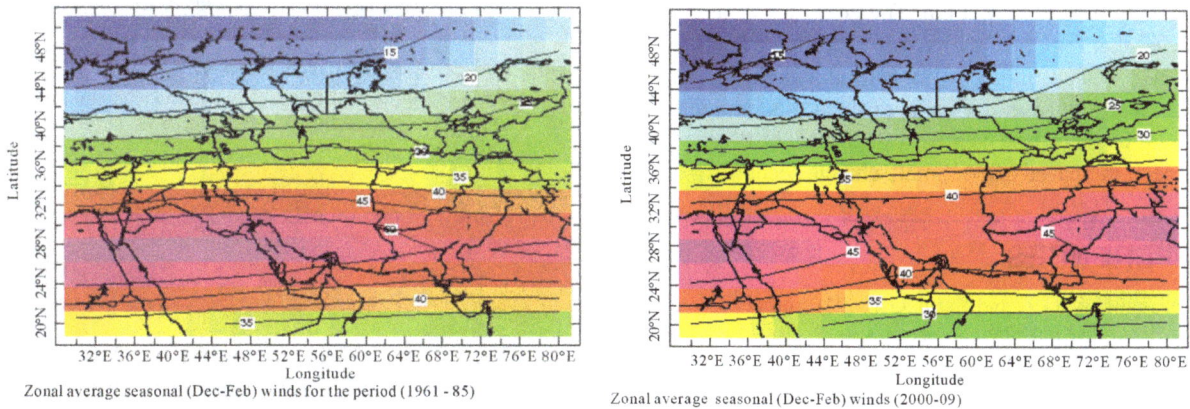

Figure 14. Comparison of zonal average seasonal (December-February) winds (m/sec) of (a) before and (b) after climate change scenario.

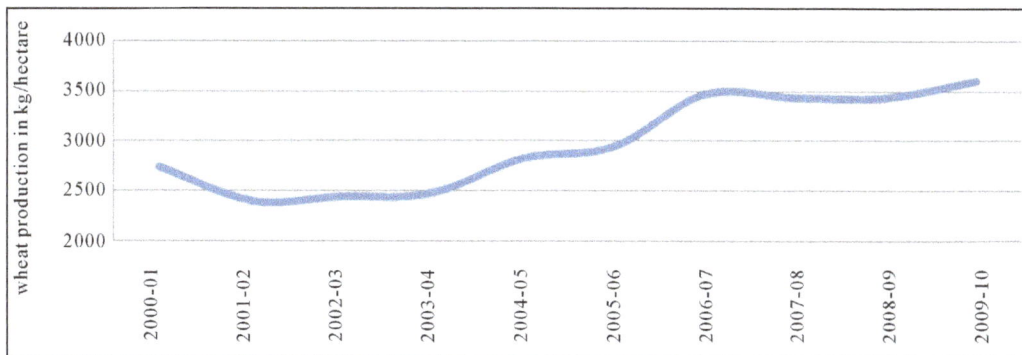

Figure 15. Time series of annual wheat production in Sindh during the last decade (2000-2001 to 2009-2010) in kg per hectare.

prominent over northern parts including glaciers. Increase in temperature caused increase in evaporation rate which increased specific humidity in the atmosphere. As a result, precipitation in the region increased without transport of moisture from surroundings. Therefore, increase in temperature is indirectly associated with increase in precipitation in these regions. This implies that as a whole mass deposited on the glacier in this region is going to decrease. Comparison in average seasonal temperature showed prominent increase in temperature over southern parts rather than northern parts of the Pakistan.

Monthly precipitation has increased (decreased) over southern Pakistan during December (January and February). However, increasing trend in seasonal precipitation is prominent over Northern and Central Pakistan. Precipitation over glacier showed increasing trend during January. Average seasonal precipitation showed increasing trend over glacier. In addition shift in average seasonal precipitation towards South Pakistan is also prominent which improved the requirement of water for agriculture land. This shift in precipitation over agriculture land is directly linked with increase in crop yields in the region. Increased in precipitation is associated with the variation in geo-potential heights at 500 - 700 mb. The variation in geo-potential height is directly linked with zonal winds at 200 mb. The zonal winds at 200 mb showed that region of maximum winds have been moved towards west, centered over central Pakistan after climate change scenario. This increase in precipitation over agriculture land has positive effect on it and crop yield has been increased after climate change scenarios.

REFERENCES

[1]   Chaudhry, Q.Z. and Rasul, G. (2004) Agroclimatic classification of Pakistan. *Science Vision*, **9**, 59.

[2]   Dowswell, C. (1989) Wheat research and development in Pakistan. Pakistan Agriculture Research Council, Collaboration Program.

[3]   Go, P. (2008) Economic survey of Pakistan (2007-08). Ministry of Finance, Government of Pakistan, Pakistan.

[4]   Harry, M.K. and Thomas E.D. (1993) Agricultural dimensions of global climate change. St. Lucie Press, Delary Beach.

[5]   Rasul, G., Dahe, Q. and Chaudhry, Q.Z. (2008) Global warming and melting glaciers along southern slopes of HKH ranges. *Pakistan Journal of Meteorology*, **5**, 14 p.

[6]   Kazmi, D.H. and Rasul, G. (2009) Early yield assessment of wheat on meteorological basis for Potohar region, Pakistan. *Journal of Meteorology*, **6**, 73.

[7]   Pratley, J. (2003) Principles of field crop production. Oxford University Press, Oxford.

[8]   Hussain, S.S. and Mudasser, M. (2004) Prospects for wheat production under changing climate in mountain areas of Pakistan—An econometric analysis. *Econpapers*, **94**, 494-501.

[9]   Elshamy, M.E., Wheater, H.S., Gedney, N. and Huntingford, C. (2006) Evaluation of the rainfall component of a weather generator for climate impact studies. *Journal of Hydrology*, **326**, 1-24.

[10] Hennessy, K.J., Gregory, J.M. and Mitchell, J.F.B. (1997) Changes in daily precipitation under enhanced greenhouse conditions. *Climate Dynamics*, **13**, 667-680.

[11] Fowler, A.M. and Hennessy, K.J. (1995) Potential impacts of global warming on the frequency and magnitude of heavy precipitation. *Natural Hazards*, **11**, 283-303.

[12] Chaudhry, Q.Z., Sheikh, M.M., Bari, A. and Hayat, A. (2001) History's worst drought conditions prevailed over Pakistan.